Test-Taker's Checklist for the
Advanced Placement Examination in Biology

This test-taker's checklist provides a quick survey of procedures to be followed when studying for the Advanced Placement Examination in Biology. The checklist highlights special study techniques that, if followed, will help you improve your examination score.

However, this checklist is not a substitute for reading and studying Chapters 1–3 in *Barron's AP® Biology*. Rather, it serves as a reminder of the study skills that will sharpen your thinking and performance on the examination. The checklist is a study organizer.

Read the checklist through. Then go back and check off the items that you can answer affirmatively. Finally, attend to the items that you did not check. It will probably take a good deal of your time to go back over content material or word review, but the extra time spent in studying will be of benefit to you.

Study Preparation

Are you familiar with the format of the Advanced Placement Examination in Biology?

_____ 120 Multiple-choice questions—1 hour and 30 minutes

_____ 4 Mandatory, Free Response (essay) questions—1 hour and 30 minutes

Are you able to approach the multiple-choice questions confidently because you

_____ read thoroughly, quickly, accurately?

_____ have familiarized yourself with the language of biology? (See the vocabulary lists in Chapters 4–14 and the glossary in *Barron's AP® Biology*.)

_____ can derive meaning from tightly constructed test questions?

_____ have committed to memory the facts of biology?

_____ are able to use the facts of biology in problem-solving situations?

_____ are flexible in thought?

_____ can use mathematical concepts to solve biological problems?

_____ are able to think in broad principles of biology?

_____ have used all the short-answer questions in *Barron's AP® Biology* for practice?

_____ can select the best answer from the choices given in each question?

_____ used the subject-area evaluation grid to determine your subject-matter strengths and weaknesses?

_____ have used Chapters 4–14 in *Barron's AP® Biology* to review modern concepts of biology?

Are you able to write point-accruing answers to essay questions because you

_____ can apportion time for each essay question?

_____ can differentiate the meanings of directional words that begin essay questions? (See page 62 in *Barron's AP® Biology*.)

_____ have improved your skill in understanding and using words?

_____ have read your biology textbook assignments thoroughly, looking up the meanings of unfamiliar words?

_____ have performed and understood the laboratory exercises required by the course of study in AP Biology?

_____ are able to write grammatically correct sentences?

_____ are able to present ideas in an orderly and precise manner?

_____ have learned the facts of biology and can apply them to problem-solving situations?

_____ have developed the technique of citing examples to support statements?

_____ have prepared written answers to questions in the model AP examinations that appear in *Barron's AP® Biology?*

_____ have prepared written answers for questions that appear at the ends of chapters in textbooks?

_____ recheck content items of which you are not as knowledgeable as you might be?

_____ are aware that your written answers to essay questions must communicate information to examination readers?

Supplies and Materials

Are you prepared to take the following materials to the examination in Advanced Placement biology?

_____ Four well-sharpened number 2 pencils

_____ A fresh ballpoint pen

_____ A clean, serviceable eraser

_____ A wristwatch with an easy-to-read dial face

_____ Your AP identification number (if you are taking more than one AP examination in the same period).

Additional Examination Pointers

Do you know that you should

_____ not waste time pondering over multiple-choice selections?

_____ read essay questions through thoroughly and quickly before starting to write?

_____ note the starting and finishing time of each essay in the margin of your booklet?

_____ leave time to reread each essay answer and make minor corrections on it?

_____ write essay questions in pencil (but legibly!) to allow for easy correction of answers?

BARRON'S

HOW TO PREPARE FOR THE

AP®

BIOLOGY

ADVANCED PLACEMENT TEST IN BIOLOGY

6TH EDITION

BARRON'S

HOW TO PREPARE FOR THE

AP®

BIOLOGY

ADVANCED PLACEMENT TEST IN BIOLOGY

6TH EDITION

Gabrielle I. Edwards
Former Assistant Principal, Supervisor of Sciences
Franklin D. Roosevelt High School, Brooklyn, N.Y.

and

Marion Cimmino
Lecturer in Ecology
Arizona Sonora Living Desert Museum, Tucson, Arizona
Former Teacher, Biology and Laboratory Techniques
Franklin D. Roosevelt High School, Brooklyn, N.Y.

BARRON'S

All inquiries should be addressed to:
Barron's Educational Series, Inc.
250 Wireless Boulevard
Hauppauge, New York 11788
http://www.barronseduc.com

Library of Congress Catalog Card No. 99-86952

International Standard Book No. 0-7641-1375-5

Library of Congress Cataloging-in-Publication Data

Edwards, Gabrielle I.
  AP biology: advanced placement test in
biology / Gabrielle I. Edwards and Marion Cimmino.—
6th ed.
     p.   cm.
  Includes bibliographical references (p.).
  ISBN 0-7641-1375-5
  1. Biology—Examinations, questions, etc.
I. Cimmino, Marion. II. Title.

QH316 .E38 2001
570'.76—dc21   99-86952

Contents

Preface

The purpose of this book is to help you organize your study activities so that you can take the Advanced Placement Examination in Biology successfully. This examination measures your knowledge of first-year college-level biology. *Barron's How To Pass The Examination in Advanced Placement Biology* provides the tools and techniques for study.

Some students prepare for this examination in a special tutorial or by independent study. Most high school students who take the examination, however, are enrolled in an intensive regular course that requires a full academic year. An Advanced Placement Biology course takes more time, requires more work, explores biology in greater depth, and offers more opportunity for individual progress and accomplishment.

To derive the full benefit from your advanced placement course or from this book, you should take the College Board Advanced Placement Examination in Biology. The structure of the examination and its depth and scope are explained in this book. Students who receive a sufficiently high mark on the examination are given recognition by more than 3,100 colleges and universities that grant them (1) advanced placement standing, which excuses them from taking the college's general biology course, or (2) college credit for the first year of biology, or (3) both.

The benefits of taking Advanced Placement Biology are many. Here are four of the advantages: (1) Colleges are interested in students who are willing to do the hard work required for success in an Advanced Placement Program. You will increase your chance of admission to the college of your choice because you will have demonstrated your ability to do college-level work. (2) It may be possible to save about $3,000 in college tuition by earning a year's college credit in biology while still in high school. (3) You will gain additional time for advanced electives in college. (4) You will benefit from becoming better acquainted with the fascinating and challenging field of modern biology.

Modern biology is best characterized as an explosion of new information. Cell and molecular biology, genetics, neural physiology, developmental biology, immunology, and environmental science represent some of the areas of study that have been revolutionized during the second half of the twentieth century. Modern methods of microscopy, elucidation of the role of DNA, the use of chemistry and physics in the solving of biology problems, and the advances in the use of technology have radically changed the depth and scope of biology. Biology is no longer one discipline. It is many disciplines, each with its store of new information.

The course in Advanced Placement Biology gives high school students the opportunity to accept the challenge of learning at the first year college level while still in high school.

Students who have the motivation and ability to do the hard work required in Advanced Placement Biology are very special people. These are the individuals who are sufficiently disciplined and capable of doing well in all their studies.

Introduction

General Information about the Advanced Placement Examination in Biology

The Advanced Placement Examination in Biology is offered to high school students during the month of May by the College Board. Its purpose is to provide an opportunity for students to demonstrate college-level achievement in biology. The examination is geared to the introductory course in college biology as described in the College Board publication *Advanced Placement Course Description/Biology*. The examination is not recommended for those who have not studied biology extensively beyond the high school level. Students who have completed the Advanced Placement Biology course of study set forth by the College Board, however, are encouraged to study for and take the examination.

Schools that participate in the Advanced Placement Program administer the examination to their own students. Students from nonparticipating schools are referred to locations where the examination is given. A fee is charged, which helps cover the cost of preparation and administration of the test.

Organization of the Examination

The Advanced Placement Examination in Biology consists of two parts, designated Section I and Section II. The test taker is given 1 hour and 30 minutes for each part. The total examination time is 3 hours. The content of the examination is drawn from topics in the suggested course outline, *Advanced Placement Course Description/Biology*.

Section I consists of 120 objective multiple-choice questions. Within the 1 hour and 30 minute time allowance, the test taker is expected to answer as many of these questions as possible. *There is a penalty of ¼ point for each wrong answer.*

Section II consists of four mandatory free-response essay questions. There being no choice, you must answer each of these questions. You are advised to allot 22 minutes of time for each question.

Question Content and Direction

Section I

The multiple-choice questions that comprise Section I are drawn from the entire course. This section is a general survey of the facts and concepts covered in a college-level modern biology course. *Section I counts for 60 percent of the total examination score.*

It is important for you to be well versed in each of the major areas presented in the *Advanced Placement Course Description/Biology*. You must develop the ability to read multiple-choice questions accurately and quickly. Knowing the facts of biology well and understanding how to use them will enable you to make appropriate answer choices.

Adopt the study habit of reading the end-of-chapter questions in your biology textbooks. As you finish reading a chapter, answer all the questions. Actively review and study the material that you do not know. To do well on Section I, you must not only know the facts of modern biology but also be able to apply them to problem solving and to data analysis. Chapter 1 in this book provides specific tips on how to deal with multiple-choice questions.

Section II

Section II, the free-response portion of the examination, affords students the opportunity to demonstrate their grasp of the principles of college-level biology. Free-response questions require essay writing, problem solving, organization of material, and other skills.

The free-response questions in Section II are drawn from the three broad areas of biology set forth in the *Advanced Placement Course Description/Biology*. One essay question is taken from Area I—Molecules and Cells, and another from Area II—Heredity and Evolution. One free-response question is derived from Area III—Organisms and Populations. One free-response question will be derived from the laboratory objectives. Each of the broad areas of study encompasses a great deal of material. An essay question can focus on any of the subsections. Within a time frame of 22 minutes, you must recall facts, relate them to a central theme, organize an intelligent answer, and produce a concise but comprehensive essay. *Section II counts for 40 percent of the total examination score.* Chapter 2 of this book provides tips on the proper approach to essay questions and analyzes sample answers.

Laboratory Experiences

An important aspect of the AP Biology course is laboratory experience. In college-level courses approximately one third of the credit is drawn from laboratory work. It is expected that an AP Biology course will emphasize laboratory work to the same degree. Work in the laboratory offers students the opportunity to develop important skills: detailed observation, accurate recording, experimental design, manual dexterity, data interpretation, statistical analysis, and the skilled use of technical equipment.

One of the four free-response questions will be designed to test students' analytical and reasoning skills. It is reasonable to expect that laboratory experiences will be reflected in a question of this type.

Here are the twelve laboratory topics on which examination questions may be based:

1. Diffusion and Osmosis
2. Enzyme Catalysis
3. Mitosis and Meiosis
4. Plant Pigments and Photosynthesis
5. Cell Respiration
6. Molecular Biology
7. Genetics of Organisms—Statistical Analysis Section

8. Population Genetics and Evolution
9. Transpiration
10. Physiology of the Circulatory System
11. Animal Behavior
12. Dissolved Oxygen and Aquatic Primary Productivity

The Meaning of the Scores

AP Biology Examinations are graded from 1 to 5 as indicated below:

Extremely well qualified	5
Well qualified	4
Qualified	3
Possibly qualified	2
No recommendations	1

Students who attain scores of 3, 4, or 5 demonstrate an above-average mastery of the course in college-level general biology. However, there is no uniformity among colleges in regard to the scores that are accepted for credit. *Advanced Placement Course Description/Biology* lists the colleges and universities that normally use Advanced Placement Examination grades, and you should inquire how the colleges that you are considering regard AP test grades. It has been the experience of the authors that a pupil receiving a grade of 5 or 4 is given advanced placement credit in biology by most colleges and universities. A score of 3 may or may not be given credit, depending on the college.

References

For additional information about the Advanced Placement Program and Advanced Placement Biology, you may wish to obtain the publications listed below:

General Information

"Challenge Yourself through AP" (brochure free of charge)
"A Guide to the Advanced Placement Program" (booklet free of charge)
"College and University Guide to the Advanced Placement Program" ($10.00)
"AP Bulletin for Students and Parents" (booklet free of charge)
"A Study of AP Students in High Schools with Large Minority Populations" (booklet free of charge)

Advanced Placement Biology

"Advanced Placement Program Course Description (Biology) May 2000, May 2001" ($12.00)
"The 1999 Advanced Placement Examination in Biology" ($20.00)
"1997 AP Biology: Free-Response Questions" ($5.00)
"AP Biology: Five-Year Set of Free-Response Questions" ($5.00)
"AP Biology Laboratory Manual for Students" (Edition D), 1997 ($15.00)
"AP Biology Teleconferencing Video" ($15.00)

References for Teachers

"Advanced Placement Biology Teacher's Laboratory Manual" (Edition D) ($15.00)
"Teacher's Guide to the Advanced Placement Course in Biology" ($12.00)
"AP Biology Teleconference Video" ($15.00)
"Secondary School Guide to the AP Program" ($10.00)
"College Explorer Plus" ($195.00) Recommended for the college or high school library. This is an IBM-compatible software program that enables students and teachers to research, review, and compare AP policies at more than 3,000 colleges.

Information about publications may be obtained from:
  College Board Publications
  Two College Way
  P.O. Box 1100
  Forrester Center, WV 25438-4100
  For customer service call 1-800-323-7155 or
                          1-888-225-5427

Information about testing and/or financial aid can be obtained from:
  Corporate Headquarters
  Educational Testing Service
  Rosedale Road
  Princeton, NJ 08541
  E-mail: etsinfo@ets.org
  http://www.collegeboard.org/html/communications000.html

PART ONE

Diagnose Your Problem

Model AP Biology Examination No. 1

Model Advanced Placement Biology Examination No. 1 is presented early in this book so that it can serve as a diagnostic test. In other words, it can help you discover strong and weak spots in your grasp of the subject. After you have taken the test, you may then feel ready for the next part of the book, Part Two—Preparing for the Examination. There you will find specific suggestions on how to approach multiple-choice and free-response questions and how to improve your answers to both types.

Parts Three, Four, and Five of the book can be helpful to you in improving your knowledge and understanding of biology concepts. Then, when you take Model Advanced Placement Biology Examination No. 2, located in Part Six, you will be able to see how much progress you have made since you took Model Examination No. 1.

The questions presented in this test have *not* appeared on any recent AP Biology Examinations. Section I of AP Biology Examinations is closed to all except test candidates, a procedure that is necessary if meaningful national norms are to be established and maintained.

However, the simulated questions presented here are varied in content and touch on a number of biological topics. Some questions require creative thinking, others, straight recall.

It is suggested that you adhere to the time limits set for both sections, follow the tips given for successful examination taking, and plunge right in.

The answers to the questions follow the exam.

Section I Multiple-Choice Questions

120 questions
1 hour and 30 minutes

Directions: Each of the questions or incomplete statements in this section is followed by five suggested answers or completions. Select the ONE that is BEST in each case.

1. The greatest source of the earth's energy is
   (A) stored in nutrient molecules
   (B) contained in fossil fuels
   (C) released in nuclear reactions
   (D) provided by the sun
   (E) provided by green plants

2. Most of the ATP made during aerobic cellular respiration is produced in the
   (A) ribosome
   (B) chloroplast
   (C) mitochondrion
   (D) cytoplasm
   (E) Golgi apparatus

3. The triplet of nucleotides of tRNA that "recognizes" a similar unit of nucleotides in mRNA is the
   (A) cistron
   (B) codon
   (C) anticodon
   (D) anticistron
   (E) peripheral

4. Which of the following sequences shows the order in which new substrate products are formed during the Krebs cycle?
   (A) Oxaloacetate, citrate, fumarate, isocitrate, ketoglutarate
   (B) Fumarate, isocitrate, ketoglutarate, citrate, oxaloacetate
   (C) Citrate, isocitrate, ketoglutarate, fumarate, oxaloacetate
   (D) Isocitrate, oxaloacetate, fumarate, ketoglutarate, citrate
   (E) Fumaric acid, ketoglutaric acid, oxaloacetic acid, isocitric acid, citric acid.

5. A modern procedure for counting human chromosomes was developed by
   (A) Lysenko and Lederberg
   (B) Tjio and Levan
   (C) Beadle and Tatum
   (D) Hardy and Weinberg
   (E) Lederberg and Zinder

6. The stems of six cocklebur plants were grafted together in a row and numbered 1 to 6. Cocklebur is a short-day plant. One leaf of plant number 6, which was at the right-hand end of the line, was given the appropriate period of darkness. Under these experimental conditions, you would expect
   (A) flowering in none of the plants
   (B) flowering only in plant number 6
   (C) flowering in plants numbered 4 to 6
   (D) flowering in all the plants
   (E) flowering in plant number 5

7. The major difference between the insect tracheal system and most other types of respiratory systems is
   (A) insects do not ventilate through their tracheal systems
   (B) insects do not remove CO_2 through their tracheal systems
   (C) insects exchange water and CO_2 through their tracheal systems
   (D) tracheal systems do not rely on the blood to transport oxygen to the tissues
   (E) the cell membranes of insects need not be moist for O_2 transport

8. The disorder of a system is measured by its
   (A) unbound energy
   (B) bound energy
   (C) entropy
   (D) energy
   (E) enteron

9. According to the first Law of Thermodynamics, energy may be converted from one form to another. However, the total energy of a closed system is always
   (A) increasing
   (B) decreasing
   (C) either increasing or decreasing
   (D) constant
   (E) variable at a given moment

10. The Hardy-Weinberg equilibrium principle supports all of the following statements EXCEPT
   (A) the proportion of alleles in a population tends to remain stable
   (B) the number of dominant and recessive genes for a given trait in a gene pool of a large population can be determined mathematically
   (C) for a given trait, the frequency of dominant alleles is greater than that of recessive alleles
   (D) for a given trait, the genotype of the heterozygous population can be determined mathematically
   (E) the genetic alleles for taste polymorphism reveal that there are more nontasters than there are tasters

11. The combination of a base and a sugar is known as a
   (A) molecular unit
   (B) nucleoside
   (C) nucleotide
   (D) protein base
   (E) base pair

12. Which of the following would be most likely to increase competition among the members of the vole population in a given area?
   (A) An epidemic of rabies within the vole population
   (B) An increase in the number of hawk predators
   (C) An increase in the reproduction of voles
   (D) An increase in temperature
   (E) An increase in the food supply

13. Genetic drift, which can lead to the loss of certain genes by chance, is particularly significant
   (A) in determining the fitness of a species
   (B) in the evolution of small populations
   (C) in determining the gene pool of a large population
   (D) in accounting for the total gene pool of a species
   (E) in determining the survival of mutations

14. Of the following, the carbohydrate that is NOT present in animal cells and tissues, except for rare exceptions, is
   (A) glucose
   (B) glycogen
   (C) fructose
   (D) cellulose
   (E) lactose

15. Tightly folded regions of the protein albumin indicate the presence of
    (A) amino acids with six-carbon rings
    (B) a denaturing agent
    (C) numerous hydrophobic amino acids
    (D) the sulfur-containing amino acid cysteine
    (E) a flattened, uncoiled alpha helix

16. Which molecule is most likely to have the greatest number of isomers?
    (A) Glucose
    (B) Valine
    (C) Glycerol
    (D) Steroid
    (E) Testosterone

17. Type O blood is known as the universal donor because its erythrocytes lack both antigens A and B and will not be agglutinated by any type of plasma into which they are introduced. It is true that
    (A) type O plasma contains anti-A and anti-B agglutinins
    (B) type O red blood cells contain isoagglutinogen A and isoagglutinogen B
    (C) individuals with type O blood can be transfused with blood that is genotypically I^AI^B
    (D) there is a genetic advantage to having type O blood
    (E) type O blood is determined by the presence of three alleles

18. The trees on a certain Caribbean island are misshapen, having no limb growth on the side of the tree battered constantly by the wind. The peculiar shape of these trees is caused by a factor in the ecosystem categorized as
    (A) adiabatic
    (B) abomasum
    (C) abiotic
    (D) abscissant
    (E) anaerobic

19. A true statement about the cells of the pericyle is that they
    (A) do not differentiate
    (B) form root hairs
    (C) control apical dominance
    (D) contain starch-storing plastids
    (E) none of the above

20. A person afflicted with the HIV virus contracts opportunistic diseases because
    (A) the immune system cannot make antibodies
    (B) phagocytic blood cells lose their ability to engulf microorganisms
    (C) T-cell lymphocytes lose the ability to reproduce
    (D) wild microphages consume white blood cells
    (E) bone marrow cells fail to reproduce

Questions 21–23

Questions 21–23 refer to the diagram below of the human digestive system.

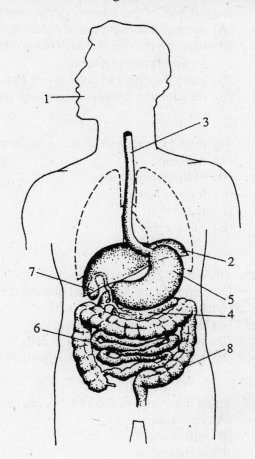

21. Which organs produce enzymes that digest protein?
    (A) 1 and 2 only
    (B) 5 and 6 only
    (C) 4, 5, and 6
    (D) 1, 3, and 7
    (E) 2, 3, and 8

22. Which organs do NOT secrete digestive enzymes?
    (A) 2 only
    (B) 1 and 5 only
    (C) 2 and 8 only
    (D) 2, 3, and 8
    (E) 4, 5, and 6

23. The accessory organs in the digestive system are
    (A) 2, 4, and 8
    (B) 1, 4, and 7
    (C) 5, 6, and 8
    (D) 1, 5, and 6
    (E) 2, 5, and 8

24. These materials are assembled at your laboratory station: frozen spinach, methyl alcohol, a hot plate, filter paper strips, a beaker, a large test tube with a cork, petroleum ether, a metric ruler, and a capillary tube. It is most likely that the goal of this laboratory exercise is to
    (A) determine the amount of chlorophyll in spinach leaves
    (B) measure the rate at which chlorophyll dissolves in a solvent
    (C) separate plant pigments
    (D) study the effect of chlorophyll in a closed system
    (E) measure the dissolving power of methyl alcohol

25. The main excretory structure in an earthworm is the
    (A) nephron
    (B) Malpighian tubules
    (C) loop of Henle
    (D) flame cell
    (E) metanephridium

26. The cuticle of plant cells
    (A) supports the xylem and phloem tubules
    (B) is secreted by the cell wall
    (C) is specialized for reproduction
    (D) waterproofs the epidermal cells
    (E) concentrates light for biochemical use

27. Small molecules of naked RNA are known as
    (A) episomes
    (B) viroids
    (C) retroviruses
    (D) transposons
    (E) centromeres

28. In humans, excess glucose is stored as a polysaccharide known as
    (A) glycerol
    (B) ecdysone
    (C) cellulose
    (D) glycogen
    (E) chitin

29. In many instances animal cells cannot synthesize part of carrier coenzyme molecules and must ingest these molecules as
    (A) hormones
    (B) polysaccharides
    (C) minerals
    (D) vitamins
    (E) amino acids

30. If the nucleoli in a eukaryotic cell were destroyed, the cellular activity that would be most affected is
    (A) the formation of the nuclear lamina
    (B) the appearance of the middle lamella
    (C) the organization of flagella
    (D) the synthesis of ribosomes
    (E) the storage of RNA

31. In female mammals, one X chromosome in each cell becomes inactivated during embryonic development. Which of the following conditions is caused by X-chromosome inactivation?
    (A) Sterility in mules
    (B) Patchwork coat of calico cats
    (C) Kwashiorkor in African populations
    (D) Hoof and mouth disease in cattle
    (E) Brachiation in monkeys

32. Of the following, which group is IMPROPERLY matched to its formula?
    (A) Hydroxyl—OH
    (B) Amine—NH_2
    (C) Sulfhydryl—SH
    (D) Phosphate—OPO_3H_2
    (E) Carbonyl—COOH

33. During daylight hours, phytoplankton are usually concentrated
    (A) in the limnetic zone
    (B) in the profundal zone
    (C) in the aphotic zone
    (D) below the thermocline
    (E) throughout the thermocline

34. The microtuble-organizing center of animal and plant cells is known as the
    (A) nucleosome
    (B) centrosome
    (C) centriole
    (D) nucleolus
    (E) centromere

35. The same alleles may have different effects on offspring depending on whether they reach the zygote by way of the egg or the sperm. This phenomenon may be explained by a process called
    (A) extranuclear inheritance
    (B) genomic imprinting
    (C) alteration of generations
    (D) sex linkage
    (E) reciprocal transformation

36. Bird migration is probably initiated by the effect of
    (A) changes in light intensity on wing muscles
    (B) changes in duration of darkness on endocrine glands
    (C) the direction of the prevailing winds
    (D) the increased temperature of the springtime sun
    (E) the availability of food

37. In general, parasites tend to
    (A) become more virulent as they live within the host
    (B) completely destroy the host
    (C) become deactivated as they live within the host
    (D) be only mildly pathogenic
    (E) require large amounts of oxygen

38. The tickbird-rhinoceros association is an example of
    (A) facultative commensalism
    (B) obligatory commensalism
    (C) obligatory mutualism
    (D) facultative mutualism
    (E) obligatory parasitism

39. The phylogeny of an organism refers to its
    (A) type of nutrition
    (B) habitat preference
    (C) evolutionary history
    (D) genetic composition
    (E) position in the food chain

40. A common feature of raw liver and raw white potato is that both contain an enzyme that
    (A) converts sugar to glycogen
    (B) takes part in oxidation activities
    (C) changes the color of Lugol's solution from amber to purple
    (D) converts polysaccharides into cellulose
    (E) changes lactate to pyruvate during anaerobic respiration

41. A compound unrelated to the substrate that binds to and alters a group in the active site is usually
    (A) a promoter
    (B) a terminator
    (C) an inhibitor
    (D) an activator
    (E) a producer

42. A cross between a horse and a zebra results in a sterile zebroid. The reason for sterility of this hybrid is BEST explained by
    (A) the failure of the chromosomes of the parent species to align properly during meiosis
    (B) the appearance of the fragile X chromosome in the zygote of the zebroid
    (C) incomplete fertilization caused by a time lag of the sperm penetrating the ovum
    (D) species incompatibility of endocrine secretions
    (E) the increased rate of nondisjunction during mitosis

43. Water is useful to living systems. Of the following, which is NOT a characteristic of water?
    (A) An excellent circulating medium
    (B) Easily compressed when squeezed
    (C) Almost a universal solvent
    (D) Relatively stable
    (E) Able to absorb a great deal of heat

44. Plastids bear a striking similarity to mitochondria in that plastids
    (A) store carbohydrates, fats, and proteins
    (B) synthesize green, red, and yellow pigments
    (C) are found in photosynthetic organisms
    (D) contain DNA, RNA, and ribosomes
    (E) serve the same function as mitochondria

45. During high light intensity, carotenoid pigments assist the process of photosynthesis by
    (A) reducing $NADP^+$ to $NADPH + H^+$
    (B) storing light energy
    (C) activating the phycobillins
    (D) slowing down the motion of excited electrons
    (E) preventing the destruction of chlorophyll *a*

46. The evolution of the Protista required the development of
    (A) a discrete nucleus
    (B) free ribosomes
    (C) functioning mitochondria
    (D) membrane bound organelles
    (E) an activated Golgi body

47. The theory that great disasters serve to maintain a population and its food supply balance was initially proposed by
    (A) Darwin
    (B) Wallace
    (C) Hardy and Weinberg
    (D) Malthus
    (E) Lyell

48. It is currently believed that the appearance of a new human species is becoming less and less likely and that the genetic evolution of *Homo sapiens* will progress very slowly in the future. Of the following, which statement supports this theory?
    (A) Modern science has made it possible to eliminate human mutations.
    (B) Modern drug therapy has decreased the incidence of human disease.
    (C) The male population has been destroyed by war.
    (D) Modern means of travel and communication have made geographic isolation a rarity.
    (E) Darwin's postulates do not apply to human beings.

49. During the breeding and egg-laying season, female red-cockaded woodpeckers store bone fragments instead of seeds in the barks of trees where they nest. From time to time, they retrieve these bones, peck at them, and ingest bone flakes. The most likely effect of this feeding practice is to
    (A) increase the number of eggs produced
    (B) add calcium supplements to the birds' diet
    (C) hasten the rate of egg production
    (D) strengthen the bones of the developing embryos
    (E) shorten the period of gestation

50. Many skeletal characteristics of primates indicate that they are basically adapted to living
    (A) in the air
    (B) in trees
    (C) in water
    (D) in swamps
    (E) on the ground

51. The upthrusting of a mountain barrier
    (A) is an instance of diastrophism
    (B) is an example of gradation
    (C) stimulates desert formation of the ocean
    (D) has no effect on the local climate
    (E) has never occurred

52. As a contributor to the many organic compounds found in animals, water is the
    (A) only source of both hydrogen and oxygen
    (B) major source of hydrogen and one of the sources for oxygen
    (C) major source for trace minerals
    (D) chief source of nitrogen, hydrogen, and oxygen
    (E) exclusive source of oxygen

53. Degradative methods to elucidate the structure of an enzyme do not offer final results. The investigator must
    (A) be able to duplicate his or her techniques
    (B) synthesize the molecule and show the same enzymatic activity
    (C) demonstrate the component fractions by means of chromotography
    (D) show how substrate molecules are deactivated
    (E) be able to convert The Active enzyme to its inactive form through operation of the lock-and-key theory

54. If the steps by which a virion is replicated are designated as follows: 1—synthesis of viral proteins; 2—fusion of virion envelope with cell membrane; 3—assembly of proteins; 4—removal of capsid; 5—release of virus from cell; 6—replication of viral RNA, which of the following represents the correct sequence?
    (A) 4—2—1—6—3—5
    (B) 6—4—1—3—5—2
    (C) 2—6—4—5—1—3
    (D) 5—2—1—4—6—3
    (E) 2—4—6—1—3—5

55. Which of the following contributes LEAST to speciation?
    (A) Sexual reproduction
    (B) Asexual reproduction
    (C) Selection
    (D) Variation
    (E) Isolation

56. Which theory explains evolution as a result of long, stable periods of time interrupted by geologically brief periods of major change?
    (A) Punctuated equilibrium
    (B) Gradualism
    (C) Natural selection
    (D) Reproductive isolation
    (E) Use and disuse

57. Hard pesticides and many metallic compounds are disrupting ecosystems because organisms in successive trophic levels of food pyramids
    (A) metabolize them rapidly
    (B) utilize them as coenzymes
    (C) excrete them as nitrogenous wastes
    (D) incorporate them into the ornithine cycle
    (E) concentrate them in their tissues

58. Organisms generally referred to as animals are NOT
    (A) heterotrophic
    (B) holotrophic
    (C) holophytic
    (D) autotrophic
    (E) parasitic

59. Several parts of the body are involved in the transmission of a stimulus. Which of the following represents the correct sequence as a stimulus is carried along the reflex pathway?
    (A) Spinal cord, sense organ, afferent neuron, efferent neuron, muscle/gland
    (B) Muscle/gland, efferent neuron, spinal cord, afferent neuron, sense organ
    (C) Sense organ, afferent neuron, spinal cord, efferent neuron, muscle/gland
    (D) Afferent neuron, sense organ, efferent neuron, muscle/gland, spinal cord
    (E) Efferent neuron, muscle/gland, sense organ, afferent neuron, spinal cord

60. A comparative study of intestines reveals that the LONGEST intestines are found in
    (A) vegetarians
    (B) snakes
    (C) herbivores
    (D) carnivores
    (E) omnivores

61. Plants without cambium are predominantly
    (A) deciduous
    (B) nonflowering
    (C) annuals
    (D) perennials
    (E) biennials

62. The apical root meristem differs from the apical stem meristem in that the former
    (A) has a higher rate of respiration
    (B) contains root hairs
    (C) has a high rate of cell division
    (D) produces the root cap
    (E) is converted into cambium

63. The person responsible for focusing attention on the study of "inborn errors of metabolism" was
    (A) Tatum
    (B) Beadle
    (C) Kornberg
    (D) Nirenberg
    (E) Garrod

64. Dikaryotic cells are characteristic of some classes of fungi. These cells have
    (A) a small nucleus and a lot of cytoplasm
    (B) doubly perforated cell membranes
    (C) a micronucleus and a macronucleus
    (D) nuclei of different genetic strains
    (E) numerous nucleoli

65. A protein may be denatured by
    (A) heat or heavy metals
    (B) action of enzymes
    (C) hydrolysis
    (D) dehydration synthesis
    (E) assimilation

66. A bacteriophage attaches to a host cell and injects into the cell its
    (A) protein coat
    (B) tail
    (C) DNA
    (D) RNA
    (E) DNA or RNA, depending upon the phage

Questions 67–69

Questions 67–69 are based on the diagram below, which shows the stages occurring during the process of cleavage, not necessarily in the correct order.

Diagram for questions 67–69

Morula
1

Early embryo
2

Zygote
3

Early
gastrula
4

2-cell stage
5

Blastula
6

67. In which of the following are the stages of cleavage listed in correct sequence?
(A) 1—2—3—4—5—6
(B) 5—3—1—4—6—2
(C) 3—5—1—6—4—2
(D) 1—3—5—6—4—2
(E) 3—1—5—2—6—4

68. Differentiation occurs during stages
(A) 2 and 4
(B) 1 and 3
(C) 3 and 5
(D) 4 and 6
(E) 3 and 2

69. The first stage of cleavage is
(A) 6
(B) 5
(C) 4
(D) 3
(E) 2

70. The only living monotremes are
    (A) echidna and tree shrew
    (B) platypus and kangaroo
    (C) spiny anteater and platypus
    (D) kangaroo and tree shrew
    (E) spiny anteater and kangaroo

71. Heredity of pneumococcus cells can be altered by adding DNA from other cells. This process is known as
    (A) induction
    (B) replication
    (C) mutation
    (D) duplication
    (E) transformation

72. Which of the following is a true statement about mitosis?
    (A) There is no duplication of chromosomes.
    (B) Two nuclear divisions occur during anaphase.
    (C) Neither synapsis nor crossing-over takes place.
    (D) Four identical daughter cells are produced.
    (E) Cytokinesis precedes nuclear division.

73. Information is transported from DNA molecules to proteins by way of
    (A) transferases
    (B) tRNA molecules
    (C) mRNA molecules
    (D) plasmagenes
    (E) ribosomal RNA molecules

74. Examination through the microscope of cells in the apical meristem of an onion root tip would enable you to see
    (A) reproducing chloroplasts
    (B) developing cork cells
    (C) crossed-over tetrads
    (D) mature root hair cells
    (E) stages of mitosis

75. Which of the following is true of the cells that form as a result of normal cleavage?
    (A) They are multinucleate.
    (B) They are polyploid.
    (C) They contain only one member of each homologous pair of chromosomes.
    (D) They are produced by synapsis and nondisjunction of chromosomes.
    (E) They have identical genetic material.

76. A gene pool is
    (A) the number of recessive genes available in an ecosystem
    (B) the sum total of all the genes of an individual
    (C) the genes that are available to a community
    (D) the sum total of all the genes of all the individuals in the population
    (E) the sum total of all the mutated genes of the various species that share an ecosystem

77. Inducible enzymes are those that
    (A) are not always present in the cellular fluid
    (B) activate substrate molecules
    (C) induce the formation of mRNA
    (D) are present in large amounts in the ribosomes
    (E) control genetic induction

78. The sandworm, *Nereis*, is classified as an annelid because it possesses
    (A) setae, a proboscis, closed circulatory system, anterior ganglion
    (B) parapodia, paired nephridia in each body segment, open circulatory system
    (C) dorsal ganglion, segmented body, nerve net, open circulatory system
    (D) paired nephridia in each body segment, ventral nerve cord, closed circulatory system
    (E) parapodia, nonsegmented body, nerve ring, mouth-anus digestive tract

79. A chemical reaction occurs spontaneously only if the amount of the free energy of the products is
    (A) the same as that of the reactants
    (B) more than that of the reactants
    (C) less than that of the reactants
    (D) twice as much as that of the reactants
    (E) augmented by an increase in activation energy

80. Translation, transcription, translocation, and termination are biochemical processes necessary for protein synthesis. Of the following, which is a TRUE statement about one of these processes?
    (A) Translation is the synthesis of tRNA and takes place just outside of the nuclear envelope.
    (B) Transcription is the synthesis of RNA taking place in the nucleus under the direction of DNA.
    (C) Translocation is a step in polypeptide building that occurs through a passive transport mechanism provided by the ribosome.
    (D) Termination is the final stage in transcription during which mRNA molecules flow into the cytoplasm.
    (E) None of the above.

81. Cytochrome is an electron carrier in the respiratory chain of all of the following EXCEPT
    (A) embryo cells
    (B) plants
    (C) animals
    (D) eukaryotic cells
    (E) prokaryotic cells

82. Which of the following are constituents of the cell membrane?

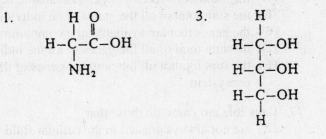

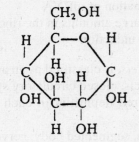

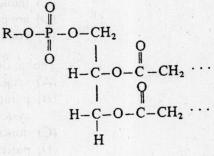

(A) 1 and 3
(B) 1 and 4
(C) 1 and 2
(D) 2 only
(E) 4 only

83. The gastric mill is the part of the stomach used for
(A) storage
(B) digestion
(C) excretion
(D) absorption
(E) grinding

84. Branching evolution, producing many branching lines of descent from a single ancestral source, is called
(A) divergent evolution
(B) convergent evolution
(C) straight-line evolution
(D) macroevolution
(E) adaptive evolution

85. The organelles most closely associated with the intracellular digestion of damaged cell components are
(A) peroxisomes
(B) lysosomes
(C) glyoxysomes
(D) dictyosomes
(E) mesosomes

86. The ability of RNA molecules to cross the nuclear membrane and enter the cytoplasm is best explained by
    (A) the spontaneous movement of molecules due to a concentration gradient
    (B) the ease in which they dissolve in water
    (C) facilitated diffusion during which RNA links to a carrier molecule
    (D) pores in the nuclear membrane that permit the passage of certain nucleotide sequences
    (E) transport vesicles that carry them across the membrane

87. The revolving-door hypothesis of the transport of material into cells can best be explained as
    (A) pinocytosis
    (B) osmosis
    (C) diffusion
    (D) facilitated transport
    (E) active transport

88. The osmotic pressure of a solution is proportional to the number of solute particles present per unit of volume. Which molecule is responsible for maintaining osmotic pressure in a guard cell?
    (A) Protein
    (B) Glucose
    (C) Linoleic acid
    (D) Phospholipid
    (E) Polysaccharide

89. Every multicellular living organism begins life as
    (A) a prokaryote
    (B) an embryo
    (C) a seed
    (D) a trophoblast
    (E) a single cell

90. The significance of the trochophore larva is that it
    (A) shows an evolutionary relationship between mollusks and annelids
    (B) develops spontaneously into an arthropod
    (C) links echinoderms and chordates
    (D) lacks a true mesoderm
    (E) has become extinct because of decreased adaptive potential

91. Mutations that can revert are known as
    (A) changelings
    (B) reversions
    (C) point mutations
    (D) recessives
    (E) lethals

92. During the photosystem II phase of photosynthesis, there is noncyclic transfer of electrons. A certain weed killer is able to block the transfer of electrons from plastoquinone to cytochrome b preventing the accumulation of H^+ inside the thylakoids. The plant dies because
    (A) light no longer strikes photosystem I
    (B) ATP production stops
    (C) O_2 becomes highly reactive
    (D) hydrogen peroxide is formed
    (E) light no longer strikes photosystem II

93. Viruses are
    (A) obligatory ectoparasites
    (B) facultative ectoparasites
    (C) facultative endoparasites
    (D) obligatory endoparasites
    (E) none of the above

No clue!

Directions: Each set of lettered items below refers to the numbered statements that follow. Select the one lettered choice that best fits each statement. A choice may be used once, more than once, or not at all.

Questions 94–96

    (A) T lineage precursor cell
    (B) Multipotential stem cell
    (C) B lineage precursor cell
    (D) Myeloid stem cell
    (E) Inducer T cell

94. The development of cells from this pathway takes place in the bone marrow.

A

95. This gives rise to memory and plasma cells.

D

96. This is the precursor cell of eosinophils, erythrocytes, and platelets.

E

Questions 97–98

T-A
G-C

    (A) Codon: AUG   Amino acid: Methionine
    (B) Codon: GUC   Amino acid: Valine
    (C) Codon: AAG   Amino acid: Lysine
    (D) Codon: UGU   Amino acid: Cysteine
    (E) Codon: CCA   Amino acid: Proline

T-U

97. The DNA code is TAC.   *AUG*

A

98. The anticodon is UUC.   *AAG*

C

Questions 99–101

    (A) Tropical rain forest
    (B) Temperate rain forest
    (C) Savanna
    (D) Taiga
    (E) Tundra

99. This biome is grassland with small trees and shrubs.

    C

100. In this biome, epiphytes grow in abundance.

    B

101. This biome has the greatest diversity of species.

    A

Questions 102–105

    (A) Pleiotropy
    (B) Polygenic inheritance
    (C) Multiple alleles
    (D) Sex linkage
    (E) Linked genes

102. This is the mode of inheritance of the ABO blood group.

    E

103. The absence of an enzyme causes three phenotypic defects.

    D

104. This is the mode of inheritance of hemophilia.

    D

105. This mode of inheritance is heterozygous for sickle-cell anemia and resistance to malaria.

    A

Questions 106–110

    (A) Endocrine system — Hormones
    (B) Nervous system
    (C) Both systems
    (D) Neither system

106. Homeostasis is maintained.

    A

107. Responses are of short duration.

    B

108. Chemical messengers are secreted.

    A

109. The plasma transport system is utilized.

    C

110. DNA replication is controlled.

    D

Directions: The groups of questions that follow are each based on a laboratory situation or an experiment. Read the description of the situation or experiment first. Then choose the one best answer to each question following it.

Questions 111–113

The Russian biologist G. F. Gause performed a series of experiments that involved the interaction of two protozoan species, *Paramecium caudatum* and *Didinium nasutum*. The latter feed on paramecia. In one experiment, didinia were introduced into a thriving paramecium culture. The figure shows the population sizes of both types of protozoa during a 16-day period.

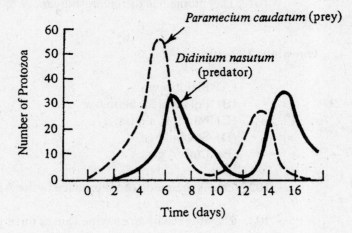

111. The figure indicates that there is an increase in the number of *P. caudatum* until
    (A) *D. nasutum* are introduced into the culture
    (B) *D. nasutum* markedly increase in number
    (C) *D. nasutum* outnumber the *P. caudatum*
    (D) *D. nasutum* begin to feed
    (E) *D. nasutum* reach old age

112. The nutritional relationship demonstrated by these experiments is properly labeled
    (A) parasite-host
    (B) symbiosis-commensalism
    (C) mutualism-commensalism
    (D) carnivore-omnivore
    (E) predator-prey

113. The figure shows that the number of didinia declined. The explanation for this is that
    (A) the paramecia were stronger than the didinia
    (B) the paramecia became immune to the didinia
    (C) the paramecia learned to avoid the didinia
    (D) as the number of paramecia declined, less food was available for the didinia
    (E) a biological conditioning took place

Questions 114–117

Questions 114–117 are based on the figure below, dealing with the seasonal fluctuations in the abundance of diatoms in the North Atlantic (solid black line-diatoms; dotted line-light intensity; dashed line-nitrates and phosphates):

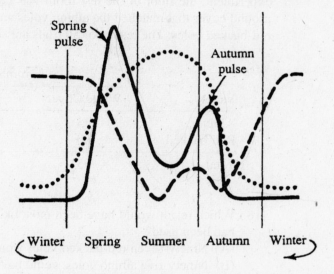

114. Which factor probably contributes to the summer decline of diatoms?
    (A) Decreased concentration of nitrates and phosphates
    (B) Increased concentration of nitrates and phosphates
    (C) Increased intensity of light
    (D) Decreased intensity of light
    (E) Increase of temperature

115. Which is probably the principal source of nitrates and phosphates?
    (A) The water cycle
    (B) Nitrogen fixation of lighting
    (C) Bacterial decay on the ocean bottom
    (D) Changes in environmental temperature
    (E) Changes in light intensity

116. A probable reason why the autumn pulse is not as great as the spring pulse is that
    (A) diatoms undergo metamorphosis in the autumn
    (B) temperature and light intensity decrease in the autumn
    (C) carnivorous animals increase in the autumn
    (D) the diatoms of the spring have used most of the proteins from the environment
    (E) bacteria of decay increase in the autumn

117. The low level of the diatom population in winter results partially from the fact that
    (A) low temperatures slow down metabolism
    (B) photosynthesis occurs only in summer
    (C) diatoms live in water
    (D) diatoms migrate to warm climates during the winter
    (E) in winter, there is a decrease in available nitrates and phosphates

Questions 118–120

Two varieties of the same species of voles (meadow mice), albino and red-backed, were used in an experiment. Both varieties were subjected to the predation of a hawk, under controlled laboratory conditions. During the experiment, the floor of the test room was covered on alternate days with white ground cover that matched the albino voles and red-brown cover that matched the red-backed voles. The results of 50 trials are shown in the following tabulation.

Variety	Number of Voles Captured		
	White Cover	Red-Brown Cover	Total
Albino	35	57	92
Red-backed	60	40	100
Total	95	97	192

118. Which result would have been most likely if only red-brown floor covering had been used?
    (A) Ninety-seven red-backed voles would have survived.
    (B) Ninety-five albino voles would have survived.
    (C) The survival rate of the albino voles would have decreased markedly.
    (D) A greater number of red-backed voles would not have survived.
    (E) There would have been no change in the results.

119. The results of this experiment lend credence to the concept of
    (A) the female of the species being more deadly than the male
    (B) use and disuse of organs and tissues
    (C) overproduction as a means of species perpetuation
    (D) adaptation permitting the survival of the species
    (E) spontaneous generation of lethal mutations

120. The purpose of the experiment was to
    (A) measure the visual acuity of hawks
    (B) decrease the vole population
    (C) compare the agility of the several varieties of voles
    (D) determine the basic intelligence of hawks and voles
    (E) compare the effects of protective coloration

Section II Free-Response #1

4 questions
1 hour and 30 minutes

Directions: Answer all four questions

1. Suppose that you have isolated an extract from a tissue and have found that the extract speeds up the rate of a particular reaction. What kind of information would you need to demonstrate that the substance responsible for increasing the rate of this reaction is an enzyme? Explain how this information would show that the catalytic effect is due to an enzyme.

2. The purpose of the immune system is to protect the body against disease caused by invading microorganisms and foreign proteins. In biochemical reactions characterized as the *immune response*, antigens, antibodies, and cells have specific roles.
    a. Describe the general function of antigens, antibodies, macrophages, and lymphocytes.
    b. Describe the structure of an antibody.
    c. Identify and describe the role of effector cells in the immune response.

3. When cells are crushed by centrifugation, the cell contents (organelles and inclusions) spill out into the lysing fluid and separate into various fractions. The heavier organelles form the bottom layers; the lighter ones form upper fractions (layers). Polyribosomes are heavier than the individual ribosomes that are bound to the membranes of the endoplasmic reticulum. Therefore discrete ribosomes and polyribosomes form different fractions. Discuss the differences in function of polyribosomes and bound ribosomes.

4. Self-regulatory, or homeostatic, feedback mechanisms are present in the respiratory, vascular, and endocrine systems of vertebrates. Describe one such feedback system, discussing evidence that indicates that the feedback occurs.

swers to Model AP Biology Examination No. 1
Section I

1. D	31. B	61. C	91. C
2. C	32. E	62. D	92. B
3. C	33. A	63. E	93. D
4. C	34. B	64. D	94. D
5. B	35. B	65. A	95. C
6. D	36. B	66. E	96. D
7. D	37. D	67. C	97. A
8. C	38. D	68. A	98. C
9. D	39. C	69. D	99. C
10. C	40. B	70. C	100. A
11. B	41. C	71. E	101. A
12. C	42. A	72. C	102. C
13. B	43. B	73. C	103. A
14. D	44. D	74. E	104. D
15. D	45. E	75. E	105. A
16. A	46. D	76. D	106. C
17. A	47. D	77. A	107. B
18. C	48. D	78. D	108. C
19. A	49. B	79. C	109. A
20. C	50. B	80. B	110. D
21. C	51. A	81. E	111. B
22. D	52. B	82. B	112. E
23. B	53. B	83. E	113. D
24. C	54. E	84. A	114. A
25. E	55. B	85. B	115. C
26. D	56. A	86. D	116. B
27. B	57. E	87. E	117. A
28. D	58. D	88. B	118. C
29. D	59. C	89. E	119. D
30. D	60. E	90. A	120. E

Missed 56

53%

$\dfrac{64}{120}$

Explanations to Model AP Biology Examination No. 1
Section I

1. D The sun's energy is the most abundant source of energy on earth. Green plants are able to trap this energy during photosynthesis, the food-making process.

2. C The mitochondrion is the organelle that serves as the site of cellular respiration.

3. C The portion of the tRNA molecule that contains bases complementary to the nucleotides of mRNA is called an anticodon.

4. C You know that the cycle begins with citric acid. (C) is the only choice in which citric acid comes first; therefore it must be the correct response. Notice that isocitric acid is next in this sequence—an additional clue that choice (C) is correct.

5. B A modern procedure for counting human chromosomes was developed by J. H. Tjio and A. Levan in 1956. They devised a technique for keeping human cells alive in tissue culture while treating them with special chemicals that cause the chromosomes to swell and spread out. Instead of the chromosomes being bunched and tangled as in the cell nucleus, they become separated for easy microscopic observation.

6. D The hormone named florigen produced by cocklebur plant number 6 in response to the appropriate period of darkness diffused across the grafted stems and stimulated flowering in all six plants.

7. D The tracheal system is a series of tubes through which gases pass into and out of the tissues of the insect's body.

8. C Entropy is the measure of the degree of disorder in a system where spontaneous reactions are occurring.

9. D Within a system, energy may be changed from one form to another, but the total amount of energy remains constant because no energy is lost.

10. C Gene frequencies within a population are quantitated and explained by the Hardy-Weinberg principle. However, the theory does not propose that one type of gene is more prevalent than another.

11. B A nucleoside is a part of a nucleotide, consisting of a sugar in combination with a base.

12. C An increase in the population of voles would increase the competition for food and space.

13. B In a small population, the loss of genes may become a critical factor. Since the number of individuals possessing a particular gene is small, a shift in the gene frequencies may eventually result in the loss of that gene.

14. D Cellulose is a polysaccharide that is synthesized by plants and used in cell wall formation.

15. D Dense coiling and folding proteins such as enzymes and albumin is brought about by interactions between amino acid R groups, such as the disulfide linkages that form between molecules of the sulfur-containing amino acid cysteine.

16. A Isomers are compounds that have the same chemical formulas but different structural ones. Glucose has several isomers: fructose, galactose, mannose.

17. A Persons with type O blood have neither antigen A nor B on the blood cells. The plasma, however, contains antibodies A and B. Because type O red blood cells lack antigens, these cells can be transfused into persons having type A, B, or AB blood without causing clumping (agglutination) of recipient cells.

18. C Abiotic factors are the physical, nonliving forces in the environment that affect living things.

19. A The cells in the pericycle remain meristemic and do not differentiate. They are capable of division as needed.

20. C The HIV virus attacks the T-cell lympocytes. Healthy T cells differentiate into T helper cells and T suppressor cells. The T helper cells encourage the work of the B lymphocyte. When the T cells cannot reproduce themselves, the immune system breaks down, leaving the body a ready target for opportunistic diseases.

21. C The pancreas secretes trypsin and chymotrypsin into the small intestine; these enzymes convert peptones into amino acids. In the stomach, pepsin converts proteins into peptones. Peptidases secreted by the small intestine digest peptones into amino acids.

22. D The liver secretes bile, which is an emulsifier, not an enzyme. Neither the esophagus nor the large intestine secretes enzymes.

23. B Accessory organs of digestion lie outside the alimentary canal. The liver, the pancreas, and the gall bladder are accessory organs.

24. C The purpose of this exercise is to separate the pigments in chlorophyll, which will diffuse from the spinach leaves in warm methyl alcohol. A concentrated drop of chlorophyll placed at the end of a filter paper strip, just barely allowed to touch the petroleum ether solvent, and then suspended in the test tube constitutes the process of protein separation known as chromatography.

25. E The metanephridium removes metabolic wastes in the earthworm.

26. D Water loss by transpiration is reduced by the cuticle that surrounds epidermal cells.

27. B Viroids, plant pathogens, are very small molecules of naked RNA. They are only several hundred nucleotides in length.

28. D Glycogen is a polysaccharide formed by the dehydration synthesis of excess glucose molecules. The reaction takes place in the liver.

29. D Some vitamins are coenzymes. This means that these molecules can donate or receive groups of electrons from a substrate while it is catalyzed by specific enzymes.

30. D Ribosomes are synthesized and assembled in the nucleolus. The nuclear lamina is the inner layer of the nuclear envelope. The middle lamella is a partition that divides newly formed plant cells. RNA is not stored by the nucleolus.

31. B In the female calico cat, the gene controlling hair color is carried on the X chromosome; one allele controls black hair and another, orange hair. A male cat (XY) can inherit one of these alleles but not both. A female calico cat is heterozygous, inheriting one gene for black fur and another for orange. However, during cleavage in the female, one of the pair of X chromosomes is randomly inactivated in each embryonic cell. In cats hybrid for hair color, cells making up the early embryo will carry an X chromosome for either black or orange hair. The patchwork coat of the calico cat is the result of a population of cells having a gene for black

color and a population of cells carrying the gene for orange hair color.

32. E COOH is the carboxyl group. The carbonyl group consists of a carbon atom joined to an oxygen atom by a double bond: $C=O$.

33. A Plankton are composed of plants, animals, and protists, namely, phytoplankton and zooplankton. Phytoplankton remain at the surface of ocean water, where they utilize the sun's energy for photosynthesis. In so doing, the phytoplankton provide oxygen and food for the zooplankton.

34. B The centrosome is the microtubule-organizing center of the cell. In many cells, microtubules grow out from a centrosome. The centrosome is located near the nucleus.

35. B Genomic imprinting is the process in which a particular gene is expressed only if it comes from the egg cell (maternal imprinting) or from the sperm (paternal imprinting). If a child inherits defective chromosome number 15 from the father, the result is Prader-Willi syndrome. If defective chromosome number 15 is inherited from the mother, the child inherits Angelman syndrome. These are two different disorders manifested in different ways.

36. B The migration of birds is dependent on a cycle referred to as the "biological clock." Migrating birds are responsive to the angles at which the sun's rays strike the earth and, perhaps, to periods of light and dark. A whole series of metabolic events, including hormone secretions, are initiated by diverse light patterns which then stimulate migrating behavior.

37. D In most cases, parasites do not kill the host immediately. (Of course, people do die from bacterial, viral, or worm infection.) It is not biologically sound for the parasite to kill the host. The parasite would then lack a home and a source of food.

38. D The tickbird and the rhinoceros can survive without each other. However, when in close association, they are mutually helpful.

39. C The phylogeny of an organism is its evolutionary history.

40. B The enzyme catalase is present in both liver and potato and functions in the oxidation activities in cells.

41. C A molecule that binds to the active site of an enzyme prevents enzyme activity. This is known as inhibition. The inhibiting molecule must be similar enough in structure to fit the enzyme's configuration in order to prevent the normal substrate from joining with the active site.

42. A When the horse and zebra mate, a sterile zebroid is produced. Sterility in the hybrid results from incompatibility of the chromosomes inherited from two different species. During meiosis, incompatible chromosomes fail to line up and segregate normally, forming abnormal, inviable gametes.

43. B Put some water into a plastic bag and then squeeze it. Note that it is not easily compressed.

44. D Plastids, like mitochondria, contain DNA, RNA, and ribosomes and are therefore able to replicate themselves. Choices A, B, and C refer only to plastids. Choice E is incorrect.

45. E During periods of high light intensity, O_2 molecules become partially reduced and form highly reactive radicals. These radicals can destroy chlorophyll *a*, but they are made inactive as they bind to the double bonds in carotenoid molecules.

46. D Protists differ from the Monera in that the former have membrane-bound organelles. The living plasma membranes that separate organelles from the cytoplasm carry out biochemical functions that are necessary to the life of the protist. The membranes are an evolutionary advancement.

47. D Thomas Malthus recognized that once the carrying capacity of an area was exceeded, organisms would die of starvation. Darwin's theory of evolution based on natural selection may have been influenced by the Malthusian concept.

48. D Speciation is made possible by the separation of gene pools of two or more populations. Population separation is affected by geographic barriers. Human beings are able to transcend land and water barriers, reducing the chance for further speciation of *Homo sapiens.*

49. B Bone contains the minerals calcium and phosphorus. This inborn feeding pattern enables the woodpeckers to replace the calcium used in shell formation.

50. B Characteristics of the primate that encourage a tree-dwelling existence are long arms, grasping fingers, and the opposable thumb.

51. A Diastrophism is a geological term applicable to the building up of land masses.

52. B Since there is no uncombined hydrogen in the atmosphere, plants obtain this element from water during the light reaction of photosynthesis.

53. B An investigator must be able to synthesize an enzymatic molecule and demonstrate its normal function to ensure the correctness of the structural formula he or she has derived.

54. E This is the correct sequence in which a virus particle is replicated.

55. B Asexual reproduction produces organisms that are genetically identical to the parent. Speciation is the result of genetic variability.

56. A Punctuated equilibrium is a theory of evolution that proposes that species are stable and remain unchanged for long periods of time. Speciation occurs when the stability is interrupted by brief periods of major changes.

57. E Each trophic level feeds on the one below it. Insecticides consumed by the herbivores will ultimately become concentrated in the tissues of the carnivores.

58. D Autotrophs are "self-feeders," such as green plants, that produce their food (organic nutrients) from inorganic materials.

59. C A stimulus is picked up by a sense organ and travels along an afferent neuron to the spinal cord. It leaves the spinal cord by way of an efferent neuron and ends in a muscle or a gland.

60. E An omnivore must have meat-digesting and plant-digesting enzymes. The longer length of the intestine provides for a greater number of intestinal glands that produce enzymes. The longer length of the intestine also provides for a greater surface area for absorption.

61. C Cambium is undifferentiated tissue required for secondary growth. Plants that complete the life cycle in one year (i.e., annuals) have no need for cambium.

62. D Meristem is undifferentiated tissue located at the tips of plant structures. Meristem at the ends of roots differentiates into cells of the root cap.

63. E Sir Archibald Garrod explained the cause for the genetic disorder alkaptonuria. He said that the disorder was caused by the lack of an enzyme. Nobel Prize winner Arthur Kornberg studied the chemistry of DNA synthesis and synthesized DNA for the first time. Marshal W. Nirenberg and Philip Leder determined the significance of tRNA. The one gene–one enzyme theory was formulated by George W. Beadle and Edward L. Tatum.

64. D In the club fungi, hyphae of different mating types fuse forming mycelia that are dikaryotic. Most of the life history of the mushrooms and its relatives is spent in the dikaryotic stage.

65. A The denaturing of a protein means that the bonds which maintain the coiled shape are broken. Heat and chemicals cause the chain to unfold, thus changing the configuration of the molecule. A denatured enzyme is useless.

66. E Viral DNA or RNA can replicate only within a living cell.

67. C The correct sequence is as follows: zygote, 2-cell stage, morula, blastula, gastrula, early embryo.

68. A Differentiation, the process in which the primary germ layers begin to develop into body tissues and organs, begins at the gastrula stage and continues throughout embryo development.

69. D The fertilized egg, called the zygote, represents the first stage of cleavage.

70. C Monotremes are egg-laying mammals. The duck-billed platypus living only in Australia and the spiny anteater (echidna) found in Australia and New Guinea are the only two living species of monotremes. Although egg laying, the mothers suckle the hatched young with milk from mammary glands.

71. E Transformation is a phenomenon in which certain species of bacteria, such as pneumococcus, when grown in the presence of killed cells, acquire some of the genetic characteristics of these strains.

72. C Neither synapsis, the pairing of homologous chromosomes, nor crossing-over, in which there is exchange of genetic information, takes place during mitosis. Synapsis and crossing-over occur during meiosis.

73. C An mRNA molecule, which is a replicated portion of a DNA molecule, acts as a template for ribosome synthesis of proteins.

74. E The apical meristem is a region of dividing cells. Meristemic cells are undifferentiated and divide to produce new cells near the growing regions of plants. Microscopic examination of this region would enable you to witness the stages of mitosis.

75. E Since cleavage is a series of mitotic cell divisions, all of the resulting cells are genetically alike.

76. D A gene pool contains all of the genes of a given population. All the genes in a gene pool are capable of being passed from one generation to the next.

77. A Inducers activate the genes that are necessary to produce enzymes. This type of enzyme is said to be inducible, not a regular constituent of the cellular fluid.

78. D Annelids are characterized by a segmented body that is bilaterally symmetrical, paired excretory organs (nephridea) in each segment, a closed circulatory system and a ventral nerve cord.

79. C Spontaneous chemical reactions occur when the change in free energy is negative. Such a reaction is exergonic. Energy is released. Thus the free energy of the products is less than that of the reactants.

80. B Transcription is the production of RNA under the direction and control of DNA taking place in the nucleus of the cell. A gene's specific sequence of DNA nucleotides serves as a template for assembling a complementary sequence of RNA nucleotides resulting in mRNA. Translation is the decoding of mRNA into a sequence of amino acids to form a polypeptide. Translocation is the movement of tRNA molecules from the A site to the P site on the ribosome. Termination is the final stage of translation in the synthesis of proteins by the ribosome.

81. E Cytochromes are respiratory enzymes in the membranes of the mitochondria. The cytochromes are not present in prokaryotic cells because of the absence of mitochondria.

82. B Cell membranes are composed of proteins (1) and phospholipids (4).

83. E Gastric mills are adaptations in the stomach of arthropods and are used for grinding.

84. A Evolution is the descent from a common ancestor. Divergent evolution is a branching out of species from a single source.

85. B Lysosomes are membrane-bound sacs containing hydrolytic enzymes. These enzymes digest large molecules not needed by the cell, nutrient molecules, and damaged cell components.

86. D Pores in the nuclear membrane that measure about 9 nm across permit the passage of RNA molecules if these molecules are arranged in specific nucleotide sequences.

87. E The revolving-door concept implies that there is an expenditure of energy as in active transport. When Na$^+$ ions leak out of a cell, K$^+$ ions leak into it. Energy must be expended by the cell to force these ions across the cell membrane against the concentration gradient.

88. B If you have looked at guard cells under a microscope, you will have seen the number of chloroplasts contained in them. Chloroplasts indicate photosynthetic activity. During the day, guard cells are filled with glucose—molecules responsible for maintaining the osmotic pressure of the cells.

89. E Fertilization—the union of egg and sperm—results in a single cell, the *zygote.*

90. A The trochophore larva of the mollusk is very similar to the larva of the marine annelid.

91. C A point mutation involves the substitution of one base for another in a nucleotide. It is very possible for this substitution to revert to the original base sequence.

92. B The energy stored in the H$^+$ thylakoid reservoir is used by enzymes to join ADP with the inorganic phosphate ion (P^1) to form ATP. The energy from ATP molecules is used during carbon fixation.

93. D A parasite is an organism that lives on or inside of a host organism and often does harm to it. Virus particles are obligatory endoparasites because they can reproduce only within a living cell, taking the DNA of the host cell and using it for the replication of viral DNA.

94. D The development of cells in the myeloid pathway takes place in the bone marrow. The multipotential stem cells give rise to lymphocytes, which develop in the thymus or in various lymphoid organs.

95. C Plasma cells and memory cells develop from a B lineage precursor cell in the spleen or in lymphoid organs.

96. D The myeloid stem cell is the precursor for red and white blood cells and for platelets. Eosinophils are a type of white blood cell; erythrocytes are red blood cells.

97. A The mRNA codon is AUG. Its DNA code must be TAC.

98. C The mRNA codon is AAG. The anticodon must be UUC. In RNA, thymine is replaced by uracil.

99. C A savanna is a grassland with small trees and shrubs. Savannas are located in subtropical and tropical regions.

100. A Epiphytes such as orchids and bromeliads grow on the surfaces of other trees. Epiphytes nourish themselves.

101. A A tropical rain forest has a greater diversity of species of plants and animals than any other community on earth. There are about 300 species of birds in about 2.5 acres (1 hectare) and about 30 species of frogs.

102. C *I^a*, *I^b*, and *i* are the three alleles for blood type ABO. The term *multiple alleles* indicates that more than one allele determines blood type ABO.

103. A *Pleiotropy* refers to the multiple effects that lack of an enzyme may have on the phenotype of an organism. The absence of the enzyme phenylalanine hydroxylase causes three major defects in a child: mental retardation, abnormal body growth, and pale skin pigmentation.

104. D The mutant gene that causes hemophilia is carried on the X chromosome. A male receiving a defective gene inherits hemophilia because there is no corresponding gene for blood clotting on the Y chromosome. Genes carried on the sex chromosomes are said to be sex-linked.

105. A The gene that causes cells to sickle has one bad effect and one good effect. Sickle cells can carry only a reduced amount of oxygen, which has a deleterious effect on the body. However, a person with one abnormal gene for defective hemoglobin develops resistance to malaria. Pleiotropy describes the multiple effects of a single gene.

106. C Both the endocrine system and the nervous system maintain homeostasis, that is, steady-state control of physiological reactions.

107. B Nerve responses are of relatively short duration.

108. C Both the endocrine and the nervous systems secrete chemical messengers. The endocrine glands secrete hormones; the nerve endings, neurotransmitters.

109. A Hormones manufactured by the endocrine system travel to target organs through the plasma transport system.

110. D Neither the endocrine nor the nervous system controls DNA replication.

111. B The population of *P. caudatum* is directly influenced by the presence of *D. nasutum,* as shown on the graph. There is a decline in the population of paramecia as the number of *D. nasutum* increases.

112. E This is definitely a predator-prey *(D. nasutum-P. caudatum)* relationship. Interestingly enough, if all of the paramecia disappear, the didinia will die for lack of food.

113. D As the didinia preyed on the paramecia, the number of paramecia decreased. Then, as less food became available, the number of didinia also declined.

114. A According to the graph, nitrates and phosphates are depleted by the summer growth.

115. C The recycling of materials to produce nitrates and phosphates is the function of organisms of decay.

116. B According to the graph, light intensity is a factor in autumn growth. It is to be assumed that temperature is also an important factor, since the data indicate the seasonal aspect of the cycle.

117. A There is an optimum temperature for metabolic activities. Low temperatures slow down the speed of chemical reactions.

118. C Albino voles stand out against a brown background. The increased visibility would have made the voles easy prey for hawks. Predation decreases population.

119. D Adaptations encourage the survival of the species in a given environment.

120. E The effects of protective coloration, as related to predation, can be evaluated by this experiment because two varieties of voles were subjected to contrasting backgrounds.

Answers to Model AP Biology Examination No. 1
Section II

1. To prove that an extract is an enzyme, the following procedures would have to be used.

Purification of extract reveals that the substance is protein in nature and has a three-dimensional structure. Purification is accomplished by salting out. The solubility of the protein is reduced by the addition of a high concentration of any salt. The extract is repeatedly subjected to the salting-out process until one fraction, which retains enzymatic activity, is obtained. The salt may be removed by dialysis. In this process crystals are obtained. These crystals may be purified further by chromatography and electophoresis.

X-ray diffraction permits examination or demonstration of the three-dimensional structure of the protein molecule. It also permits demonstration of the active site that determines specificity. A beam of X-rays strikes the protein crystal. A crystal is an external manifestation of the internal geometric arrangement of atoms. Part of the beam passes through the crystal, and part of the beam is scattered. The scattered beam impinges on a photographic film. The darker spots on the film represent a higher intensity of X-rays. The intensity of each spot is related to the position of various atoms in the crystals. By using a mathematical relationship involving sine and cosine (Fourier series), a three-dimensional picture is constructed.

A substrate can be added to the protein crystals. The substrate will diffuse into the crystals. The crystals can then be subjected to X-ray diffraction. The pattern of X-ray diffraction can then be compared to the original pattern made by the pure crystal. This method allows for the identification of the active site. The substrate will penetrate the active site.

The specificity of the enzyme can be determined through the addition of other molecules to the purified fraction. The lack of biochemical reactions would indicate the inability of the enzyme to catalyze the reaction.

The reusability of the enzyme can be determined as follows. After the purified extract interacts with the specific substrate molecules, the protein can be separated and purified by following the techniques already indicated above. The crystals can be added to a solution of substrate molecules and observations will indicate if the protein molecules are still capable of catalytic activity.

2. a. An *antigen* is any substance that evokes a specific immune response. Antigens are usually proteins, carbohydrates, or glycoproteins. Inert molecules such as carbon particles and fats are not antigenic. *Antibodies* are glycoproteins (molecules) produced by B-cell lymphocytes in response to invading microorganisms or molecules (antigens). Antibodies are specific, binding only to antigens that have their same molecular structure to form an antigen-antibody complex. *Macrophages* are phagocytic cells that circulate throughout the body. When a macrophage encounters a bacterium coated with antibody, the macrophage ingests the bacterium. *Lymphocytes* are produced in the blood-making tissues in the bone marrow. Lymphocytes that mature in the thymus form T cells. The function of T cells is to kill virus-infected cells.

b. An antibody is composed of four polypeptide chains connected to each other by disulfide. Two of the chains are small and are called light chains. The larger chains are called the heavy chains. A light chain is connected to a heavy chain at the amino terminal of the heavy chain called the Fab region. The Fc region is the carboxyl terminals of the heavy chains connected to each other. There are two kinds of light chains and five kinds of heavy chains. The five classes of antibody molecules are IgM, IgG, IgA, IgE, and IgD.

c. The effector cells in the immune system are composed of antibody-producing B cells and two kinds of T cells. The T helper cells stimulate the B cells to produce antibodies under certain conditions. When an antigen displayed on the surface of a macrophage is recognized as "self" by a T helper cell, the T cell becomes activated, divides, and produces more helper cells that stimulate the B cells to reproduce and thus produce more antibodies. The T cytoxic cells kill cells that are infected with viruses.

3. Polyribosomes consist of several ribosomes, numbering from five in some cells to as many as forty in others. Polyribosomes are bound together by an mRNA molecule and resemble beads on a sting. The various ribosomes in the polyribosome group read the mRNA molecule. Each ribosome moves along the entire length of the cistron of mRNA, from the beginning codon to the terminator codon. Then each ribosome drops off. At the same time, other ribosomes gather at the initiator codon and then travel along the mRNA. The various ribosomes in a polyribosome produce individual copies of the same polypeptide at the same time.

Bound ribosomes are attached to one side of the membrane of rough endoplasmic reticula. Polypeptides produced by free ribosomes are released into the cytoplasm. Polypeptides produced by bound ribosomes are moved across the ER membrane and released in the channel formed by the membranes. Polypeptides produced by ribosomes bound to the ER are either incorporated into the cell membrane, secreted from the cell, or become part of the hydrolytic enzymes found within a lysosome.

4. Feedback is a mechanism that contributes to homeostatic control of cellular activities. Glycolysis, the Krebs citric acid cycle, and the respiratory chain are controlled by allosteric monitoring by some of the enzymes involved. Allostery can be explained simply. When some products that are produced later in the respiratory cycle increase in concentration, this increase in concentration suppresses the activity of some enzymes that function earlier in the respiratory process. For example, the enzyme phosphofructokinase controls the phosphorylation of fructose-6-phosphate to the diphosphate. When a very little ATP is produced during anaerobic respiration, phosphofructokinase will work to speed up the phosphorylation of fructose-6-phosphate. When the amount of ATP is increased, the enzyme does not work.

A feedback mechanism regulates heart rate. During times of exercise, the right atrium becomes filled with a great deal of blood. The heart wall is stretched. Stretch receptors in the wall of the atrium send excitatory messages to the accelerating center in the medulla. Now the medulla sends more impulses to the

S-A node causing the heart to beat faster. Increased heart rate increases the blood pressure. Pressure receptors in the aorta respond to this increase of pressure by sending impulses to the inhibition center in the medulla. The medulla, in turn, sends inhibiting impulses to the S-A node. As a result, heart rate is slowed down. An increase of carbon dioxide in the blood stimulates the excitatory center in the medulla in much the same way as the stretch receptors.

Feedback control is demonstrated by the interaction between the anterior pituitary and the other endocrine glands. For example, thyrotrophic hormone is secreted by the pituitary gland when there is too little thyroxin in the blood. Thyrotrophic hormone stimulates the thyroid gland to release thyroxin. When the level of thyroxin in the blood increases, the pituitary is inhibited from releasing more thyrotropin. Thus the pituitary exerts influence over the thyroid and the thyroid exerts influence over the pituitary. The message from the pituitary to the thyroid stimulates; the message from the thyroid to the pituitary inhibits. This kind of relationship exists between the pituitary and other endocrine glands such as the adrenal cortex and gonads.

Evaluation of Model AP Test No. 1 Results

Below is a breakdown of the test questions by subject area. Place a check mark (✓) in the box below each question answered correctly, and write the total number correct for that area in the space provided at the right. Then figure out what percentage you answered correctly in that area. If, for example, you answered 10 correctly in the Cells section, then you had 66⅔% or ⅔, correct.

Finally, add up the numbers of correct answers for all the areas to get your overall score. Then determine your overall percent correct. You should have a minimum of 90 correct answers, or ⁹⁰/₁₂₀, or 75% correct. Concentrate your review and test preparation on the areas where your scores were poorest.

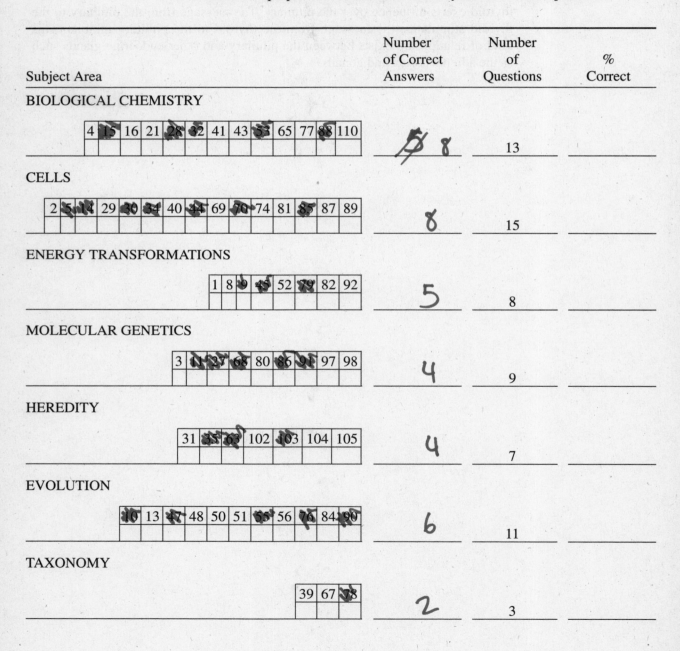

Subject Area	Number of Correct Answers	Number of Questions	% Correct
BIOLOGICAL CHEMISTRY 4 15 16 21 28 32 41 43 53 65 77 88 110	8	13	
CELLS 2 5 14 29 30 34 40 44 69 70 74 81 85 87 89	8	15	
ENERGY TRANSFORMATIONS 1 8 9 45 52 79 82 92	5	8	
MOLECULAR GENETICS 3 11 27 68 80 86 91 97 98	4	9	
HEREDITY 31 35 63 102 103 104 105	4	7	
EVOLUTION 10 13 47 48 50 51 55 56 76 84 90	6	11	
TAXONOMY 39 67 78	2	3	

MONERA, PROTISTA, AND FUNGI

22	46	54	64	66	93	111	112	113

8 _____ 9 _____

PLANTS

6	19	23	33	61	62

2 _____ 6 _____

ANIMALS

9	17	20	24	25	26	36	42	49	58	59	60	71	72	73

75	83	94	95	96	106	107	108	109	118	119	120

11 _____ 27 _____

ECOLOGY

12	18	37	38	57	99	100	101	114	115	116	117

8 _____ 12 _____

Total number of correct answers _____ 120 _____

PART TWO

Preparing for the Examination

CHAPTER ONE
How to Approach the Multiple-Choice Questions

Multiple-choice questions are called "objective" because the candidate's opinion is not asked for. Everyone taking the examination is expected to select the correct answer for each test item from the five choices given. For each multiple-choice question there is only *one* correct answer; there are no allowances for individual differences or personal bias. The answer that you choose is recorded by blackening the appropriate space on the separate answer sheet. There is a penalty of $1/4$ point for each wrong choice.

Although they may appear to be easy, multiple-choice questions are often difficult. You, the test taker, have limits set both in time and in content. You are given 1 hour and 30 minutes to answer 120 questions. Therefore you have $3/4$ minute for each question. You must be a thorough and rapid reader, and you must know the facts of biology. This test is designed to separate the superior students from the average ones. Remember that biology experts select the content; test-making experts refine the language of the question.

As you begin the study of Advanced Placement Biology, keep in mind that you must develop the skill of test taking in order to be successful. These skills are not acquired overnight but must be developed day by day. To do well on this part of the examination, you will have to:

1. Read proficiently—thoroughly, accurately, and quickly.
2. Become intimately familiar with biology terms.
3. Understand and obtain meaning from the tightly constructed questions of a college-level examination.
4. Have the facts of biology at your fingertips.
5. Use facts in problem-solving situations.
6. Use facts to make generalizations.
7. Exhibit flexibility in thought.
8. Be able to identify and recall the broad principles involved.
9. Use mathematical concepts to solve problems.
10. Use the tools of physics and chemistry to help solve biology problems.

The only way you can become skillful in answering objective questions is by consistent hard work. You must read your assignments with care and interest. The work that you do in the laboratory will also contribute to your reading. The notes you take must be useful for study.

Let us now look at some objective-type questions in order to determine the best way to answer them. Remember that you have 1 hour and 30 minutes in which to complete 120 questions. Keep these guidelines in mind:

- Read thoroughly, quickly, and accurately.
- Know the facts of biology.
- Select the BEST choice.

Questions of Fact

1. In which organ does meiosis take place?
   - (A) Lung
   - (B) Kidney
   - (C) Heart
   - (D) Ovary
   - (E) Ureter

2. Which prefix indicates the smallest unit of measure?
   - (A) Centi
   - (B) Milli
   - (C) Micro
   - (D) Nano
   - (E) Pico

3. A capsid is the external coat of a (an)
   - (A) firefly
   - (B) antibody
   - (C) leukocyte
   - (D) ribosome
   - (E) virus

4. A high content of digestive enzymes is contained in
   - (A) lysosomes
   - (B) pinocytic vesicles
   - (C) ribosomes
   - (D) chloroplasts
   - (E) nucleoli

5. Ligaments of the knee are composed of
   - (A) squamous epithelial tissue
   - (B) elastic cartilage tissue
   - (C) smooth muscle fibers
   - (D) striated muscle fibers
   - (E) elastic connective tissue

6. The nucleolus functions in the
   - (A) provision of energy to the cell
   - (B) synthesis of ribosomes
   - (C) synthesis of DNA
   - (D) secretion of enzymes
   - (E) manufacture of lipids

7. The organisms belonging to Class Hirudinea are
   (A) spiders
   (B) sea stars
   (C) leeches
   (D) tunicates
   (E) dragonflies

8. The medulla oblongata is a structure in the
   (A) kidney
   (B) brain
   (C) adrenal gland
   (D) eustachian tube
   (E) colon

9. Which of the following plants lack a vascular system?
   (A) Bryophytes
   (B) Club mosses
   (C) Ferns
   (D) Horsetails
   (E) Conifers

10. The area where a freshwater stream or river merges with the ocean is called a (an)
    (A) intertidal zone
    (B) biome
    (C) pond
    (D) pelagic zone
    (E) estuary

Answers for Questions 1–10

| 1. D | 3. E | 5. E | 7. C | 9. A |
| 2. E | 4. A | 6. B | 8. B | 10. E |

Analysis

The objective questions thus far presented encourage speed in reading and answering. Note that these questions to test your knowledge of biological facts are deliberately written with short stems and, with one exception, choices that are limited to one or two words or very short phrases. Questions of this type are interspersed throughout Section I, allowing the test taker a "breather," or mental relaxation, between questions that require thought and problem solving. But don't be fooled by short, well-constructed objective questions. You must have the facts at your command to answer these questions correctly in as short a time as possible.

Some students make the mistake of believing they can understand biology without memorizing certain basic facts. Actually, you must have a thorough background in biology in order to score well on the examination. If you have to puzzle over each of the short fact questions, you will lose valuable time; you will also be at a loss in problem solving and in answering questions of practical application.

Read and answer the four short questions of fact that follow.

---

11. A "brush border" enzyme is
    (A) enterokinase
    (B) nuclease
    (C) trypsin
    (D) sucrase
    (E) lipase

12. Examples of Ascomycetes are
    (A) yeasts
    (B) toadstools
    (C) rusts
    (D) smuts
    (E) downy mildews

13. The forests of the Mesozoic era were composed mostly of
    (A) angiosperms
    (B) conifers and cycads
    (C) large tree ferns
    (D) giant mosses and ferns
    (E) lichens and mosses

14. *Nereis* reproduces by means of
    (A) sporulation
    (B) binary fission
    (C) eggs and sperms
    (D) parthenogensis
    (E) budding

Answers for Questions 11–14

11. D   12. A   13. B   14. C

Analysis

These four questions illustrate several points. Although they are easy to read, are they easy to answer? You not only need to know facts but also must have at your command the specific vocabulary of biology.

Let us look closely at Questions 11 through 14. Question 11 is a question of fact: either you know the answer or you don't. Sucrase is the *only* correct answer. It is the only enzyme in the list that is bound in the epithelial lining or "brush border" of the small intestine. The other enzymes work within the lumen of the small intestine. Question 12 demands much more than just recall. Not only are you expected to recognize the class name Ascomycetes, but also you must be able to identify yeasts as the only example of this group. The toadstools, rusts, and smuts belong to the class Basidiomycetes and the downy mildews to the Phycomycetes. Question 13 is a "give away" *only* if you know that conifers are cone bearers and cycads are large tropical plants that have fernlike leaves. Can you just envision a Mesozoic forest? Question 14 requires that you identify *Nereis* as a sand worm or annelid. You can then select the correct answer—eggs and sperm—without hesitation.

A point worthy of emphasis is that a knowledge of the vocabulary of biology is essential. The Glossary at the back of this book will help you in this respect.

Questions of Comprehension

The approaches to multiple-choice questions differ. Some questions require just fast recall in order to make a comparison or judgment. Other questions that are more lengthy and involved demand more attentive reading. While you are getting the sense of such a question, you must also think through the choices offered and then select the *one* correct answer. Some of the key words or phrases used in comprehension questions serve as qualifiers: *how much, which, which is greater than, which is less than, which includes the other, to what extent, all of the following except.* Familiarize yourself with this type of question. You must be flexible in thought and sharp in comprehending quickly.

Study the questions that follow and their answers.

---

15. If the enzyme reverse transcriptase were denatured, which of the following events could NOT take place?
    (A) The release of carbon dioxide by yeast
    (B) The uptake of radioactive potassium by plant root hairs
    (C) Transcription of mRNA from a DNA template
    (D) Transcription of RNA from DNA by retroviruses
    (E) Transposition of the double helix

16. Experiments with the plant hormone phytochrome reveal that it
    (A) controls growth by activating auxins
    (B) informs the plant that light is present
    (C) signals the germination of seeds
    (D) regulates the ripening of fruit
    (E) prevents out-of-season leaf drop

17. Which is TRUE of a gene that is recessive?
    (A) It will usually code for hybrid vigor.
    (B) It more readily disappears from gene pools than its dominant allele.
    (C) It will appear more frequently in progeny than its dominant allele.
    (D) Its phenotypic effect will be apparent in the heterozygous condition.
    (E) Its phenotypic effect will be apparent in the homozygous condition.

18. The genetic basis for sickle-cell anemia has been traced to a change in a single nucleotide in a gene that codes for one of the proteins of hemoglobin. The amino acid, valine, is substituted for glutamic acid. Based on the information provided herewith, this type of mutation is BEST classified as a
    (A) base-pair transcription
    (B) point mutation
    (C) nonsense mutation
    (D) frame-shift mutation
    (E) mutagen

19. Of the following pairs, which CORRECTLY belong to the Kingdom Protista?
    (A) Cyanobacteria and water molds
    (B) Amoebas and mycoplasmas
    (C) Ciliates and golden algae
    (D) Diatoms and bryophytes
    (E) Spirochetes and chytrids

20. The genome in a prokaryotic cell differs from the genome in a eukaryotic cell in that
    (A) it is composed of RNA instead of DNA
    (B) it is not enclosed in a nuclear envelope
    (C) it has more protein associated with its DNA
    (D) it consists of a single strand of DNA
    (E) its ribosomes are smaller and function uniquely

21. In a favorable environment, a single bacterium will give rise to a colony of offspring very quickly as the result of
    (A) endospore formation
    (B) conjugation
    (C) transduction
    (D) binary fission
    (E) transformation

22. In which of the following functions is a fibrous root system MOST LIKELY to be more efficient than a taproot system?
    (A) Elongation of the root cap
    (B) Transport of fluids through the xylem and phloem
    (C) Anchorage in the soil
    (D) Food and water storage
    (E) Absorption of water and minerals from the soil

23. The return of salmon to their streams of origin to spawn is an example of
    (A) conditioned reflex
    (B) associative learning
    (C) olfactory imprinting
    (D) habit formation
    (E) operant conditioning

24. A scientist studying a virus particle notes that its protein coat has knobs that project outward, the inner core contains two RNA molecules, and each RNA has one molecule of reverse transcriptase attached to it. This virus is correctly classified as a
    (A) herpes virus
    (B) papilloma virus
    (C) polio virus
    (D) retrovirus
    (E) poxvirus

25. Which of the following is a TRUE statement about gray matter?
    (A) It is composed of cell bodies of neurons.
    (B) It is made up of several fat protective layers.
    (C) It is found only in the brain.
    (D) It is positioned outside the spinal cord.
    (E) It is located in the ventricles of the vertebrate brain.

26. Which of the following terms includes the others?
    (A) Disaccharide
    (B) Starch
    (C) Polysaccharide
    (D) Carbohydrate
    (E) Monosaccharide

27. A characteristic common to all algae is that they
    (A) reproduce with flagellated gametes
    (B) contain chlorophyll
    (C) form multicellular colonies
    (D) are prokaryotic
    (E) are heterotrophic

28. Of the following, which statement BEST characterizes mycoplasmas?
    (A) They are gram-negative bacteria that resemble spirochetes.
    (B) They are host cells to a number of plant and animal parasites.
    (C) Mycoplasma genomes code for at least 2,000 proteins.
    (D) The lack of cell walls makes them resistant to penicillin.
    (E) They are used commercially to make cheese and some fermented products.

29. Target cells are affected by hormones in all of the following ways EXCEPT
    (A) changes in cell permeability
    (B) increase in cyclic AMP concentration
    (C) changes in genome composition
    (D) changes in rate of metabolism
    (E) synthesis of different mRNA and proteins

Answers for
Questions 15–29

15. D	18. B	21. D	24. D	27. B
16. B	19. C	22. E	25. A	28. D
17. E	20. B	23. C	26. D	29. C

Analysis

Now that you have studied Questions 15–29, let us discuss the way in which you determine the answers to comprehension questions. First, read the question thoroughly, quickly, and accurately. Second, look for words that signal. For example, in Question 16 you are asked to identify the function of phytochrome. You should know that the prefix "phyto" refers to light and that "chrome" means color. Thus phytochrome has to do with colored light. Item B is the only choice that mentions light. This should bring to mind photoperiodism in plants and thus the function of phytochrome.

Now look at Question 18. You are asked to identify a mutation type. The information given is that valine is substituted for glutamic acid in the nucleotide of the gene that codes for one of the proteins of hemoglobin. A change in a gene is called a mutation. A chemical change in one nucleotide (a base-pair substitution) is a point mutation.

In Question 24, "reverse transcriptase" signals Item D as the correct choice. All retroviruses contain this enzyme. Note that the other choices are viruses that cause specific diseases.

Signal words or phrases help you to determine correct answers for comprehension questions. In Question 21, "favorable environment" and "quickly" serve as phrase and word signals. In an optimum environment of pH, temperature, nutrition, and salt concentrations, bacteria will reproduce "quickly." Choices B, C, and E are methods of genetic recombination that occur in the prokaryotes. Endospores are resistant cells that form in unfavorable environments. Binary fission is a method of asexual reproduction in which bacterial cells reproduce quickly forming a colony of cells.

Questions Requiring the Interpretation of Data

Interpretation questions present a set of facts or a series of conditions in a number of ways: written discourse, tables (word list or numerical notation), or diagrams. From the facts or experimental conditions given, you are called upon to demonstrate an ability to understand the meaning of the material.

Questions of interpretation have certain characteristics. They usually deal with experimental data and involve basic biological facts. The data are set forth in any of the ways mentioned above. The presentation of the data is necessarily more lengthy than in the usual multiple-choice question. By using the data, you must answer four to five multiple-choice questions that require knowledge of a given area of biology. In addition, these questions test your ability to draw correct conclusions based upon biological fact and experimental data.

Let us analyze some such questions.

THE DATA FOR QUESTIONS 30–34:

A normal female from a laboratory culture of *Mormoniella vitripennis*, a type of wasp, was mated with a male that was captured from the natural environment. All of the offspring from this mating had red eyes except one male, which had white eyes. This white-eyed male was used in a series of matings as shown below.

P_1 White-eyed male × Red-eyed female sisters
  Results: 65 males, red-eyed
        70 females, peach-eyed
F_1 Red-eyed males (65) × Peach-eyed females (70)
F_2 Results: 25% males, white-eyed
        25% males, red-eyed
        25% females, peach-eyed
        25% females, red-eyed

30. The best explanation for the appearance of a white-eyed male in the P_1 generation is as
  (A) an expression of a recessive allele
  (B) a manifestation of a mutation
  (C) a result of artificial laboratory media on developing larvae
  (D) a consequence of mating a laboratory wasp with a wild wasp
  (E) a culturing of offspring under laboratory conditions that are more restrictive than the natural environment

31. On the assumption that a single point on a chromosome determines the eye color, the gene seems to function as
  (A) a recessive allele
  (B) a dominant allele
  (C) an incompletely dominant allele
  (D) a translocation
  (E) a trisomic anomaly

32. The data reveal that the gene for this characteristic is physically located on the chromosome designated as
  (A) 15
  (B) 21
  (C) 1
  (D) X
  (E) Y

33. The best explanation for the fact that the F_1 males do not have white eyes is that the
    (A) gene for white eyes is latent in males
    (B) effect of the gene is diminished by an autosome
    (C) gene for white eyes, in males, is masked by the gene for red eyes
    (D) white-eyed trait is expressed in alternate generations
    (E) gene for white eyes is not present in F_1 males

34. In order to determine whether a white-eyed female can be obtained, an investigator would have to cross a
    (A) white-eyed male with a peach-eyed female
    (B) normal male with a peach-eyed female
    (C) red-eyed male with a peach-eyed female
    (D) white-eyed male with a red-eyed female
    (E) red-eyed male with a red-eyed female

Answers for Questions 30–34

| 30. B | 31. C | 32. D | 33. E | 34. A |

Analysis

These are fairly easy questions requiring a knowledge of the principles of simple Mendelian inheritance and the modifying exceptions of blending inheritance and sex-linkage. The data present you with necessary information: Two red-eyed wasps are crossed and produce normal offspring with the exception of one male that is white-eyed. Where did this white-eyed male come from? The data tell us that a cross between the white-eyed male and red-eyed sisters yields some females that are peach-eyed. This immediately indicates that blending inheritance has occurred between the two opposite alleles. If red × white results in peach eyes in females, then the original white-eyed male must represent a mutation in the gene for eye color. There is direct indication that eye color in the wasps is sex-linked, because white eyes are present only in the males. By following the reasoning presented, you should be able to answer the questions accompanying the problem.

Note: Data are useful only if you know the underlying biological principles.

Let us study another type of interpretation question—one that presents data numerically.

THE DATA FOR QUESTIONS 35–37:

A corn plant that has obtained full growth was subjected to chemical analysis for the purpose of determining its normal chemical composition.

Overall Composition of the Corn Plant

Water	79.7%
Organic matter	
Carbohydrate	17.2%
Fat	0.5%
Protein	1.8%
Total organic matter	19.5%
Mineral elements	0.8%
Total	100.0%

Elemental Composition of the Corn Plant

Carbon	44.58%
Oxygen	43.79%
Hydrogen	6.26%
Total C, O, H	94.63%
Nitrogen	1.43%
Potassium	1.62%
Phosphorus	0.25%
Calcium	0.59%
Magnesium	0.44%
Iron	0.10%
Sulfur	0.05%
Chlorine	0.20%
Sodium	0.15%
Silicone	0.54%
Total	100.0%

35. What percentage of the dry matter in the plant had its origin from the products of photosynthesis?
    (A)   0.8%
    (B)   19.5%
    (C)   20.3%
    (D)   94.63%
    (E) 100.0%

36. As the plant stood in the field, what percentage of its dry weight was due to inorganic salts?
    (A)   0.8%
    (B) 19.5%
    (C)   0.15%
    (D)   0.95%
    (E) 20.3%

37. One should infer from the elemental composition of the corn plant that
    (A) the source of the mineral elements in unimportant
    (B) although in small quantities, trace elements are important
    (C) sodium is of very little use to the plant
    (D) the rate of photosynthesis depends on the percentage of nitrogen present
    (E) substitutions can be made for iron in plant cells

Answers for
Questions 35–37

35. B   36. A   37. B

Analysis

To answer these questions, the first concept that you should recall is that carbohydrates, fats, and proteins have their "origin from the products of photosynthesis." The tabular material shows that these organic compounds make up 19.5% of the overall composition of the corn plant. The mineral elements in the inorganic salts are shown to comprise 0.8%. Although trace elements such as sulfur and iron are present in only minute amounts, they are vital to the proper growth of the plant.

Note: Refer solely to the data given in the problem.

Questions Requiring Mathematical Computation

The kinds of questions that lend themselves best to mathematical computation are those that are related to population genetics. You must familiarize yourself with the Hardy-Weinberg formula:

$$p^2 + 2pq + q^2 = 1$$

where p = one allele and q = the other allele, and 1 = the total gene pool for a given trait.

Brush up on basic mathematical computations so that you are able to apply the algebraic equation quickly and correctly.

THE PROBLEM FOR QUESTIONS 38–40:

It is known that vestigial wings in the parasitic wasp *Mormoniella vitripennis* are determined by a single recessive gene. In a selected population, the statistical figure of 10^{-4} of the adults were found to have vestigial wings.

38. According to the Hardy-Weinberg formula, what is the frequency of the recessive allele in the population?
   (A) 1.00
   (B) 0.99
   (C) 0.01
   (D) 0.0001
   (E) 1.99

39. Before natural selection can occur, what frequency of dominant alleles should be expected in the next generation?
   (A) 1.00
   (B) 0.99
   (C) 0.01
   (D) 0.198
   (E) 0.0198

40. The frequency of heterozygotes that will exist in the early stages of the next generation is
   (A) 1.00
   (B) 0.99
   (C) 0.01
   (D) 0.198
   (E) 0.0198

Answers for Questions 38–40

38. C   39. B   40. E

Analysis

Let us begin with the formula $p^2 + 2pq + q^2 = 1$. According to the information given, $q^2 = 10^{-4}$. Therefore q = the square root of 10^{-4} or $(10^{-4})^{1/2} = 0.01$. Once we know the value of q, we can compute the value of p:

$$p = 1 - 0.01 = 0.99.$$

Knowing the values of both p and q permits us to find the value

$$2pq = 2(0.99 \times 0.01) = 0.0198.$$

Note: Brush up on basic mathematical computations so you are able to solve algebraic equations quickly and correctly.

Handling Numbers and Quantitative Relationships

The modern biologist makes extensive use of numbers to explain biological phenomena and to express experimental data as accurately as possible. Therefore, numbers are used to measure and to count. You are urged to spend time each day learning the metric system of measurement (see pages 56–60 and Appendix A) and using metric units for problem solving.

Today's biologists quantify information. For example, some writers describe the cell membrane as being about 10 nanometers thick; others say that it is about 70 angstrom units (Å) thick. What is the relationship of these units to each other and to other metric units of measure? Would it help if you were given the information that 1 Å is equal to 10^{-8} cm? This introduces another problem of numbers.

In addition to measuring in metrics, you must gain facility in counting by using a system based on exponents. What does it mean when a biologist states that the mass of a proton is 1.67×10^{-24} grams? Is this a small number? If so, how small? Let us now direct our thoughts toward understanding and using numbers.

Exponents

Scientists use a counting system based on exponents. From your work in mathematics, you know that an exponent raises a number to a power. A power tells us how many times a certain number is to be multiplied by itself. For example, $2^2 = 2 \times 2$; $2^3 = 2 \times 2 \times 2$; $2^4 = 2 \times 2 \times 2 \times 2$. In biology it is common practice to use a counting system based on 10. Therefore we shall concern ourselves here with exponents to the base of 10, the decimal system.

A *positive exponent* moves the decimal point a corresponding number of places to the right. Expressing this in another way, we can say that the exponent is the number of zeros between 1 and the decimal point. The exponent indicates the number of times that 10 is to be multiplied, as shown in Table 1.1.

TABLE 1.1 POSITIVE POWERS OF 10

$$10^1 = 10$$
$$10^2 = 100$$
$$10^3 = 1{,}000$$
$$10^4 = 10{,}000$$
$$10^5 = 100{,}000$$

A *negative exponent* moves the decimal point a corresponding number of places to the left:

$$10^{-1} \text{ means } 0.1 \text{ or } \frac{1}{10}.$$

In other words, a negative exponent indicates that 1 is divided by a 10 with the corresponding positive exponent. Thus

$$10^{-2} = \frac{1}{10^2} = \frac{1}{100}.$$

In decimal form, this is written as 0.01. Therefore,

$$10^{-2} = \frac{1}{10^2} = 0.01.$$

The number of zeros to the right of the decimal point is always one less than the exponential number. For example,

$$10^{-4} = \frac{1}{10^4} = \frac{1}{10,000} = 0.0001.$$

Notice that a negative 4 exponent produces three zeros between the decimal point and the 1. A number with a negative exponent is always less than 1. See Table 1.2.

<u>TABLE 1.2   NEGATIVE POWERS OF 10</u>

$$10^{-1} = \frac{1}{10} = 0.1$$

$$10^{-2} = \frac{1}{100} = 0.01$$

$$10^{-3} = \frac{1}{1,000} = 0.001$$

$$10^{-4} = \frac{1}{10,000} = 0.0001$$

$$10^{-5} = \frac{1}{100,000} = 0.00001$$

Note: $10^0 = 1.$

Very large or very small numbers can be handled more efficiently and meaningfully when exponents are used. The number 32,000 is better written as 3.2×10^4. Since 3.2 is a number between 1 and 10, it has greater significance to us than does the number 32. We tend to comprehend small numbers better. In an earlier example the weight of a proton was given as 1.67×10^{-24}. Would this number have more meaning if written with 23 zeros? Try it. Write the following numbers in exponential form:

41. 6,900,000   _____

42. 0.000008   _____

43. 0.0000000245   _____

44. 4,500,000,000   _____

45. 0.000000001   _____

Answers for Questions 41–45

41. 6.9×10^6 42. 8×10^{-6} 43. 2.45×10^{-8}

44. 4.5×10^9 45. 1×10^{-9}

The Metric System

The metric system of measurement is based on multiples of 10. The unit of *length* is the meter. The unit of *mass* is the gram. The unit of *volume* is the liter. With each of these units the prefix that is used indicates the multiple(s) of the unit. The most commonly used prefixes are shown in Table 1.3.

TABLE 1.3   THE METRIC SYSTEM PREFIXES

Prefix Indicating More than the Unit	Prefix Indicating Less than the Unit
kilo (k) $= 10^3 = 1000$	deci (d) $= 10^{-1} = 0.1$
hecto (h) $= 10^2 = 100$	centi (c) $= 10^{-2} = 0.01$
deka (da) $= 10^1 = 10$	milli (m) $= 10^{-3} = 0.001$
	micro (μ) $= 10^{-6} = 0.000001$

Table 1.4 shows metric measures of *length*. Note that in biology you will use the millimeter and the centimeter quite frequently. Microscopic objects viewed under the light microscope are measured in micrometers, sometimes referred to as microns. Cellular structures as seen under the electron microscope are measured in angstrom units and nanometers.

TABLE 1.4   METRIC MEASURES OF LENGTH

Unit	Abbreviation	Equal to	Base 10
1 kilometer[a]	km	1000 m	10^3 m
1 meter[b]	m	0.001 km	10^{-3} km
1 centimeter	cm	0.01 m	10^{-2} m
1 millimeter	mm	0.001 m	10^{-3} m
1 micrometer	μm	0.001 mm	10^{-3} mm
		1/millionth m	10^{-4} cm
			10^{-6} m
1 nanometer	nm	0.001 micron	10^{-9} m
1 millimicron	mμ	1/billionth m	
1 angstron unit[c]	Å	0.1 nm	10^{-10} m
		1/10,000 μ	10^{-8} cm
		1/10 billionth m	

[a] English equivalent is 0.62 mile.
[b] English equivalent is 39.37 inches.
[c] Although being phased out, "angstrom unit" is still found in biological literature.

The metric unit of *mass* is the gram. One gram is about the mass of a green pea. In laboratory work that you do in school, you will measure things in terms of grams and milligrams. You may even use kilograms for large organisms. Experimental biologists engaged in biochemical work often measure in micrograms and nanograms. Tables 1.5 to 1.7 provide information on metric measures of mass, volume, and temperature.

TABLE 1.5   METRIC MEASURES OF MASS

Unit	Abbreviation	Equal to	Base 10
1 kilogram[a]	kg	1000 g	10^3 g
1 gram	g	1/1000 kg	10^{-3} kg
1 milligram	mg	1/1000 g	10^{-3} g
1 microgram	μg	1/1000 mg	10^{-3} mg
		1/millionth g	10^{-6} g
1 nanogram	ng	1/1000 μg	10^{-3} μg
		1/billionth g	10^{-9} g
1 picogram	pg	1/trillionth g	10^{-12} g

[a] English equivalent is 2.2 pounds.

TABLE 1.6   METRIC MEASURES OF VOLUME

Unit	Abbreviation	Equal to	Base 10
1 liter[a]	l	1000 ml	10^3 ml
1 milliliter	ml	1/1000 l	10^{-3} l
1 cubic centimeter	cc or cm^3	1 cc of H_2O at 4°C weighs 1 g	
1 microliter	μl	1/1000 ml	10^{-3} ml
		1/millionth l	10^{-6} l
		1 μl of H_2O at 4°C weighs 1 mg	

[a] English equivalent is 1.06 quarts.

TABLE 1.7   THERMOMETER REFERENCE POINTS

Reference Point	Degrees C	Degrees F
Freezing point of water	0	32
Boiling point of water	100	212
Standard room temperature	20	68
Standard body temperature	37	98.6

The metric unit of *volume* is the liter. No doubt you already know that the milliliter is almost identical to the cubic centimeter. Research scientists measure in microliters as well. Table 1.6 provides information on metric measures of volume. Note that Appendix A provides conversion tables that indicate how to change measures from the English system to the metric system.

Included in the metric system is the Celsius scale for measuring temperature. Study Figure 1.1, which compares the Celsius (centigrade) scale with the Fahrenheit. Note that the Celsius degree is relatively larger than the Fahrenheit degree. To compute degrees Fahrenheit, multiply the degrees Celsius by 1.8 (or $^9/_5$) and add 32. To compute degrees Celsius, subtract 32 from the degrees Fahrenheit and divide by 1.8 (or multiply by $^5/_9$).

$$C = \frac{(F - 32)}{1.8}$$

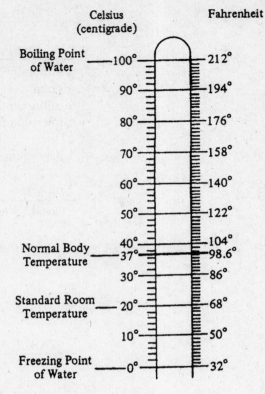

Figure 1.1.   Comparison of Fahrenheit and Celsius thermometers

$$F = (C \times 1.8) + 32$$

Convert to the unit indicated.

46. 1 kilogram   = _____ gram(s)

47. 50 dekaliters   = _____ liter(s)

48. 43 centigrams   = _____ gram(s)

49. 210 dekameters   = _____ meter(s)

50. 2 hectograms   = _____ grams(s)

51. 35 dekaliters   = _____ liter(s)

52. 31 kiloliters   = _____ liters (s)

53. 5 kiloliters   = _____ liter(s)

54. 1 nanogram   = _____ microgram(s)

55. 10 hectograms   = _____ gram(s)

56. 5 decigrams   = _____ gram(s)

57. 42 centiliters   = _____ liter(s)

58. 35 millimeters   = _____ meter(s)

59. 102 deciliters   = _____ liter(s)

60. 18 milliliters   = _____ liter(s)

61. 15 centimeters   = _____ meter(s)

62. 1 centigram   = _____ gram(s)

63. 85 decimeters   = _____ meter(s)

64. 1,000 milligrams = _____ gram(s)

65. 0.5 deciliter   = _____ liter(s)

Select the correct answer.

66. A scientist expresses 0.00001 as
     (A) 10^4
     (B) 10^{-4}
     (C) 10^5
     (D) 10^{-5}

67. 10^4 is equal to
     (A) 400,000
     (B) 40,000
     (C) 10,000
     (D) 100,000

68. 10^{-7} is equal to
     (A) 0.0000001
     (B) 0.00000001
     (C) 0.000001
     (D) $1/10^{-7}$

69. A scientist records 5,643,000 as
     (A) 56.43×10^6
     (B) 5.643×10^6
     (C) 5.643×10^8
     (D) 5.643×10^7

70. A nanometer is equal to
     (A) 10^{-3} millimeter
     (B) 10^{-3} angstrom unit
     (C) 10^{-3} micrometer
     (D) 10^{-3} meter

71. How many millimeters are in a centimeter?
    (A) 5
    (B) 1000
    (C) 100
    (D) 10

72. An angstrom unit is equal to
    (A) 10^{-10} cm
    (B) 10^{-4} μm
    (C) 10^{-3} mm
    (D) 10^{-9} m

73. The boiling point of water on the Celsius scale is
    (A) 0°C
    (B) 32°C
    (C) 100°C
    (D) 212°C

74. One microliter of water at 4°C weighs
    (A) 1 microgram
    (B) 1 milligram
    (C) 1 decigram
    (D) 1 centigram

75. 100 ml of water is approximately equal to
    (A) 0.1 m³
    (B) 1.0 ml³
    (C) 1 liter
    (D) 1 quart

Answers for
Questions 46–75

46. 1,000	52. 31,000	58. 0.035	64. 1.0	70. C
47. 500	53. 5,000	59. 10.2	65. 0.05	71. D
48. 0.43	54. 10−3	60. 0.018	66. D	72. B
49. 2,100	55. 1,000	61. 0.15	67. C	73. B
50. 200	56. 0.5	62. 0.01	68. A	74. A
51. 350	57. 0.42	63. 8.5	69. B	75. A

CHAPTER TWO
How to Approach the Free-Response Question

A free-response question is an essay question—an expository or writing question that requires the examination candidate to give an answer in declarative sentences. Although the free-response question deals with a specific segment of content, the test taker's answer should indicate an in-depth knowledge of the given topic. Unlike multiple-choice questions, the essay question evokes free response. This means that the test candidate is expected to bring individuality to the organization and expression of the answer. It is reasonable to assume that the responses to a given essay question will be as individual as the persons answering the question. Each writer will offer a presentation differing in approach to the problem, writing style, word usage, and content emphasis.

No matter how writing styles differ, however, the facts of biology remain the same. Free-response questions may ask you to give an explanation or a description or a discussion of a specific area of content. In any case, a knowledge of the **facts** of biology is necessary in order for you to write point-scoring answers.

Preparation for the Free-Response Question

Preparation for free-response questions logically begins at the start of your course in AP Biology and continues throughout the two semesters of work. You should become familiar with certain techniques that will help you to improve your skill in answering these questions.

Keep in mind what is required of you: you are to answer four essay questions in 1 hour and 30 minutes. One question will be taken from the content area Molecules and Cells (Chapters 4–6), and another question will be from the content area Genetics and Evolution (Chapters 7–9). Two questions will be based on the content area Organisms and Populations (Chapters 10–14). In addition, laboratory experiences will be incorporated into these questions.

To score well on Section II, you must:

1. Know the facts of biology thoroughly.
2. Be able to approach biology content on several levels: biochemical, genetic, evolutionary, ecological, morphological, anatomical.
3. Develop the ability to read and absorb meaning quickly and accurately.
4. Be able to formulate answers rapidly, taking cues from key words, ideas, and concepts.
5. Be able to express in writing what you wish to say.
6. Apportion your time so that you make the best use of each 22-minute time segment.

Learning the Terminology

All free-response questions use directional words that tell a candidate what to do. It is not unusual for a person to lose credit on an answer because of lack of understanding of a key directional word. We have studied many free-response questions and are listing below the directional words (and their meanings) that you must know before answering such questions. We suggest that you study this list, paying attention to subtle differences in meaning.

Cite   substantiate a statement of fact by an example or documentation of proof.

Compare   select characteristics of organisms or biological events and state the similarities and differences between the items.

Contrast   emphasize the dissimilarity in characteristics of organisms or biological events.

Construct   build a written "model," showing with concrete examples how a biological phenomenon can be accomplished.

Criticize   analyze a situation, using examples and other proof, to show why a situation could not be as stated.

Define   state the meaning in precise terminology. A definition must answer the question, "What is it?" Be specific in your statement. For example: "Photosynthesis is a series of biochemical processes in which...."

Delimit   set the limits of or mark the boundaries of a problem. You may be given a very broad problem from which you must identify and delimit smaller problems that contribute to the whole.

Describe   list the characteristics (shape, color, texture, size) of an organism or the sequence of activities in a biological event.

Diagram   make a line drawing or chart. (In the AP Biology Examinations, it has been customary for the candidate to answer questions based on a diagram that is presented.)

Discuss   present information that treats all sides of a question. You must give reasons from biological evidence for taking a given position.

Enumerate   list items, usually in outline form.

Evaluate   use data or illustrative examples to substantiate or disprove the problem situation that is given. An evaluation must not be based on one's "gut" feeling or a matter of hearsay or conjecture. To assess a biological situation, one must deal in facts and figures from which conclusions can be inferred.

Explain   answer the questions, "How does it work?" and "What does it do?" In a series of well-developed statements, you must clearly spell out your interpretation, based upon fact, of a biological event.

Illustrate   make a statement more specific and precise by citing evidence, giving an example, or drawing a diagram.

Indicate   point out the steps that affect a biological process, or show how a phenomenon occurs.

Interpret   give meaning to experimental data by translating and interpreting experimental observations into written discourse, using illustrative material.

Justify   give proof of an event, an occurrence, or a process.

List   make a list of the items asked for. The list may be a series of words, phrases, or sentences. However, if phrases or sentences are used, each should be sequentially lettered or numbered.

Outline   summarize a topic, relating major points to subordinate points by means of Roman and Arabic numerals, capital and small letters, and indentations.

Prove   show that something is true by presenting specific facts, data, and examples to substantiate the statement(s) given.

Relate   establish the cause-and-effect relationship of one item to another. By using concrete examples, show how one process or event is connected to or associated with another process or event.

Review   present a general account or survey of the topic given. Be careful of this word since it is not defined here as a critical discussion, as in English courses.

Specify   state explicitly as a condition for the result mentioned.

State   present a statement of fact without citing illustrative material or proof.

Summarize   give the main points of an event, process, or hypothesis.

Trace   describe the sequence of a process. In a historical sense, "trace" means to describe an evolutionary sequence.

Apportioning Time

Become familiar with the pattern. You must answer four mandatory questions in 1 hour and 30 minutes. Each question has equal weight.

Use the first 5 minutes of your total examination time to read *all* of the questions. If you can plunge into Question 1 or 2, do so. If you cannot, select one of the others for your beginning. Number your answer as the question is numbered.

As you reread the question you choose to answer first, jot down key words along the margin of your test paper. Write the number of the question in the box at the top. Pay attention to the directional words. Use 1 minute to get your thoughts together and then begin to write. Write in short, declarative sentences. Substantiate statements of fact with pertinent examples. As you write, think of accruing points: $1/2$ point for each fact; $1/2$ point for each example. You should aim to amass 10–16 points per essay.

Write continuously for 16 minutes. Since you have used 6 minutes for preliminary reading and decision making, use the remaining 2 minutes to reread and correct what you have written thus far. Now go on to the next question.

Use 2 minutes to reread and analyze the question and to jot down key words. Write for 16 minutes. Review and correct your answer for 2 minutes. Go on to the next question following this pattern. Complete all four questions.

By the time you have finished answering the fourth essay question, you should have about 5 minutes left before the end of the examination. Reread your entire paper. Add additional facts and examples where you can. Empty sentences do not gain points. Only facts and examples can be counted.

Analysis and Dissection of a Free-Response Question

We will use the topic of excretion to illustrate how free-response questions with various approaches may be derived from one content area.

Sample 1:
Comparative
Morphology

Describe the process of osmoregulation in three types of invertebrate excretory systems. Use examples of organisms that live (a) in fresh water, (b) in moist soil, and (c) on dry land.

Analysis

You must grasp immediately what you are asked to do. The directional word *describe* points out the approach to formulating your answer. The word *osmoregulation* tells you what you are to describe in terms of invertebrate excretory systems. Examples (a), (b), and (c) set boundaries to your description.

Dissection

In the time frame of 20 minutes, write the answer to this question. Remember to use short, declarative sentences. The maximum credit that you can earn for this question is 10 points. Allow $1/2$ point for identifying each organism, $1/2$ point for naming the excretory system, and 2 points for description. For these 2 points, describe the excretory structures (for 1 point) and tell what they do (for 1 point).

As you begin, place your watch onto the table where you can see the time at a glance. Note on your paper the time when you begin, the time when you should stop writing, and the end of the time period—for example; 4:00, 4:20, 4:25. Be sure to do this. It will enable you to allocate your time better.

Answer the question as directed. Do not include your opinion or evaluation. You are asked to *describe* and to *give examples* that will illustrate your description. For an item analysis of responses to this free-response question, see Table 2.1. The underlined items indicate where points are assigned.

TABLE 2.1   ITEM ANALYSIS FOR RESPONSES
TO SAMPLE 1 FREE-RESPONSE QUESTION

Describe the processes of osmoregulation in three types of invertebrate excretory systems. . . .	
Explanation	Point*
Osmoregulation is the removal of excess water by excretory systems of organisms that live in hypoosmotic environments.	$1/2$
(a) The excretory system of freshwater <u>flatworms</u> is the <u>flame cell system</u>.	$1/2$ $1/2$
This is a <u>branched system of tubules</u> that extend through the body of the flatworm. At the tip of each of the small tubules is a rounded <u>flame cell</u>. Water bathing the animal tissues diffuses into the tubular system through the flame cells. Beating <u>cilia in the flame cell push the fluid</u> into excretory ducts that empty to the external environment. The <u>excreted fluid is very dilute</u>. The flame cell excretory system keeps an <u>osmotic balance between the internal environment</u> of the flatworm and its external hypoosmotic external environment.	$1/2$ $1/2$ $1/2$ $1/2$ $1/2$
(b) <u>Earthworms</u> live in moist soil from which a great deal of water enters the body. The excretory system consists of tubules, known as <u>metanephridia</u>, with internal openings that collect body fluids.	$1/2$ $1/2$

TABLE 2.1 (Continued)

Explanation	Point*
The metanephridia occur in pairs in each of the earthworm's segments. The tubules are surrounded by capillaries. <u>Water collected from the body cavity enters the ciliated openings, called nephrostomes.</u> As water passes through the tubules, <u>salt is pumped out</u> and reabsorbed into the blood. Much dilute urine leaves the body by way of external openings called <u>nephridiopores.</u> The osmotic balance is thus maintained.	$^1/_2$ $^1/_2$ $^1/_2$ $^1/_2$
(c) <u>Grasshoppers</u> are examples of invertebrates that live on dry land. The excretory system consists of a system of tubules known as <u>Malpighian tubules.</u>	$^1/_2$ $^1/_2$
<u>Fluid from the body cavity is pumped into the tubules by transport epithelium.</u> Salts and <u>nitrogenous wastes</u> are also pumped into the tubules. The water, salts, and nitrogenous wastes are <u>emptied into the midgut and transported</u> to the hindgut. The <u>salts are then pumped back</u> into the body fluids by transport epithelium in the rectum. <u>Water follows these salts by osmosis.</u> The nitrogenous wastes are excreted as dry matter. <u>Osmoregulation</u> of the system occurs, preventing the <u>organism from losing water.</u>	$^1/_2$ $^1/_2$ $^1/_2$ $^1/_2$ $^1/_2$ $^1/_2$

Point Distribution. By using the system of $^1/_2$ point per fact, we have accrued 11 points for this question. Assigning $^1/_2$ point for each statement of fact or example is a good rule of thumb that enables you to estimate how much is expected of you for each question.

Sample 2: Comparative Physiology

Justify this statement: The nitrogen excretory product is correlated with the supply of water available to the animal.

Analysis

You are asked to *justify* or *prove* that the statement is true. In other words, the type of nitrogenous waste excreted is related to the water environment of the organism. Think in terms of the nitrogenous wastes excreted from the animal body. Immediately you should think of ammonia, urea, and uric acid and of the relationship of these compounds to water availability. No other nitrogenous metabolic wastes are excreted.

Dissection

In a time frame of 20 minutes, write the answer to this question. Assume that the total maximum number of credits for this question is 10. The answer to the question requires a *survey* of the animal kingdom—water-dwelling and land-dwelling organisms and *appropriate examples*.

TABLE 2.2 ITEM ANALYSIS FOR RESPONSE TO SAMPLE 2 FREE-RESPONSE QUESTION

The nitrogen excretory product is correlated with the supply of water....

Explanation	Point*
The following animals excrete ammonia:	
Most water-dwelling animals secrete nitrogenous waste in the form of ammonia. This molecule is small and	$1/2$
easily diffuses across permeable membranes.	$1/2$
Most of the aquatic soft-bodied invertebrates excrete	$1/2$
ammonia. It diffuses across the entire body surface	
into the surrounding water. Ammonia is toxic in	$1/2$
concentrated solutions and must be excreted in	$1/2$
lots of water.	
Amphibian larvae (tadpoles) excrete nitrogenous waste	$1/2$
in the form of ammonia. These organisms live in	
water until they metamorphose to the adult form.	$1/2$
In freshwater fish, the ammonium ion (NH_4^+)	$1/2$
diffuses across the epithelium of the gills into the	$1/2$
surrounding water.	
The following animals excrete uric acid:	
Reptiles, birds, and insects secrete their nitrogenous	$1/2$
wastes as uric acid. These animals live on land,	$1/2$
where conservation of water is necessary for life.	$1/2$
Uric acid is an almost solid (sometimes crystalline)	
nitrogenous waste from which most of the water	$1/2$
has been removed. It conserves water and	$1/2$
reduces toxicity.	$1/2$
The following animals excrete urea:	
Marine fish, adult amphibians, and mammals secrete	$1/2$
nitrogenous waste in the form of urea.	$1/2$
Marine fish lose water through the gills because the	$1/2$
body fluids are less concentrated than the seawater.	$1/2$
Thus fish living in the sea have excretory mechanisms	
for conserving water. The gills pump out salt. The	$1/2$
kidneys convert ammonia to urea.	
Amphibians that undergo metamorphosis change from	
excreting ammonia to urea.	$1/2$
In mammals, ammonia is converted to urea in the liver.	$1/2$
Urea can be stored in higher concentrations in a	
body where the water supply is limited.	$1/2$

*Point Distribution. With the $1/2$-point method, this question can accrue $11\frac{1}{2}$ points. You need a total of 10 points for a perfect score. The $1/2$-point system forces you to put down on paper, in an organized manner, as many facts as possible. Remember to use short, declarative sentences.

**Sample 3:
Biochemistry**

Describe the cyclical process of urea formation. Explain why urea formation is beneficial even though it requires much expenditure of energy.

Analysis

You are asked to give a *description* of the process through which urea is formed. Since this is a biochemical process, you are expected to *list the steps* necessary to form urea. The word "cyclical" in the question means that you must show a recycling of the process. You are also asked to *explain* why a process requiring so much energy is beneficial to the organism. Think about the low toxicity of urea. How does this benefit an organism such as yourself?

Dissection

In the time frame of 20 minutes, you are going to think, compose an answer, and then write it. The "describe" portion of this question is restrictive. You must know where to begin; a logical starting point is toxic ammonia. You must also know the facts of urea formation. Use the $^1/_2$-point method for assigning credits.

TABLE 2.3   ITEM ANALYSIS FOR RESPONSES TO SAMPLE 2 FREE-RESPONSE QUESTION

Describe the cyclical process of urea formation. Explain why urea formation is beneficial. . . .

Explanation	Point*
<u>Urea formation</u> takes place in the <u>liver</u>.	$^1/_2$
<u>Ammonia</u> + CO_2 + amino groups energized by <u>ATP</u> in the presence of glutamic acid form <u>carbamyl phosphate</u>.	$^1/_2$
Ornithine + carbamyl phosphate form <u>citrulline</u>.	$^1/_2$
Aspartic acid + ATP + citrulline form <u>argininosuccinic acid</u>.	$^1/_2$
Argininosuccinic acid splits into <u>arginine and fumaric acid</u>.	$^1/_2$
Fumaric acid is transformed back into <u>aspartic acid</u>.	$^1/_2$
Some arginine is converted into <u>urea</u>.	$^1/_2$
Some arginine is converted into <u>protein</u>.	$^1/_2$
Some arginine is converted into <u>ornithine, which initiates the cycle</u>.	$^1/_2$
Hence the <u>name ornithine cycle</u>.	$^1/_2$
The cycle is powered by <u>energy from ATP</u>.	$^1/_2$
It is worth using this energy to form urea, a <u>less-toxic nitrogenous waste</u>. It can be <u>stored in high concentrations</u> in the body of an animal that has a limited water supply. <u>Ammonia</u>, on the other hand, <u>destroys</u> body tissues. Urea can be <u>excreted in high concentrations</u>, thus <u>requiring less water</u>.	$^1/_2$ $^1/_2$ $^1/_2$ 1

*Point Distribution. By using the system of $^1/_2$ point per fact, we have accrued a total of 9 points for this answer. For previous questions, no mention has been made of point accrual for question organization. For a question such as this, requiring the statements of sequential biochemical events, however, a point might be given for logical organization. Remember that we are using the *half-point system*. The AP Development and Reading Committees may set the standard to grade each statement of fact as 1 point. Therefore a question such as this would reach the 10 point maximum easily.

Sample 4:
Evolution

Describe the evolutionary changes that have occurred in the vertebrate kidney. Include in your answer a brief discussion of the regulation of kidney function by hormones.

Analysis

The question asks you to *describe* changes in the vertebrate kidney over evolutionary time. A description of change involves comparison with simpler excretory systems that are functionally related to the modern vertebrate kidney. Although models of excretory mechanisms from invertebrate phyla may be appropriate to use, you will have to show a relationship of these mechanisms to the kidney of ancestral vertebrates. You are also asked to mention the *role of hormones* in your answer.

Dissection

In a time frame of 20 minutes, you are to write the answer to this question, which is restrictive. You cannot "talk around" the question and hope to earn a few points. Your response should include a description of the design and functions of the modern vertebrate kidney and a comparison with primitive types, as exemplified in developing vertebrate embryos. You are also asked to include a brief discussion of the regulation of kidney function by hormones. Ten points is the maximum credit for this answer. Your "discussion" of hormones should be a naming of them and a brief description of their functions.

TABLE 2.4   ITEM ANALYSIS FOR RESPONSES TO SAMPLE 4 FREE-RESPONSE QUESTION

Describe the evolutionary changes that have taken place in the vertebrate kidney. . . .

Explanation	Point*
The <u>functional unit</u> of the vertebrate kidney is the <u>nephron</u>.	$^1/_2$
The nephron consists of a <u>renal tubule</u> in close association with a knot of capillaries, known as the <u>glomerulus</u>.	$^1/_2$
Plasma diffuses out of the capillaries into the renal tubules, where <u>filtration of wastes</u> and <u>reabsorption of useful compounds</u> take place.	$^1/_2$
The ancestral kidney probably was a <u>segmented tubular structure</u>, extending from the coelomic cavity to the cloaca.	$^1/_2$
Each segment had a <u>ciliated</u>, funnel-shaped <u>protrusion opening into the coelom</u>.	$^1/_2$
The ciliated protrusions are known as <u>nephrostomes</u>.	$^1/_2$
The segmented, tubular structures are known as <u>metanephridia</u>.	$^1/_2$
<u>Excess water</u> emptied into the metanephridia and was <u>excreted</u> from the body of the water-dwelling organism <u>by way of the cloaca</u>.	$^1/_2$

TABLE 2.4 (Continued)

Explanation	Point*
Over evolutionary time, capillaries formed external glomeruli that protruded from the body wall into the openings of the metanephridia.	$^1/_2$
During later stages of kidney development, the renal tubule lost segmentation, while the openings disappeared.	$^1/_2$
A Bowman's capsule enclosed the glomerulus. The number of nephrons per kidney increased markedly.	$^1/_2$
There are approximately 1 million nephrons in each human kidney.	$^1/_2$
Renal tubules became structures of filtration and reabsorption.	$^1/_2$
In addition to glomeruli, another set of capillaries extends along the length of each nephron, permitting further exchange with the blood stream.	$^1/_2$
Evolutionary modifications in the kidney resulted in an organ that functions dually to conserve water and to excrete nitrogenous wastes.	$^1/_2$
In marine fish, the number of glomeruli is reduced and in some cases, glomeruli are nonexistent. Urine is formed by secretory functions of the renal tubules.	$^1/_2$
The gills excrete salts.	$^1/_2$
In the mammal kidney, the loop of Henle is a water-conserving device that permits excretion of urine that is hyperosmotic to kidney tissue.	$^1/_2$
Developing vertebrate embryos demonstrate the various stages of kidney development.	$^1/_2$
Antidiuretic hormone (ADH) is secreted by the posterior pituitary gland in the brain.	$^1/_2$
ADH increases the reabsorption of water from the urine.	$^1/_2$
Aldosterone is secreted by the cortex of the adrenal	$^1/_2$
gland. It promotes the reabsorption of sodium by the renal distal tubule.	$^1/_2$

Point Distribution. The answer to this question demonstrates your grasp of a rather specific, limited segment of biology. A total of $11^1/_2$ points has been accrued.

IN SUMMARY: To achieve your highest score on the free-response questions, you must:

- read each question thoroughly;
- know the facts, concepts, and principles of biology well;
- vary your approach appropriately to particular questions.

Be sure to study biochemical, ecological, genetic, evolutionary, morphological, and anatomical concepts.

CHAPTER THREE

How to Improve the Writing of Answers to Free-Response Questions

In Chapter 2, techniques were given for a more effective approach to answering free-response questions. The suggestions offered concerned the mechanics of response: observing directional words, point accrual, time allotment. With these items in mind, you must work to develop other dimensions of writing that will improve the content and quality of your answers, namely, vocabulary, presentation, and application.

Vocabulary

As you proceed in the AP Biology course, direct your attention to the ways in which words are used in textbook presentations, in journal articles, and in lectures. In other words, you are going to make a conscious effort to improve your vocabulary. A section of Chapter 2 was devoted to a review of directional words that occur in essay questions. In addition to these question markers, you must become familiar with language that denotes preciseness in description and presentation. The vocabulary of biology can be learned through the subject matter. However, you must be acquainted with a host of nonbiological words so you can derive full meaning from well-written essay questions. An expanded vocabulary will aid you in writing answers of more acceptable quality. Just as a person's thoughts are limited by lack of language, so are they expanded by increased word power.

The following question appeared on an AP Biology Examination.

> All living cells exploit their environment for energy and for molecular components in order to maintain their internal environments. Describe the roles of several different membrane systems in these activities.

This, seemingly, is a straightforward question requiring information that is basic to the study of biology. However, some students did not get the full import of what the question asks. Specifically, words such as "exploit," "molecular components," and "membrane systems" did not convey to these students precise and complete conceptual meanings. Before reading further, do the following exercises:

1. Explain the meaning of the word "exploit" as it appears in context in the question above.
2. What is the difference between a molecule and a molecular component?
3. What relationship exists between a membrane and a membrane system?

How do your answers compare with these?

Exploit means to utilize toward one's own advantage. Therefore, cells "raid" the environment for energy necessary for maintaining homeostasis.

A *molecular component* is a part of a molecule; one of the elemental units from which a molecule is built. A molecule is the smallest part of a compound that retains the properties of the compound. Therefore a molecular component is not a molecule.

A *membrane system* indicates that a series of membranes or membranous components are utilized to effect some special activity. The term "membrane system" implies something more than a membrane. There is the implication of structure that accomplishes specific function.

You can estimate your understanding of this question by completing the two exercises that follow:

1. Use your own words to explain what this question asks you to do. Write out this explanation.
2. Within a time allotment of 20 minutes, write a complete answer to the question. Try to aim for 10 credits based on $1/2$ point for each statement of fact or substantiating example. For an item response to the question, see Table 3.1.

TABLE 3.1   ANSWER TO SAMPLE FREE-RESPONSE QUESTION

Several different membrane systems use the cell's energy and molecular components to support the biochemical activities of the cell. A description follows:

Membrane	Point*
The plasma membrane regulates the entrance and exit of solvents, gases, ions, and organic molecules.	$1/2$
This is accomplished by diffusion and osmosis.	$1/2$
Diffusion is the movement of molecules along a gradient from high to low concentration.	$1/2$
Osmosis is the movement of water across a membrane.	$1/2$
The arrangement of phospholipids and proteins makes possible the passive movement.	$1/2$
Proteins form pores that aid the movement of ions.	$1/2$
Neutral gases and water are helped to cross the membrane by its hydrophobic and hydrophilic regions.	$1/2$
All of the above transport methods are classified as passive transport because they do not require the cell's energy.	$1/2$

TABLE 3.1 (Continued)

Membrane	Point*
Active transport is the movement of molecules against the gradient from low to high concentration.	$^1/_2$
The sodium-potassium pump mechanism, which is attributed to the arrangement of proteins in the plasma membrane, requires an expenditure of ATP.	$^1/_2$
Active transport is associated with polarity across the plasma membrane, which becomes operational during muscle contraction and nerve impulse movement.	$^1/_2$
The mitochondrial membrane is essential to the production of ATP.	$^1/_2$
The electron transport chain is an integral part of the protein arrangement in the membrane system of mitochondria.	$^1/_2$
The proteins are arranged in a specific way to carry electrons to oxygen.	$^1/_2$
For every two electrons that move through the chain, three molecules of ATP are produced.	$^1/_2$
The chloroplast membrane is essential to the biochemical activity of photosynthesis that takes place in the chloroplast.	$^1/_2$
The protein arrangement in the membrane permits the chloroplast to utilize sunlight for the production of carbohydrate and ATP.	$^1/_2$
The endoplasmic reticulum is a series of membranes that separate the cell into various compartments, making possible a number of biochemical activities.	$^1/_2$
The concentrations of certain molecules differ in the endoplasmic reticulum and in the cytoplasm. Thus a concentration gradient is established between these two structures.	$^1/_2$
Likewise an ionic gradient is established between the membranes of the endoplasmic reticulum and the cytoplasm.	$^1/_2$

*There are 10 points on which to score in this question.

No one can anticipate the specific nonbiological terms that test writers use in structuring questions. Theirs is the task to choose words that are precise and pertinent. It is impossible for your AP Biology teacher to drill you in all general terms that might appear on the examination. You must accept the responsibility of vocabulary building by the means available to you; these include extensive reading, active learning of new words, and incorporation of these words into your speaking and writing vocabulary. It is recommended that you obtain a copy of a vocabulary improvement book. Diligent study of its contents will reap dividends in your improved ability to understand and to communicate.

Presentation

Answering essay questions is not creative writing. The test taker does not invent fictional situations but is required to demonstrate knowledge of factual material substantiated by appropriate examples.

Too many students experience difficulty in recording facts on paper because of deficient vocabulary and lack of formal drilling in sentence construction. A most important aspect of the presentation of information is the careful learning of content facts. Command of vocabulary, knowledge of the rules of sentence structure, and fact facility are the ingredients necessary to produce a well-written paper.

"Presentation" means the orderliness of thought that must be conveyed by the written response. A satisfactory answer to an essay question can be judged on such criteria as the following:

1. Are you answering the question asked?
2. Are your ideas presented clearly and logically?
3. Is there evidence that you understand the basic principles of biology?
4. Do you apply these biological principles intelligently in problem-solving situations?
5. Have you shown evidence of more than surface mastery of the content?
6. Are the examples you cited appropriate and illustrative?

Application

The problem situations in biology with which you may be confronted on an AP examination are limitless. Any repetition of essay questions is unlikely because the subject matter of biology is versatile and lends itself to varying treatment. However, each biological discipline has a core of basic information that is fleshed out by pertinent examples and by exceptions to the rules. It is not enough for you to learn the facts of biology; you must also know how to apply these facts to the solutions of problems. This is what is meant by application.

It is generally accepted that young people learn well from each other and from example. By using an essay question that appeared in an AP Biology Examination, we present an analysis of the answers of three students, together with model answers to each part of the question with point values noted in brackets in the answers.

Sample Essay
Question

> Regulation of biological systems is commonly achieved by means of feedback control. In each of the following systems, describe how feedback control is used for regulation, and give a specific example for each system.
>
> (1) The size of a population.
> (2) The rate of a physiological process.
> (3) The rate of an enzyme reaction.

Student A Response

Part (1) Feedback
Control and
Population Size

The size of a population is regulated by a feedback system. The system is actually the death and birth of the members of a population. If the individuals would only be born and never dies there would be no feedback; and if the individuals only dies and were never born, there would be no feedback mainly because eventually there would be no one to feed back to. They would go extinct. Death provides a feedback which is relatively stable if the environment of the population is not disturbed greatly. The actual feedback is the animal being born—lives and eats and removes wastes—dies—decomposes—and this decomposition can either provide food to other animals or be utilized back in the earth to help grow plants and trees. Many minerals are left by the decomposition of a body.

Analysis of
Student A Response

Student A does not understand the concept of feedback as a biological phenomenon. An individual is not born in response to another's dying. Neither does death take place as the result of birth. Not only is there a deficiency in the understanding of a biological principle, but also there is poor mastery of words. Something does not "go extinct." The thinking is imprecise and disorderly. (The underlined items indicate questionable word usage and weak grammatical construction.)

The rating of this answer might be as follows: $^1/_2$ point for reference to death and birth of members of a population.

Student B Response

Part (1) Feedback
Control and
Population Size

Feedback control plays a basic and fundamental role in regulating the size of a population. Food, its amount and availability, has a major role in determining the size of a population. The more food is available in a certain community, the larger would be the population. But this increase in number of individuals does not continue indefinitely for the simple reason that as soon as the number of individuals increases in a certain environment, the amount of food that each individual in this population can obtain decreases, which in turn limits the number of individuals that can successfully survive and reproduce in this community. Thus the amount of food has a negative feedback control on the size of any population. The space available for a certain population in which this population can grow and survive has also a feedback control on the size of that population.

In a forest, for instance, the number of individuals of a certain species of tree that would be found is determined, in part, by the space available in which these trees can grow. Thus the number of trees in this community is checked by the space available for the growth of these trees.

Analysis of
Student B Response

Student B has attempted to show the operation of feedback control between population and available food. The point is made that population size will not increase indefinitely because the situation will reverse itself, that is, the available food will not be able to support the increased population. The student cited another example of feedback in terms of tree growth in the forest. However, this idea was not developed.

The weakness in this answer is that the information presented is too general; specific examples are omitted.

This answer might be scored as follows: $^1/_2$ point for including food as a feedback mechanism; $^1/_2$ point for the explanation of population versus food; $^1/_2$ point for the example of trees and space.

Student C Response

Part (1) Feedback Control and Population Size

Regulation of biological systems is commonly achieved by feedback control. Feedback controls normally regulate many systems on a collective and also on an individual basis. One such feedback system is on population control. This system is operated by the populations of prey and predators, keeping their respective populations fairly constant. As the prey organisms begin to increase in size, the predators begin feeding on them more and more. This prompts more reproduction among the predators.

As the predators continue to eat and as their population rises, the prey population begins to decrease. Now the predators do not have enough food and their numbers begin to decline. As this happens, the prey population again begins to increase due to less predation, and the cycle begins again. The cycle is exhibited by the arctic blue fox and the arctic hare, and it keeps the populations in equilibrium.

Analysis of Student C Response

First sentence wastes time. It does not accomplish points. Second sentence does not gain points. It is too general to have meaning.

However, much of this answer deals in specifics. The regulation of the sizes of predator and prey populations is achieved by means of feedback control, as exemplified by the arctic hare and arctic fox. "Size" in the fifth sentence refers to size of population, not size of individual.

The answer might be scored as follows: $\frac{1}{2}$ point for the selection of prey and predator populations; $\frac{1}{2}$ point for stating that the size of prey population affects size of predator population; $\frac{1}{2}$ point for stating that the increase in predator population decreases prey population; $\frac{1}{2}$ point for stating the cyclical nature of population control; $\frac{1}{2}$ point for the example of fox and hare.

This is a much better answer in terms of written expression and content. However, there are weaknesses in sentence structure and the presentation of ideas.

Model Response X

Part (1) Feedback Control and Population Size

Population density studies of the snowshoe hare and its predator [$\frac{1}{2}$ point], the lynx, show that as the density of the prey species increases, the density of the species preying on it increases also [$\frac{1}{2}$ point]. An increase of the predators causes increased predation [$\frac{1}{2}$ point], resulting in a fall in the prey population. As the prey population decreases, so does the predator species [$\frac{1}{2}$ point]. The conditions described are a series of density fluctuations: an example of natural feedback [$\frac{1}{2}$ point].

Analysis of Model Response X

Note that the answer gets to the point. There are no wasted sentences. The response rates a total of $2\frac{1}{2}$ points.

Model Response Y

Part (1) Feedback Control and Population Size

The deer population on the Kaibab Plateau numbered about 4,000 animals [$\frac{1}{2}$ point]. The predation of pumas and wolves kept the population size of deer in check [$\frac{1}{2}$ point]. When people killed the predators [$\frac{1}{2}$ point], the size of the deer population rose to 100,000. The vegetation of the Plateau was stripped as the result of this increase in the deer population [$\frac{1}{2}$ point]. The carrying capacity of the Plateau was destroyed [$\frac{1}{2}$ point], and starvation killed more than 50% of the deer [$\frac{1}{2}$ point]. Destruction of the predators disrupted a stable deer population [$\frac{1}{2}$ point]. Increase in the number of deer denuded their feeding range [$\frac{1}{2}$ point].

Analysis of
Model Response Y

Each sentence is a statement of fact and earns $\frac{1}{2}$ point. The response rates a total of 4 points.

Comments on
Responses

Remember that these answers are responses to a question that has three parts. Not more than 6 minutes should be utilized in answering each part, since other parts of this question may require more extensive answers. As you write, try to direct each sentence toward point accrual or toward useful clarification.

Student A Response

Part (2) Feedback
in the Rate of
a Physiological
Process

The rate of a physiological process such as digestion works on a feedback system [$\frac{1}{2}$ point]. All movements involved in digestion require energy. Some of these are peristalsis, churning of the stomach muscle, and just general movement of any organ or pipe or tube during digestion. The process of movement requires energy which is found in the form of ATP. However, ATP can only be formed by the process of cellular respiration, and cellular respiration can only occur if there is something in the form of nutrition to keep the cells alive. Therefore digestion allows cellular respiration to occur by providing the cells with nutrition usually in the form of glucose, which is the first stage in cellular respiration, and then glucose is converted to fructose 6 . . . fructose-1, 6-phosphate . . . PGAL . . . pyruvic acid . . . Krebs cycle. Through this entire process 38 ATPs are produced which then can be used during any of the physiological processes that require movement.

Analysis of
Student A Response

A score of $\frac{1}{2}$ point was given for problem identification. However, there is no follow-through of the idea that digestion works on a feedback system. Student A used what is called the "shotgun" approach, scattering bits of information here and there. An attempt was made to show how movement of the digestive organs requires energy. However, the student failed to show a connection between feedback and any one of the ideas mentioned.

Student B Response

Part (2) Feedback
in the Rate of
a Physiological
Process

The rate at which a certain physiological process occurs in an organism is the result of a complex series of interactions and feedback controls exerted by the rest of the external and internal environment of this organism [$\frac{1}{2}$ point]. The rate at which the heart beats is an example that can be used to clarify this point [$\frac{1}{2}$ point]. Cardiac muscle by its very nature is contractile. But as everybody knows, the heart beats as a whole (the auricles at one time, the ventricles a fraction of a second later) and not in an erratic manner. What causes the heart to contract at a certain rate is determined by the interaction of two sets of nerves, one being part of the sympathetic [$\frac{1}{2}$ point] nervous system, whose final effect is to stimulate the contraction of the heart, and the other nerve being part of the parasympathetic [$\frac{1}{2}$ point] nervous system, whose final effect is to slow down the rate of contraction of the heart. These two sets of nerves act antagonistically [$\frac{1}{2}$ point] to each other. And it is the interaction of these two nerves that finally determines the rate at which the heart beats. But these two nerves in turn are controlled by other factors such as the need for a faster circulation of blood in the body. Such factors as these determine the rate of the heart beat by stimulating the sympathetic nervous system and the heartbeat center [$\frac{1}{2}$ point] in the medulla oblongata. This is usually accomplished by an increase in the concentration of CO_2 in the blood, and an increase in the temperature of the body, or the pressure that the blood exerts on the walls of the vascular system [$\frac{1}{2}$ point]. To check the rate at which the heart is pumping, e.g., from working too hard, an impulse caused by the friction of the blood rushing through the aorta stimulates the parasympathetic nerve and thus causing the heart to slow down [$^1/_2$ point]. It is through interaction and feedback such as this that homeostasis [$^1/_2$ point] is achieved in the body and at the same time, the need of the body is achieved.

Analysis of Student B Response

This student scored a total of 4$^1/_2$ points. Although the written English needs refining, this answer indicates a knowledge of feedback control as related to heartbeat.

Student C Response

Part (2) Feedback in in the Rate of a Physiological Process

Feedback in a physiological system can be seen in the blood sugar level of humans [$^1/_2$ point]. The pancreas produces insulin which is a hormone necessary in the release of sugar into the blood. The pancreas is sensitive to the concentration of sugar in the blood and when too much sugar is in the blood nervous impulses are sent to the brain which then turns off the insulin into the blood until the body has used it up to a below-normal level. At this below-normal level impulses are again sent to the brain, and then the insulin flow is turned on again. Now more glycogen is turned into sugar and blood sugar level rises again, starting the feedback system again.

Analysis of Student C Response

Some incorrect ideas are expressed in this answer, which received only $^1/_2$ point. Insulin is secreted by the islets of Langerhans in the pancreas in response to the level of sugar in the blood. Insulin is a hormone that makes cell membranes permeable to sugar. Insulin regulates blood sugar level by bringing molecules of glucose into cells. Insulin flow is not under the control of the hypothalamus of the brain.

Model Response X

Part (2) Feedback in the Rate of a Physiological Process

Feedback control regulating a physiological process is exemplified by endocrine activity [$^1/_2$ point]. Neurohormones secreted from the hypothalamus of the brain stimulate the adenohypophysis (anterior pituitary) to release its trophic hormones [$^1/_2$ point]. In turn, the trophic hormones activate certain endocrine glands that, in response, secrete hormones [$^1/_2$ point]; for example, the hypothalamus secretes thyrotrophin-release factor (TRF) [$^1/_2$ point]. The anterior lobe of the pituitary gland responds by releasing the trophic hormone thryotopin [$^1/_2$ point], which signals the thyroid gland to secrete thyroxin [$^1/_2$ point]. Body metabolism is regulated by thyroxin [$^1/_2$ point]. When the level of thyroxin in the blood is high, the hypothalamus ceases secreting TRF [$^1/_2$ point]. The adenohypophysis turns off its secretion of thyrotrophin [$^1/_2$ point]. In this way, signals to the thyroid gland stop and thyroxin synthesis and secretion cease. When the volume of thyroxin in the blood drops, the hypothalamus initiates the cycle again [$^1/_2$ point].

Analysis of Model Response X

This response received a total of 5 points. The points were awarded for identification of physiological processes, discussion of the general process, citation of specific examples, and description of feedback effect.

Model Response Y

Part (2) Feedback in the Rate of a Physiological Process

Hormones travel from the sites of secretion to target organs by way of the bloodstream [$^1/_2$ point]. When levels of hormones from endocrine glands are high, the hypothalamus of the brain ceases secreting its neurohormones [$^1/_2$ point]. Therefore the anterior pituitary is not stimulated to discharge its trophic hormones [$^1/_2$ point]. As a result of the inactivity of the adenohypophuysis, the thyroid, the gonads, and the adrenal cortex do not release their hormonal secretions [$^1/_2$ point]. The decrease of thyroxin, estrogen (or testosterone), progesterone, and corticosterone [1 $\frac{1}{2}$ points] in the blood signals the hypothalamus to resume the process of discharging the various releasing factors which trigger adenohypophysis activity. The cycle of secretion and stimulation commences [$\frac{1}{2}$ point]. The hypothalamus-hypophysis-endocrine complex is an example of negative feedback [$\frac{1}{2}$ point].

Analysis of Model Response Y

This response is a variation of the model X response and is presented for comparison. A total of 4 $\frac{1}{2}$ points was scored by this student. The points were awarded for a general description of hormonal activity, examples of specific endocrine glands and hormones, and a demonstration of feedback.

Student A Response

Part (3) Feedback Control of the Rate of Enzymatic Activity

The rate of an enzyme reaction works on a feedback system. For example, the rate at which digestive enzymes digest is dependent on the rate of the individual's metabolism. Also, the metabolism rate is measured by the rate of enzyme action. Phosphorylation is also an example of an enzyme feedback. ATP can be converted into ADP and P and back. In cellular respiration, the process of phosphorylation is needed in certain areas and then later on in other areas, the ADP + P is changed back into ATP. It is a feedback cycle.

Analysis of Student A Response

The rate at which an enzyme works may be controlled by a feedback mechanism. If so, the student should have presented a clear-cut example. Actually, a person's metabolic rate is determined by measuring carbon dioxide output. The answer was begun with a consideration of digestive enzymes and then moved rather rapidly to ADP and ATP. Poor word usage makes this response difficult to mark. Note: "Phosphorylation is also an example of an enzyme feedback." Is this sentence the result of imprecise thinking? An answer such as this deserves no credit.

Student B Response

Part (3) Feedback Control of the Rate of Enzymatic Activity

The rate of enzymatic reaction is directly proportional to the amount of enzyme and substrate available, up to a certain extent of course. Temperature plays a vital role in determining the rate of an enzymatic reaction. For a certain enzyme to function at its optimal level, a certain temperature (usually between 30–40°C) must be maintained. But if the temperature is increased more than that of 40°C for the most part what happens is that the rate of this enzyme reaction drops to a rate of near zero. This is so since enzyme being protein in nature, they tend to denature when heated, and thus lose their ability to catalyze their specific chemical reactions. So in this case temperature has a feedback control in the regulation of the rate of enzymatic reactions.

Analysis of Student B Response

Although some information is given relating to the nature of enzymes, the problem posed in the question is not fully answered. Nowhere is a link established between enzymatic activity and feedback. The concept of feedback is important in biology. As you study a physiological process, think of its control in terms of feedback, a self-governing mechanism that increases or decreases action. The written expression exhibited in this answer needs improvement. This answer earned no credit.

Student C Response

Part (3) Feedback Control of the Rate of Enzymatic Activity

Equilibrium is kept in most all body reactions since they are enzymatically controlled. A certain amount of reactants A and B are present in a cell. Now if an enzyme's job is to combine A and B into AB, it can combine A and B into AB at a very fast rate since A and B are present in large concentrations [$\frac{1}{2}$ point]. But as the concentrations of the reactants begin to drop [$\frac{1}{2}$ point] and the concentration of AB begins to rise, it is much more difficult for the enzyme to find the proper substrate since it is scarce, so rate of reaction falls. <u>Since A and B are present in low concentration inside the cell, they will diffuse into the cell and AB will diffuse out and be used up</u> [$^1/_2$ point]. As this happens, the enzyme again has a greater chance to find its substrate and thus the rate of activity will rise, starting the feedback system again [$^1/_2$ point]. This feedback system is used in most cellular reactions of simple unicellular organisms.

Analysis of Student C Response

The first sentence has substance but is poorly presented. Throughout this response, the reader has to ferret out meaning and interpret. Rewriting the underlined sentence would afford greater clarity of content: "Since A and B are in low concentration inside the cell, they will be replenished by diffusion of like molecules into the cell from the extracellular fluid." This response earned a score of 2.

Model Response X

Part (3) Feedback Control of the Rate of Enzymatic Activity

The production of inducible enzymes is dependent upon the operation of a feedback mechanism [$^1/_2$ point]. An inducible enzyme is synthesized as needed. Let us assume that a bacterial culture uses sucrose as an energy source [$^1/_2$ point] but does not contain enough enzyme to hydrolyze lactose. If this species of bacterium is put into a lactose medium, within 3 minutes, two enzymes appear, namely, permease and B-galactosidase [$^1/_2$ point]. Permease controls the active transport of lactose into the cell. B-galactosidase splits lactose into glucose and galactose [$^1/_2$ point]. As long as the bacterial culture contains lactose, the inducible enzymes will be synthesized. When these bacteria are in a lactose-free medium, the synthesis of permease and B-galactosidase will be repressed [$^1/_2$ point]. This is an example of a positive feedback mechanism which summons the synthesis of enzymes by stimulating the production of a short-lived mRNA [$^1/_2$ point].

Analysis of Model Response X

A total of 3 points was awarded for this response. The points were given for problem identification and explanation; enzyme identification; discussions of physiological processes, feedback, and enzyme repression; and elucidation of feedback mechanisms.

Model Response Y

Part (3) Feedback Control of the Rate of Enzymatic Activity

Enzymes that function in cellular respiration are set into action or inhibited from function by allosteric control [$^1/_2$ point]. Allostery is a mechanism which regulates the activity of enzymes by the binding at a site on the enzyme molecule that is not the site of catalytic activity [$^1/_2$ point]. Phosphofructokinase is an enzyme that catalyzes the conversion of fructose-6-phosphate to fructose-1,6-diphosphate [$^1/_2$ point]. As long as glycolysis proceeds resulting in a small amount of ATP, this enzyme is functional. When large amounts of ATP are produced during the Krebs cycle, phosphofructokinase is allosterically controlled [$^1/_2$ point]. ATP inhibits the activity of phosphofructokinase, a condition which stops anaerobic respiration. When the number of ATP molecules is decreased, phosphofructokinase is released and its activity resumes. Throughout the cycle feedback mechanisms direct the efficiency of enzymatic activities [$^1/_2$ point].

Analysis of Model Response Y

A total of 3 points were awarded to this response. The points were given for topic identification, definition of allostery, description of allosteric effect, statement of a specific example, demonstration of feedback control of a specific enzyme, and generalization of ideas presented.

More Practice in Answering a Free-Response Question

A teacher of AP Biology gave the class the following practice essay:

Sample Essay Question

The immune response is dependent upon the activity of the B-cell lymphocytes.

(a) Support this statement by discussing the role of the B-cell lymphocytes in the primary and secondary immune responses.
(b) Explain how the macrophages and neutrophils play supportive roles in immune response.

Give yourself 20 minutes to answer this question as completely as you can. Include 3 minutes needed to check your paper. Before you begin to write, make sure you understand what you are to do. Note these key words: *immune response, B-cell lymphocyte, macrophages, neutrophils.*

Marking Scheme

The maximum score allowable for both parts of this essay question is 12 points, apportioned as shown in Table 3.2. For each statement of fact or example, 1 point (not $^1/_2$ point as in the table) was given. (Writing for $^1/_2$ point per fact forces you to write more and, consequently, learn more.)

TABLE 3.2 MARKING SCHEME

Concept	Points	Presentation
Part (a)*		
Primary Immune Response	$^1/_2$	
Antigen recognition sites on B-cell membrane	$^1/_2$	Specify that each cell is a specialist
	$^1/_2$	State that a matching antigen activates a B cell
	$^1/_2$	Indicate the attachment of antigen to recognition site
	$^1/_2$	Indicate that growth is stimulated in the antigen-aroused cell
B cell reproduces	$^1/_2$	Note that at full growth the cell goes through mitosis, producing new cells
Differentiation	$^1/_2$	Indicate that the new cells differentiate into
	$^1/_2$	plasma cells
	$^1/_2$	memory cells
Antibody production	$^1/_2$	State that the plasma cells produce antibodies
	$^1/_2$	Antibodies immobilize cells

TABLE 3.2 (Continued)

Concept	Points	Presentation
Secondary Immune Response	$^1/_2$	
Instructions stored	$^1/_2$	Describe role of memory cells in storing immune instructions
	$^1/_2$	State that memory cells produce plasma cells during second attack
Antibody production	$^1/_2$	Plasma cells produce huge numbers of antibodies
	$^1/_2$	Logical organization of answer
*Part (b)***		
Specialized white blood cells	$^1/_2$	Identify macrophages as amoeboid white blood cells
	$^1/_2$	Identify neutrophils as smaller amoeboid cells
Phagocytosis	$^1/_2$	Both cells engulf foreign antigens
	$^1/_2$	Some macrophages remain in spleen
	$^1/_2$	Other macrophages and neutrophils patrol the body
Role of B cells	$^1/_2$	Indicate that B cells immobilize foreign antigens
	$^1/_2$	Macrophages and neutrophils phagocytize immobilized antigens
Short life span	$^1/_2$	Both cells die from bacterial toxins

*An 8-point maximum is given for Part (a) of the question
**A 4-point maximum is given for Part (b) of the question

Summary

Attention must be given to development of vocabulary, improvement of writing, and acquisition of biological facts as you prepare for the AP Biology Examination. Your paper will be graded on your ability to communicate information to others. A fact has to be stated clearly; examples must be appropriate. Practice in writing answers to sample essay questions is necessarily a continuous process. A grade of 5, or even 4 or 3, on an AP Biology Examination demands hard work.

PART THREE

Area I of Advanced Placement Biology: Molecules and Cells

CHAPTER FOUR
Biological Chemistry

Capsule Concept

This chapter provides a brief review of the basic biochemistry of the compounds that work in cells. Organic and inorganic compounds are involved in chemical reactions that sustain the life of the cell through controlled release and absorption of energy. Enzymes are organic compounds that maintain the steady state of chemical and physiological reactions. How enzymes are able to work so efficiently is reviewed in this chapter.

Vocabulary

Activation energy   the energy required to initiate a specific chemical reaction
Allostery   a change in enzyme shape due to a second binding site
Cofactors   inorganic molecules and metallic ions that assist the work of enzymes
Endergonic reactions   reactions that absorb free energy
Enzymes   biological catalysts
Exergonic reactions   reactions that release free energy
Functional groups   groups of certain atoms attached to organic molecules
Homeostasis   the steady-state control of physiological reactions
Inorganic compounds   compounds that do not contain carbon
Organic compounds   compounds of carbon

Compounds and Functional Groups

Cells are composed of inorganic and organic compounds. Most of the cell, 70–95%, is made up of water, an *inorganic* compound. *Organic* compounds are carbon compounds. Carbon atoms form the skeletal backbone upon which complex molecules are built. Four major classes of complex organic molecules are found in cells: carbohydrates, lipids, proteins, and nucleic acids (Table 4.1).

Each class of molecule contains numerous compounds that differ from each other in regard to their *functional groups*, that is, groups of certain atoms that are attached to organic molecules (Table 4.2).

TABLE 4.1   MAJOR CLASSES OF ORGANIC COMPOUNDS

Class	Percent by Dry Weight	Types	Subunit	Example
Carbohydrates	5%	Monosaccharides	Monosaccharide	Glucose
		Disaccharides	Monosaccharides	Sucrose
		Polysaccharides	Monosaccharides	Cellulose (starch)
Lipids	12%	Fats	Glycerol + 3 fatty acids	Butter
		Phospholipids	Glycerol + 2 fatty acids + phosphate	Lecithin
		Steroids	4 Rings + C skeleton	Cholesterol
Proteins	71%	Simple	Amino acids	Pepsin
		Conjugated	Amino acids + prosthetic group	Hemoglobin
Nucleic acids	7%	DNA	Nucleotides	DNA
		RNA	Nucleotides	mRNA

TABLE 4.2   FUNCTIONAL GROUPS IN ORGANIC COMPOUNDS

Group	Formula	Name of Compound
Amino	$R-N{\overset{H}{\underset{H}{}}}$	Amine
Carboxyl	$R-C{\overset{O}{\underset{OH}{}}}$	Carboxyl acid
Hydroxyl	$R-OH$	Alcohol
Carbonyl	$R-C{\overset{O}{\underset{H}{}}}$	Aldehyde
	$R-C{\overset{O}{}}-R$	Ketone
Sulfhydryl	$R-SH$	Thiol
Phosphate	$R-O-\overset{O}{\underset{O^-}{\overset{\|}{\underset{\|}{P}}}}-O^-$	Organic phosphate

Chemical Reactions

To sustain life, chemical reactions—the making and breaking of chemical bonds—must occur. For a reaction to occur, a specific amount of free energy (energy that can be used to do work) must be available. There are two types of reactions based upon free-energy changes.

Exergonic reactions release free energy; these are spontaneous reactions. Chemical reactions eventually reach equilibrium. Equilibrium reduces the free energy of a system. In living cells, metabolic pathways remove the products of a reaction, preventing equilibrium and allowing the reaction to continue.

Endergonic reactions absorb free energy. The production of carbohydrates by plants is powered by the absorption of light energy.

Both endergonic and exergonic reactions occur in cells.

Model Questions Related to Compounds and Functional Groups

1. Which term includes all the other terms?
   (A) Starch
   (B) Monosaccharide
   (C) Polysaccharide
   (D) Carbohydrate
   (E) Disaccharide

2. Which functional group is represented in the outlined portion of the organic molecule?

   (A) Carboxyl
   (B) Carbonyl
   (C) Hydroxyl
   (D) Amino
   (E) Sulfhydryl

Answers and Explanations to Model Questions: Compounds and Functional Groups

1. D A carbohydrate is a compound that contains carbon and hydrogen and oxygen in a 2:1 ratio. Carbohydrates include sugars and their polymers.

   Incorrect Choices
   A A starch is a polysaccharide.
   B A monosaccharide is a simple sugar.
   C A polysaccharide is a compound composed of hundreds of simple sugars.
   E A disaccharide consists of two monosaccharide units.

2. **A**   A carboxyl group is a COOH group, as shown in the diagram.

Incorrect Choices

B   A carbonyl group is an aldo or keto group: $R - \overset{\overset{\textstyle O}{\|}}{C} \diagdown_H$ or $R - \overset{\overset{\textstyle O}{\|}}{C} - R$

C   A hydroxyl group is an OH group: $R - OH$

D   An amino group has this formula: $R - N \diagup^H _{\diagdown H}$

E   A sulfhydryl group has this formula: $R - SH$

Acids, Bases, pH, Buffers

Water is a universal solvent and the most important compound for life. A water molecule consists of two hydrogen atoms and one oxygen atom. Sometimes a hydrogen atom shared between two water molecules shifts from one molecule to the other, forming a hydronium ion and a hydroxide ion:

$$H_2O \; + \; H_2O \; \rightleftharpoons \; \underset{\substack{\text{hydronium} \\ \text{ion}}}{H_3O^+} \; + \; \underset{\substack{\text{hydroxide} \\ \text{ion}}}{OH^-}$$

Dissociation of Water

It is easier, however, to think of the separation of one water molecule (*dissociation*) into H^+ and OH^-:

$$H_2O \; \rightleftharpoons \; \underset{\substack{\text{hydrogen} \\ \text{ion}}}{H^+} \; + \; \underset{\substack{\text{hydroxide} \\ \text{ion}}}{OH^-}$$

About one out of 554 million water molecules dissociates in pure water.

Since the dissociation of pure water produces one H^+ and one OH^-, there are equal concentrations of hydrogen ions and hydroxide ions in pure water. The concentration of each ion is expressed as 10^{-7}. This means that there is only one ten-millionth of a mole of hydrogen ions and one ten-millionth of a mole of hydroxide ions per liter of pure water.

Acids and Bases

Substances can be added to water that alter the H^+ and OH^- balance. A substance that increases the hydrogen-ion concentration of a solution is called an *acid*. Acids may alter the $H^+ - OH^-$ balance either by contributing H^+'s or by removing OH^-'s from the solution. A substance that reduces the H^+ concentration and increases the OH^- concentration is called a *base*.

pH

The concentration of H^+ and OH^- will vary in a solution depending on the presence of acids or bases. A pH scale ranging from 0 to 14 has been devised to indicate the variation of ions in solution; pH (the power of the hydrogen ion) is expressed by utilizing the mathematical device known as logarithms. Since pH equals $-\log [H^+]$ and 10^{-7} is the concentration of H^+ in water, pH $= -\log [H^+]$ or $-(-7) = 7$. But 10^{-7} is also the OH^- concentration of pure water; therefore, pH $= -\log [OH^-] = -(-7) = 7$. Consequently, 7 on a pH scale describes a neutral solution. If the H^+ concentration is increased from 10^{-7} to 10^{-3}, the pH $= -\log [H^+]$

= – (–3) = 3. The pH values of acids decrease as the concentration of H^+ increases. As H^+ increases, OH^- decreases. Conversely, the pH of a base increases as the OH^- concentration increases.

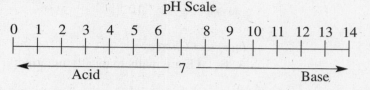

pH Scale

Each pH unit on the scale represents a tenfold difference in H^+ and OH^- concentrations. Therefore, a solution of pH 2 is 1,000 times more acidic than a solution of pH 5: 2 to 5 is an increase of 3 pH units, and $10 \times 10 \times 10 = 1,000$.

The pH levels of the internal and external environments of an organism are important to its survival. Tropical fish and many plants require specific external environmental pH values for health and growth. Biochemical pathways with their enzymatic reactions depend on specific internal pH levels. With the exception of digestive enzymes, biochemical reactions in human cells require a neutral internal pH. Body fluids, blood and urine, and the internal material of living cells are also maintained at or about a neutral pH.

Buffers

To maintain a constant neutral environment, living systems employ buffers. Buffers are chemical substances that can combine with excess H^+ or OH^-, thus removing these ions from the environment. In human red blood cells, carbonic acid (H_2CO_3) and its bicarbonate (HCO_3^-) ion are important buffers.

Action of Buffers

$$H_2CO_3 \; + \; OH^- \; \rightleftharpoons \; HCO_3^- \; + \; H_2O$$
$$HCO_3^- \; + \; H^+ \; \rightleftharpoons \; H_2CO_3$$

Model Questions Related to H^+ and OH^- Concentrations

1. In a lake, acid rain has lowered the pH to 3. What is the hydrogen ion concentration of the lake?
   (A) 3.0
   (B) 10^{-11}
   (C) 10^{-3}
   (D) 103
   (E) 3%

2. What is the hydroxide ion concentration of the lake in Question 1?
   (A) 10^{-7}
   (B) 10^{-3}
   (C) 10^{-11}
   (D) 10^{-14}
   (E) 10

Answers and Explanations to Model Questions: H^+ and OH^- Concentrations

1. C pH = –log [H^+]; therefore, 3 = –log [H^+].

   Incorrect Choices
   A, B, D, E are mathematically incorrect.

2. C The concentrations of [H⁺] and OH⁻ are constant in pure water.

The constant $= [H^+] \times [OH^-] = 10^{-7} \times 10^{-7} = 10^{-14}$

$$10^{-14} = [H^+][OH^-] = (10^{-3})(x)$$

$$x = 10^{-11}$$

Incorrect Choices
A, B, D, E are mathematically incorrect.

Enzymes

A catalyst is a substance that changes the rate of a chemical reaction without being used up, destroyed, or incorporated into the end product of the reaction. Organic catalysts are called *enzymes*. These protein molecules are vital to the orderly metabolic processes of the cell and thus are essential to the life of an organism.

The rate of all cellular reactions is controlled by enzymes. Although these enzymes are synthesized in the ribosomes, as are other protein molecules, their production is genetically controlled. The survival of an organism depends not only on the synthesis and degradation of compounds within the cells but also on the speed at which these reactions occur. Enzymes increase the rates of chemical reactions within the isothermic (constant temperature) limits of the cell. It is true that, as the temperature is increased, the movement of molecules increases. However, an increase in heat can damage the delicate structure of cell organelles. What, then, are the characteristics of enzymes that permit them to regulate the biochemical reactions at temperatures that are compatible with the life of the cell?

Enzymes in Profile

Enzymes are globular proteins ranging in molecular weight from 10,000 to several million. The smaller proteins consist of a single, folded polypeptide chain. The larger proteins may contain several chains. Enzymes have an active site that binds to substrate molecules. Associated with the active site are metallic ions, among which are zinc, magnesium, iron, copper, and cobalt. The metallic ions at the active sites help to bind the substrate molecule and/or withdraw electrons.

Traditionally, it has been believed that the active site is structured to fit the substrate molecule. This theory, proposed by Emil Fisher in 1894, was known as the *lock-and-key theory*.

More recently a new enzymatic model called the *induced-fit hypothesis* has been proposed. An enzyme is not a rigid molecule, but, rather flexible. An enzyme undergoes a *conformational change* as the substrate enters the active site. The conformational change results in the folding of the enzyme around the substrate molecule, forming a tight fit. The forces that maintain the folded structure of the enzyme and the binding of the substrate are hydrogen bonds, electrostatic attraction and repulsion of charged chemical groups, and the interaction of hydrophobic groups. Carboxypeptidase and hexokinase are two enzymes known to operate on the induced-fit model.

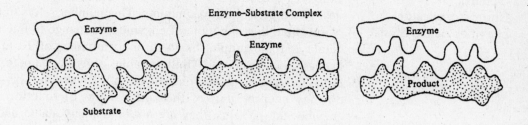

Figure 4.1   Lock-and-key theory of enzyme action

The importance of flexibility in enzyme molecules is illustrated by the property of *allostery*. Many enzyme proteins, although quite large, are not built from an enormous polypeptide chain. Rather, they are packaged together in identical subunits. One of these subunits contains the active site where substrate molecules bind. Another subunit has a regulatory, or allosteric, site where molecules that may or may not be related to reactants or products may bind. Binding at the regulatory site can deform the structure of the enzyme at the active site, misshaping the latter and rendering it incapable of binding to substrate molecules. Enzyme inhibition is the result. On the other hand, allosteric behavior can enhance the activity of the active site. In some molecules, allostery causes a proton shift, inducing in the enzyme molecule a change in acid-base balance. This change in acidity can increase the activity of the active site by changing the relative positions of the subunits.

Homeostasis is the maintenance of a steady state in the physiological environment of the cell (Figure 4.2). Enzymatic activity may result in the production of too much or too little product. Imbalances of this kind are avoided by the operation of feedback mechanisms that control the rate and quantity of work effected by enzymes. It is not unusual for an enzyme to have both an active and a regulatory site. An excess of the catalyzed product may bind to the regulatory site, deforming the molecule and thus altering the shape of the active site. This situation causes enzymatic activity to cease. If too little product is formed, the active site remains unaltered and enzymatic activity continues. The endocrine system offers an excellent example of feedback control. The level of a given hormone in the blood either stimulates or inhibits endocrine activity, depending on metabolic need.

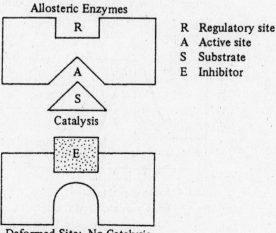

R   Regulatory site
A   Active site
S   Substrate
E   Inhibitor

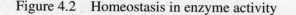

Figure 4.2   Homeostasis in enzyme activity

General Characteristics of Enzymes as Catalysts

1. Enzymes are effective in small amounts. The efficiency of an enzyme is designated by its turnover number. This number refers to the number of moles of substrate that one mole of enzyme can change in one minute. The enzyme catalase, for example, has a turnover number of 10 million. One mole of catalase will turn over 10 million moles of hydrogen peroxide per minute.
2. Usually enzymes remain unchanged in the reaction. They are not used up, nor do they become part of the end product of reaction. However, because enzymes are proteins, they are not completely stable under all conditions. A change in pH or a rise in temperature can render enzymes inactive.
3. Enzymes do not affect the equilibrium of a chemical reaction. Enzymes hasten a reaction until it reaches equilibrium and can speed up the process in either direction.
4. Enzymes are specific in their ability to act on substrate molecules. Some enzymes are highly specific for a given molecule; others are specific for a group of substances.
5. The presence or absence of *cofactors* affects enzyme activity. Cofactors are usually inorganic molecules or metals (iron, zinc, copper) that are enzyme helpers. An organic enzyme helper such as a vitamin is known as a *coenzyme*. NAD and FAD are coenzymes.
6. The presence of *inhibitor* substances affects enzyme activity. Substances that mimic a substrate are known as *competitive inhibitors*. Substances that do not bind to the active site of an enzyme but cause the enzyme to change shape are known as *noncompetitive inhibitors*.

Effect of Enzymes on Activation Energy

Biochemical reactions in cells are spontaneous, that is, given time, a biochemical reaction will take place without outside stimulus. However, the time necessary for stable molecules to react is infinite. To increase the rate of reaction between stable molecules, energy has to be added to the system. Through collisions, some molecules derive energy that enhances their ability to react. Molecules that are energy deficient must overcome the energy barrier to reactions. The higher the energy barrier, the more stable the molecule. The energy required to catapult biochemical molecules over the energy barrier is known as activation energy. Enzymes have the ability to lower the energy of activation (Figure 4.3), thus permitting reactions within cells to occur at the normal body temperature of the organism.

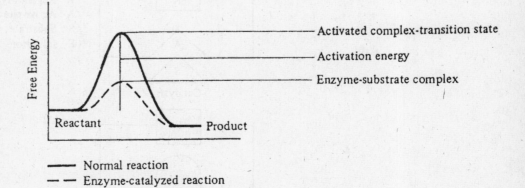

Figure 4.3   Energy of activation

Enzymes lower the activation energy in a unique way. A particular enzyme forms impermanent bonds with its substrate, orienting these molecules in spatial relationships that promote reactions. The reacting molecules are brought into contact with each other in a way that eliminates the forces of repulsion between them without weakening or disrupting bonds. The efficiency of enzymes in lowering the activation energy is illustrated in Table 4.3.

TABLE 4.3   ENZYME EFFICIENCY

Enzyme	Substrate	Activation Energy Required
Liver catalase	H_2O_2	Less than 31%
Trypsin + HCl	Casein	56%
Yeast invertase	Sucrose	Less than 50%

Factors Affecting the Rate of Enzyme Action

The rate of enzyme action varies with such factors as temperature, concentration of the substrate, concentration of the enzyme, and pH. These effects are illustrated in the graphs shown in Figure 4.4.

1. The Effect of Temperature on Enzyme Activity

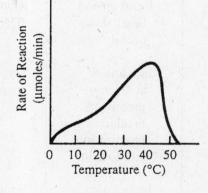

2. The Effect of Substrate Concentration on Enzyme Activity

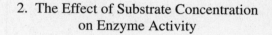
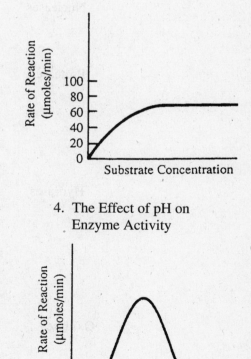

3. The Effect of Enzyme Concentration on Enzyme Activity

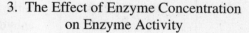
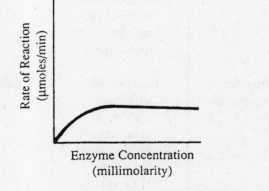

4. The Effect of pH on Enzyme Activity

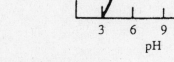

Figure 4.4   Factors affecting rate of enzyme action

Table 4.4 lists the types of enzymes, their functions, and their effects and gives examples of the various types.

TABLE 4.4   CLASSIFICATION OF ENZYMES

Enzyme Type	Function	Effects	Examples
Hydrolases	Addition or removal of water	Hydrolysis and dehydration synthesis	
Phosphatases		Add or remove phosphate groups	ATPase
Glycosidases		Combine monosaccharides; break down polysaccharides	Amylase, sucrase
Nucleases		Build up and break down nucleic acids	Ribonuclease, DNAase
Proteases		Join amino acids into proteins; break down proteins into amino acids	Trypsin, pepsin
Hydrases		Incorporate the elements of water into, or take away water from, substrate molecules	Enolase, carbonic anhydrase
Oxidases	Transfer of electrons	Reaction of substrates with molecular oxygen breaks down H_2O_2 into H_2O and O_2	Catalase

TABLE 4.4   (Continued)

Enzyme Type	Function	Effects	Examples
Dehydroge-nases	Removal of H_2 from substrate and trans-port by means of a coenzyme or other carrier groups	Joins H_2 to molecular O_2 to form water	Cytochrome oxidase
		FAD as co-enzyme	Riboflavin de-hydrogenase
		Riboflavin prosthetic groups	Cytochrome c reductase
		NAD or NADH as coenzyme	Pyridine nucleotide dehydrogenase
Transferases	Transfer of radicals		
Transaminases		Transfer amino groups	
Transphos-phorylases		Transfer phosphate groups	
Trans-methylases		Transfer methyl groups	
Transgly-cosidases		Transfer mono-saccharides	
Desmolases	Rupture of C-C bonds	Removal of carboxyl group from amino acids	Amino acid decarboxylase

Model Questions Related to Enzymes

1. All of the following statements about enzymes are true EXCEPT
   (A) high temperatures denature enzymes
   (B) enzymes are reusable
   (C) enzymes are specific
   (D) enzymes contain coenzymes
   (E) enzymes can operate only in living cells

Answer Questions 2 and 3 on the basis of the following information and on your knowledge of biology.

A student learned that biological oxidation occurs if cells contain the enzyme dehydrogenase. She wished to determine whether this enzyme is present in liver cells. She also knew that methylene blue becomes colorless when it accepts hydrogen ions.

The student set up a series of test tubes numbered 1–5. The contents of each tube (summarized in the chart below) were covered with a layer of mineral oil. Each tube was placed into a water bath at 37°C for 20 minutes.

Tube No.	Raw Liver	Boiled Liver	Distilled Water	Methylene Blue	Resulting Color
1	—	—	3 ml	2 ml	Blue
2	+	—	5 ml	—	None
3	+	—	3 ml	2 ml	None
4	—	+	3 ml	2 ml	Blue
5	—	+	5 ml	—	None

2. The student's question, "Is enzyme present in liver cells?," was answered by test tubes numbered

(A) 1 and 2
(B) 3 and 4
(C) 4 and 5
(D) 2 and 3
(E) 1 and 5

3. Each test tube was topped with a layer of mineral oil in order to

(A) ensure the sterility of the tube
(B) prevent the oxidation of methylene blue by the air
(C) maintain anaerobic conditions, permitting the continuation of dehydrogenase activity
(D) prevent the reduction of methylene blue
(E) inhibit the activity of bacteria of decay

4. Competitive inhibition of enzymes may occur when

(A) the temperature is increased beyond the optimum requirements
(B) the pH is lowered
(C) two molecularly similar compounds (one being the substrate) are present
(D) a molecule combines with the substrate blocking the enzyme activity
(E) some substance changes the configuration of the enzyme

The following diagram shows a biochemical pathway depicting the production of various amino acids by the bacterium Escherichia coli. Answer Questions 5 and 6 on the basis of the diagram.

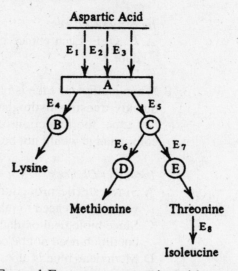

5. Three enzymes, E_1, E_2, and E_3 convert aspartic acid to product A. These enzymes are inhibited by negative feedback of lysine, methionine, and threonine, respectively. If there is an excess of lysine in the system, which statement is TRUE?
   (A) No aspartic acid is converted to product A.
   (B) The production of product A is cut by approximately one-third.
   (C) There will be an increase in all three amino acids.
   (D) No threonine or methionine will be produced.
   (E) All of product A will be converted to isoleucine.

6. If a gene mutation occurs, resulting in a defect in E_7, which of the following statements is true?
   (A) The production of product A will be cut by approximately one third.
   (B) The production of aspartic acid will be affected.
   (C) The production of lysine will cease.
   (D) The production of isoleucine may decrease.
   (E) The production of product C will be reduced.

Answers and Explanations to Model Questions: Enzymes

1. **E** One of the earliest experiments conducted on fermentation by Eduard Buchner involved the use of yeast cell extracts.

   Incorrect Choices
   A At high temperatures the internal bonds of the enzyme are disrupted, thus destroying the integrity of the molecule.
   B Once the substrate has been acted on, the enzyme falls away and is free to combine with another substrate molecule.
   C Because of the specificity loop found in enzymes due to their distribution of charges and amino acid arrangement, only molecules with the proper shape and spatial orientation can occupy the space in the loop.
   D Not all enzymes contain a coenzyme portion.

2. B In test tubes 3 and 4, all of the conditions for enzyme activity are met. The difference between these tubes is that one contains raw liver and the other boiled liver.

Incorrect Choices
A, C, D, E Each choice has something missing in the experimental procedure.

3. B When methylene blue is blue in color, it is in the oxidized state. However, it quickly transfers hydrogens to oxygen. If oxygen were not kept out of the test tube, the reduction-oxidation reaction would occur so rapidly that a color change would not be noticed.

Incorrect Choices
A Since sterile procedures were not included in the experiment, there would be no need to maintain such a condition.
C Since biological oxidation involves the presence of oxygen, an anaerobic condition need not be considered.
D Methylene blue is already reduced, as indicated by the color change.
E Since liver will not significantly decompose in the short time needed to conduct the experiment, the bacteria of decay are not a factor.

4. C Two similar molecules will have similar spatial arrangements of atoms and charges. Both will fit into the enzyme's configuration. Whether the normal substrate or its analog occupies the active site depends on the concentration of each.

Incorrect Choices
A At high temperatures, the internal bonds of the enzyme are disrupted, thus destroying the integrity of the molecule.
B The proper concentration of the hydrogen ion is essential for enzyme activity because a shift in pH may cause a change in the internal structure of the enzyme.
D In competitive inhibition, the enzyme is usually not permanently blocked because the competing substrate is worked on and new products are released. The cells suffer, not from enzyme blockage, but from the presence of the new metabolites.
E An enzyme's configuration is altered, not at the active site, but at different portions of the molecule, the regulatory site. This does not involve competition for the active site.

5. B Since it is assumed that all three enzymes are operating to capacity, cessation of the operation of one enzyme necessarily reduces the catalyzed product.

Incorrect Choices
A For the production of product A to cease completely, all three enzymes must be inoperative.
C The production of lysine will be inhibited, along with the inhibition of the other two amino acids, because there is a reduction of product A.
D The other enzyme systems are not affected by the inoperation of E_1.
E All three amino acids will be produced but in reduced amounts.

6. **D** Since E_7 is part of the pathway that leads from C to isoleucine, the effect of some change on the enzyme will occur in this pathway.

Incorrect Choices

A Only if the concentration of isoleucine increases will there be a decrease in product A. Since the type of gene mutation has not been stipulated, this assumption cannot be made.

B In the diagram there is no indication of how aspartic acid is produced. It may be obtained from the diet.

C The production of lysine involves a different enzyme.

E Refer to choice C.

CHAPTER FIVE
Cells

Capsule Concept

The cell is the basic unit of life. All of the activities necessary for life are carried on within a single unit so small that it cannot be resolved by the naked eye. In time, measured in fractions of seconds, the organelles within the cell perform biochemical activities that stagger the imagination. In this chapter, you will review the structures that compose the cell and their functions.

Vocabulary

Cell membrane   the outer living boundary of the cell, which controls movement of substances into and out of the cell

Cystol   the semifluid cytoplasm

Cytoplasm   the living material of the cell, which lies between the cell membrane and the nucleus

Endocytosis   the cellular uptake and incorporation of macromolecules into an intracellular vesicle

Eukaryotic cell   a cell that has a discrete nucleus separated from the cytoplasm by a living nuclear membrane

Nucleoid   a region in prokaryotic cells where genetic material is concentrated

Organelles   structures within the cell that carry out the work of the cell

Phagocytosis   the process by which a cell engulfs particulate matter

Pinocytosis   the process by which the cell ingests extracellular fluid and its solutes

Prokaryotic cell   a cell without a discrete nucleus

Protoplasm   the living material of the cell; its membranes and living organelles

Transport   the means through which materials enter and leave the cell

Cytological Tools

Microscopy made the science of cytology possible. The compound light microscope without oil is able to resolve cellular structures that measure at least 0.4μ. The resolving power of an oil immersion light microscope is 0.2μ. (The abbreviation μ stands for micrometer.) However, subcellular structures, or organelles, are too small to be seen through light microscopes because they are beyond the limit of resolution. In the 1950s, the electron microscope was introduced to the study of biological specimens, making visible subcellular structures.

The electron microscope shows objects with 10,000 times greater resolution than that of a light microscope. The electron microscope directs a beam of electrons through a thinly sliced specimen stained with heavy metal atoms. Two types of electron microscopes are used: the *transmission electron microscope* (TEM) and the *scanning electron microscope* (SEM). Each of these microscopes is specialized to enhance specific observations. The TEM resolves the inner structure of organelles. The SEM scans the surfaces of organelles, producing a three-dimensional image. The major disadvantage of the electron microscope is that living specimens cannot be observed. Preparation procedures kill specimen tissues.

Cytology is necessarily concerned with the function of organelles. To study the function of organelles, cytologists take cells apart in a procedure known as *cell fractionation,* which allows the isolation of cell parts. Tissues and lysing (loosening) solution are put in the ultracentrifuge, an instrument that rotates at 80,000 rpm (revolutions per minute). This force causes the membranes of tissue cells to split open, releasing their contents into the lysing fluid. Organelles form layers in the lysing fluid based on their densities. The densest organelles fall to the bottom of the centrifuge tube. Organelles of intermediate densities are distributed in zones between the top and bottom layers. The least dense organelles are in the top layers. The technique of cell fractionation enables the cytologist to remove layered zones and to study them separately, either by biochemical means or by way of the electron microscope.

Cell Structures and Functions

Protoplasm

Historically, the term *protoplasm* has been used to describe the physical substance of life. Modern biologists tend to deemphasize the importance of this term because it does not explain the intricate biochemical relationships that exist among molecules in the living material of cells. However, the term still provides a convenient means of referring to living material.

Protoplasm can be divided into structural units called cells. Whether an organism be unicellular or multicellular, certain processes take place that involve (1) utilizing food sources for energy, (2) eliminating the resulting waste, and (3) responding to various stimuli in the cell's environment.

The Cell

The *cell* is the unit of structure, the unit of function, the unit of growth, and the unit of heredity in all living things. It consists of cytoplasm and genetic material. The differences in the internal organization of cells separate cells into two types.

The *prokaryotic cell* lacks a true nucleus. There is no membrane separating the genetic material from the cytoplasm. The genetic material is concentrated in a region of the cytoplasm designated as the *nucleoid*. In contrast, the *eukaryotic cell* has a true nucleus. A nuclear membrane separates the genetic material from the cytoplasm.

Within the nucleus are materials for inheritance and cellular function. Within the cytoplasm are found both living and nonliving structures; the living structures are referred to as *organelles*. Prokaryotes lack many of these organelles. The eukaryotic cell (Figure 5.1), with its nucleus and organelles, is the basic unit of life.

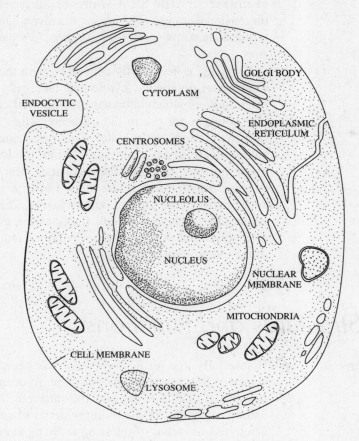

Figure 5.1   Diagram of eukaryotic cell

The Nucleus

In *eukaryotic cells*, the nucleus, together with its genetic material, is enveloped by a double membrane, the outer part of which resembles rough endoplasmic reticulum (ER). It is believed that the ER is a continuation of the nuclear membrane. Pores—structures having regulatory functions—are prominent features of the nuclear membrane.

The growth and reproduction of the cell are controlled by the nucleus, as demonstrated conclusively by laboratory research involving the marine alga *Acetabularia* and the protozoan *Amoeba proteus*. During cell division (mitosis), two cylindrical organelles called *centrioles* are present in animal cells and in the cells of some lower plants. *Asters* (centrioles with radiations) and *spindle* fibers (radiations between the two asters) consist of proteins similar to muscle actin and myosin. Structurally, centrioles resemble basal bodies that control the movements of cilia and flagella.

Present also in the nucleus is the *nucleolus*. This is where ribosomal synthesis takes place.

Cytoplasm

The cytoplasm is a semifluid medium called the *cystol*. Cytoplasm in the cell is a colloid; a colloid is a dispersion of intermediate-size particles in a medium. Part of the cytoplasm of the amoeba, for example, consists of colloid particles dispersed randomly in a fluid medium, forming a phase known as a *sol*. In another part of the same amoeba, the cytoplasm contains colloid particles that have interlocked to form an orderly network, rendering the semifluid medium more viscous. The latter is the *gel* state. These cytoplasmic changes from sol to gel and from gel to sol confer mobility on the amoeba. (This sol-gel transition is visible as one watches the flowing amoeba.) Although present in all cells, the sol and gel phases are not as dramatically demonstrated elsewhere as in the nearly transparent, single-cell amoeba.

The Cell Membrane

The cell membrane is a living boundary between the internal and the external environment of the cell. It is a gel in which the orderly arrangement of colloidal particles forms a very stable and elastic boundary. The colloidal particles in the cell membrane (and in the other cellular organelles) are *proteins* (see page 101).

Other components of cell membranes are *phosphatides* (Figure 5.2) and *phospholipids*. The latter consist of glycerol, fatty acid, and phosphoric acid molecules. When the phosphoric acid group is esterified with a compound of nitrogen, a phosphatide is formed. The phospholipid in cell membranes is, in reality, a phosphatide, although the terms are used interchangeably in various texts. The phosphatide resembles a clothespin. The head is charged and strongly *hydrophilic* (water soluble). The hydrocarbon tails are not charged and therefore are *hydrophobic* (water insoluble). Phospholipids are amphipatic molecules; that is, they have both a hydrophilic and a hydrophobic region.

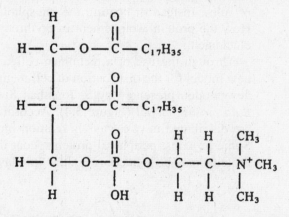

Figure 5.2   Representation of a phosphatide

According to the model proposed by Danielli, proteins and phospholipids are arranged in a specific way in cell membranes (Figure 5.3). There are two phospholipid layers coated on each side by proteins. The protein/phospholipid bilayer/protein arrangement of molecules is popularly referred to as the "butter sandwich."

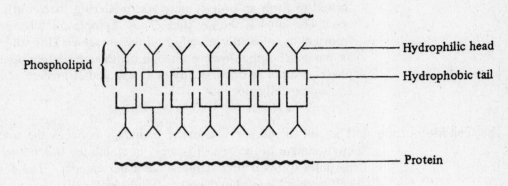

Figure 5.3   Danielli model of the cell membrane

The phospholipid bilayer resembles a double row of clothespins with the charged ends facing the protein molecules. The forked, uncharged tails face each other. When the phospholipids, such as lecithin or cephalin, are placed in a liquid medium, the hydrophobic tails remain in the air while the charged ends become immersed in water, forming a double-layered boundary (as seen in electron micrographs).

As with any simplified description, there are exceptions that complicate matters. Proteins in the membrane of red blood cells appear to follow a different set of rules. Instead of coating the phospholipid bilayer, they seem to penetrate it. How the protein arrangement in erythrocytes aids transport has not been clearly elucidated.

Through the use of a technique called *freeze-fracture electron microscopy*, a new model for the distribution of cell membrane proteins has been proposed. This new model, presented by S. Jonathan Singer and Garth Nicolson, is called the *fluid mosaic model* (Figure 5.4). According to this model, the globular proteins are distributed in two ways in relationship to the lipid bilayer of the membrane. Some proteins, peripheral proteins, coat the surface layers of the lipids. Integral proteins may be located *within* the lipid layers or may *penetrate* the lipid layers.

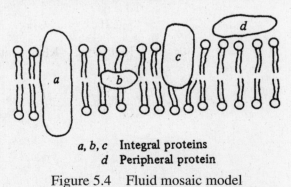

a, b, c   Integral proteins
d   Peripheral protein

Figure 5.4   Fluid mosaic model

Proteins determine the function of the cell membrane. Membrane proteins are classified according to their functions: transport, enzyme, receptor site, cell adhesion, and cytoskeletal attachment.

The Cell Membrane and Cell Transport

The cell membrane serves a regulatory function. It is selectively permeable and controls the flow of material into and out of the cell by passive and active transport. Passive transport (diffusion) depends on the molecular movement of particles along a concentration gradient; energy is not utilized. For example, osmosis is the diffusion of water through the cell membrane. Theoretically, the movement of water should continue until equilibrium is reached—in simple terms, until the number of water molecules entering a cell equals the number of water molecules leaving the cell. (Actually, osmosis depends on the number of particles in solution, or the concentration differential. Equilibrium is reached when the numbers of particles on both sides of the membrane are equal.)

Osmotic pressure is the force exerted by particles in solution against the cell membrane. In living systems, osmotic pressure is controlled by colloidal proteins that tie up free water. All cells exhibit osmotic pressure adaptions for water control of their internal environments. Because an amoeba lives in a freshwater environment, its osmotic pressure is very high; the contractile vacuole prevents rupture of the cell. A red blood cell exists in an environment in which the concentration of the internal protein is virtually the same as the concentration of the protein in its external environment; therefore the osmotic pressure is low. The exchange of materials between the blood plasma and the blood cell is due to a combination of osmotic pressure and hydrostatic pressure (blood pressure), which will be discussed later.

Table 5.1 shows the rates of penetration of certain materials through the cell membrane.

TABLE 5.1   PENETRATION RATES OF VARIOUS MATERIALS

Very Rapid	Rapid	Slow	Very Slow	No Penetration
CO_2	Water	Glucose	Strong electrolytes	Proteins
N_2		Amino acids	Acids and bases	Polypeptides
Fat solvent		Fatty acids	Disaccharides	
Alcohol Ether Chloroform				

Transport proteins are involved in the diffusion of ions and nonpolar molecules across the membrane. Because of their molecular configuration, transport proteins form channels through the membrane. There are three classes of transport proteins. Uniport carriers transport one molecule in one direction. Symport carriers transport two different molecules in the same direction. Antiport carriers transport two different molecules in different directions. *Cystinuria* is a kidney disease in which the kidney cells fail to reabsorb cystine from the urine because of a lack of transport proteins. The result is the formation of kidney stones.

Glucose Transport

As stated previously, *passive transport* does not require an expenditure of energy by the cell. Glucose does not diffuse into muscle cells freely but spreads through these cells by *carrier-facilitated diffusion*. It is believed that a special molecule (permease), located within the cell membrane, carries glucose across the membrane following the mode of diffusion from an area of high concentration to an area of low concentration. Such a mechanism of diffusion establishes an equilibrium constant and maintains a homeostatic balance.

Absorption of glucose from the intestinal tract occurs against a concentration gradient. Energy is expended by the intestinal cells. This is *active transport*. The "revolving-door" hypothesis is a means of explaining how glucose absorption occurs. Within the plasma membranes of the intestinal cells there are enzyme-like carrier proteins which have active sites that are able to "recognize" specific molecules such as glucose. Once a glucose molecule is bound to the active site of the carrier protein, the protein changes shape and rotates. In this way, the carrier protein transports the target molecule through the lipid layer to the inner surface of the plasma membrane, where the union between the glucose molecule and its carrier is terminated.

The interior of cells, in general, exhibits a negative potential as compared with the exterior. Active transport, such as the sodium-potassium pump of cell membranes of the nervous and muscle systems, utilizes this difference in potential. The cells of kidney tubules are excellent examples. For every sodium ion that is pumped out of the cell, a potassium or hydrogen ion is pumped in from the filtrate. Another example of a membrane pump is the proton pump, which transports H^+ across the membrane.

Large particles enter and leave the cell by the processes of endocytosis and exocytosis (Figure 5.5). Vesicles derived from the plasma membrane draw in or expel large particles. Phagocytosis and pinocytosis are types of endocytosis. Phagocytosis is the process by which a cell engulfs particulate matter. Pinocytosis is the process by which a cell engulfs droplets.

Extracellular fluid

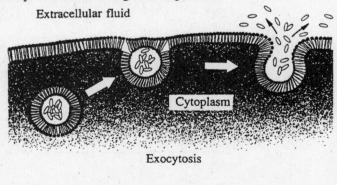

Exocytosis

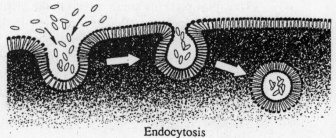

Endocytosis

Figure 5.5 Endocytosis and exocytosis

Organelles

Within the cytoplasm of eukaryotic cells is a system of internal membranes called the *endoplasmic reticulum* (ER). There are two regions of ER, which differ in structure and function—rough ER and smooth ER.

Rough ER is "rough" in appearance because it is dotted with ribosomes, the sites of protein synthesis. Proteins manufactured in the ribosomes are usually secreted by cells, but some are carried in vesicles formed by the ER to the Golgi apparatus.

Smooth ER lacks ribosomes and serves in the synthesis of lipids, in carbohydrate metabolism, and in the detoxification of drugs and poisons. Liver cells are rich in smooth ER. Muscle cells depend on Ca^{2+} pumps within the smooth ER to control the contractile process.

The *Golgi apparatus* is composed of flattened, stacked membranes. Each stack is known as a *dictyosome*. The Golgi apparatus repackages proteins; carbohydrates are added here to form glycoproteins. Glycoproteins form salivary, intestinal, and respiratory tract secretions. They are important constituents of blood proteins, hormones, enzymes, and connective tissues.

The *lysosome* is an organelle that contains powerful digestive enzymes capable of destroying the cell. Some of the lysosomal enzymes are activated through the packaging system of the Golgi apparatus. Lysosomes are found not only in protozoans but also in leukocytes (monocytes and lymphocytes), in thryoid cells (where the stored precursor of thyroxin must be digested before assembly), and in sperm and egg cells.

In hereditary fat-metabolism diseases, lysosomal enzymes necessary to catalyze lipid degradation are missing. As a result, excessive amounts of some lipids begin to accumulate in certain tissues. Lipid-storage disorders in human beings result in retardation, muscle weakness, neurological disorders, and death. Tay-Sachs disease, Fabry's disease, and Niemann-Pick disease are examples of inherited lipid metabolism diseases in children.

In plants, a structure known as the aleurone vacuole contains lysosome-like enzymes. These enzymes are responsible for hydrolyzing the endosperm into diffusable products.

The *mitochondrion* (Figure 5.6) is a membranous organelle. It consists of a smooth outer membrane and a folded inner membrane. The folds are called *cristae*. Like the cell membrane, the mitochondrion membrane is composed of proteins and phospholipids. However, the inner and outer membranes of the mitochondrion differ in proteinaceous material, thereby accounting for the dissimilar functions of portions of the membrane. The mitochondria are necessary for aerobic respiration to take place in cells.

The electron transport system is built into the membrane of the cristae. The enzymes that catalyze reactions of the Krebs cycle are located in the fluid matrix.

Figure 5.6   A mitochondrion

Other Cell Structures   Cell walls and chloroplasts are parts of plant cells. The cellulose wall provides both support and rigidity for the cell. Chloroplasts (Figure 5.7) are essential for the process of photosynthesis, which takes place in the cells of green plants.

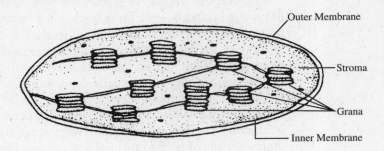

Figure 5.7   Chloroplast

Microbodies are organelles composed of membranous compartments specialized for metabolic activities. Each microbody contains specific enzymes. There are two important types of microbodies, peroxisomes and glyoxysomes.

Peroxisomes contain enzymes that transfer hydrogen to oxygen, forming hydrogen peroxide. The breakdown of fat and the detoxification of alcohol by the liver involve peroxisomes.

Glyoxysomes contain enzymes that convert fat to glucose in germinating seeds.

The *cytoskeleton* is a network of fibers found throughout the cytoplasm. It provides mechanical support and gives shape and motility to cells. There are three types of fibers.

Microtubules, the thickest fibers, which are composed of the globular protein tubulin, make up the centrioles, cilia, and flagella. Microfilaments, the thinnest fibers, are required for muscular contraction and microvilli formation. Intermediate filaments give shape to the cell and fix the organelles in the cell.

Vacuoles (food and contractile) are membranous sacs within the cell. A *tonoplast* is a membrane around a large central plant vacuole that is part of the ER. It stores inorganic ions, wastes, and toxic products and contains hydrolytic enzymes.

The various cell structures and their presence or absence in prokaryotic and eukaryotic cells are shown in Table 5.2.

TABLE 5.2   COMPARISON OF PROKARYOTIC AND EUKARYOTIC CELLS

Structure	Prokaryotic Cell	Eukaryotic Cell Animal	Plant
Plasma membrane	+	+	+
Cell wall	+	-	+
Nucleus	-	+	+
Ribosome	+	+	+
ER	-	+	+
Golgi apparatus	-	+	+
Lysosome	-	+	+
Mitochondria	-	+	+
Chloroplast	-	-	+
Vacuole	-	Small	Large
Centriole	-	+	-

1. In a paramecium, hydrolytic enzymes are found in the
   (A) lysosomes
   (B) chloroplasts
   (C) nucleus
   (D) mitochondria
   (E) ribosomes

2. Of the following, which can be seen with a light microscope?
   (A) Microtubule
   (B) Peroxisome
   (C) Ribosome
   (D) Mitochondrion
   (E) Endoplasmic reticulum

3. A structure commonly found in animal cells but rarely in plant cells is the
   (A) Golgi apparatus
   (B) centriole
   (C) nucleus
   (D) mitochondrion
   (E) endoplasmic reticulum

For Questions 4–8 select the letter of the item that BEST matches the numbered
statement.

   (A) Ribosome
   (B) Cell membrane
   (C) Mitochondrion
   (D) Centriole
   (E) Lysosome
   (F) Golgi apparatus

4. Is the site of protein synthesis

5. Houses the cytochrome system

6. Is required in animal cell division

7. Contains digestive enzymes

8. Adds carbohydrates to proteins

9. Certain types of lymphocytes in the lymph nodes ingest bacteria and debris.
   This function most likely occurs by
   (A) exocytosis
   (B) passive transport
   (C) pinocytosis
   (D) phagocytosis
   (E) facilitated transport

10. A selectively permeable dializing membrane containing specific concentrations of two solutions is immersed in a beaker containing four different solutions. The membrane is permeable to simple sugars but impermeable to disaccharides and polysaccharides. Which solute will diffuse into the cell?

    (A) Sucrose
    (B) Glucose
    (C) Fructose
    (D) Starch
    (E) Water

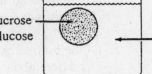

0.02M Sucrose
0.04M Glucose

0.01M Sucrose
0.02M Glucose
0.02M Fructose
0.01M Starch

Answers and Explanations to Model Questions: Cell Structures and Functions

1. A Hydrolytic enzymes are digestive enzymes. These are stored in packages called lysosomes. If the enzymes were allowed to enter the internal environment of the cell, the cell would be destroyed by self-digestion.

    Incorrect Choices
    B Chlorophyll, the pigment of green plants, is found in the chloroplast.
    C DNA is the genetic material of the cell, present in the nucleus.
    D Cytochromes are necessary for cellular respiration. They are found in the mitochondria.
    E Ribosomes are necessary for protein synthesis.

2. D The limitation of the light microscope is not its power of magnification but its power of resolution. Resolution, the ability to distinguish details, depends on the wavelength of light of the visible spectrum and the size of the aperture of the objective. The formula is:

    $$\text{Visible structure} \ = \ \frac{\text{Wavelength}}{\text{Numerical aperture}}$$

    Therefore objects that are one-half the length of the shortest wavelength of light (4,000 Å) become clear. Any object that is smaller than 2,000 Å is not distinguishable.

Mitochondrion	2,000–55,000 Å
Ribosome	200 Å
Endoplasmic reticulum	77 Å
Bacteriophage	65 Å
Tobacco mosaic virus	1.5 Å

3. B All cells contain certain structures. Plant cells usually have cell walls and chlorophyll-containing structures. On the other hand, animal cells contain centrioles and many inclusion bodies.

4. A

5. C

6. D

7. E

8. F

9. D Phagocytosis is the process by which amoeboid-like cells ingest solid material.

Incorrect Choices

A Exocytosis is the expulsion of large particles from the cell through the fusion of vesicles with the cell membrane.

B Passive transport is the movement of material along a concentration gradient across a cell membrane.

C Pinocytosis is the intake of cellular fluids and their dissolved solutes into a cell by the pinching off of the cell membrane to form intracellular vacuoles.

E Facilitated transport is the movement of material along a concentration gradient across a membrane through the action of a carrier molecule.

10. C Fructose is a simple sugar. Fructose diffuses across the membrane because of the concentration gradient.

Incorrect Choices

A Sucrose is a disaccharide.

B Glucose diffuses out of the cell into the beaker because of the concentration gradient.

D Starch is a polysaccharide.

E Water is a solvent.

Cell Division

The life cycle of a cell begins at the time when it is formed by the division of a *parent cell* and continues until the cell dies or itself divides to form new *daughter* cells. The process in which a cell makes another of its kind is called *cell division*. You know that the DNA in a cell stores the genetic information necessary to specify all the proteins it will be able to produce. In prokaryotic cells, you will recall, there is no organized nucleus. The DNA is contained in a single circular chromosome. In eukaryotic cells, the highly organized nucleus has its DNA divided among a number of chromosomes. The pattern of cell division differs in prokaryotes and in eukaryotes. However, in both nucleated and nonnucleated cells, the replication of chromosomes takes place before cell division.

The Cell Cycle

Each kind of cell has a specific life span. After reaching a certain volume, some cells go through cell division. A cell, such as yeast, may reach its critical volume in as little as two hours. The amoeba may begin to divide in a few days. Other cells have life spans of 100 days or more.

Some cells are not able to divide even after they have matured and differentiated. The red blood cell has a life span of 120 days. However, it cannot reproduce itself, neither can nerve cells nor muscular cells.

Eukaryotic cells that can divide have a specific life span knows as a *cell cycle* (Figure 5.8). Cell division involves two processes that may or may not occur together. These are nuclear division (*mitosis*) and cytoplasmic division (*cytokinesis*). Before mitosis and cytokinesis occur, a nondividing cell state or *interphase* exists. A cell cycle encompasses interphase, mitosis, and cytokinesis.

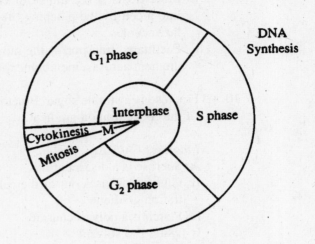

Figure 5.8. Cell cycle

During interphase, a cell grows and replicates its chromosomes. The interphase state is subdivided into three growth phases designated as G_1, S, and G_2. The G_1 phase is characterized by protein synthesis and cytoplasmic organelle production. Chromosomal replication occurs during the S (synthesis) phase. The cell prepares for mitosis during the G_2 phase. Nucleoli, extraneuclear centromeres, and asters appear.

While in the interphase stage, chromosomes are so long and thin that they appear as granules (chromatin) under the light microscope. When interphase comes to an end, the chromosomes shorten and thicken, becoming highly visible. At this time, enough chromosomal material is in the nucleus for two cells.

This orderly process that divided the chromosome equally between two daughter cells is known as mitosis. Mitosis has four stages: *prophase, metaphase, anaphase,* and *telophase* (Figure 5.9). The M stage, which includes both mitosis and cytokinesis, is usually the shortest part of the cell cycle.

The mechanism controlling the cell cycle has not yet been completely determined. However, chemical chain reactions occur that involve the interaction of cyclins and MPF.

Cyclins are proteins that help control the cell cycle. There are two types of cyclins. S-cyclin is involved in DNA synthesis. M-cyclin triggers mitosis. These cyclins activate *cdc*, a component of *MPF* (a complex of M phase promoting factor protein). cdc occurs in an inactive form called cdc_1. cdc_1 is converted into an active form called cdc_s by S-cyclin. M-cyclin converts cdc_1 into an active form called cdc_m. The activated cdc_1 and cdc_m become phosphorylating enzymes (protein kinase). Each acts on a different messenger compound. One messenger triggers DNA synthesis, and the other messenger triggers mitosis. S-cyclin is synthesized in the G_1 phase, and M-cyclin in synthesized in the G_2 phase.

Within the G_1 phase of the cell cycle there is a critical control point just prior to the S phase that determines if cell division is to proceed. This point is known as the *restriction point*. The restriction point may be considered the point of no return. If all conditions are favorable, the cell will enter the S phase. Once the S phase begins, cell division cannot be stopped.

One of the most important factors for passing the restriction point is cell size or the ratio between the cytoplasmic volume and genome size. The cell must grow to a specific but undetermined size during the G_1 phase.

If the signals are not favorable, the cell will enter a G_0 phase, a nondividing state. Most human cells are in the G_0 phase. Nerve cells, skeletal muscle cells, and the red blood cells will never divide. Other cells, such as liver cells, can be induced to reenter the cell cycle by the presence of factors due to injury.

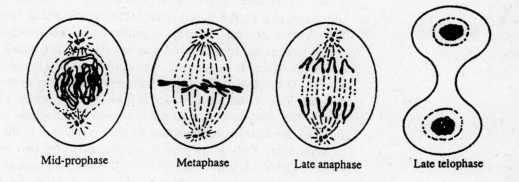

| Mid-prophase | Metaphase | Late anaphase | Late telophase |

Figure 5.9   Stages of mitosis

Cytokinesis

While changes are taking place in the nucleus, activity is occurring in the cytoplasm. There is a doubling of the molecules that make up the cytoplasm and the reassembling of cytoplasmic structures such as the endoplasmic reticulum, ribosomes, Golgi apparatus, and vacuoles. Chloroplasts and mitochondria replicate themselves. During the anaphase stage of mitosis in animal cells, a *cleavage furrow* forms that will later cut across the entire cell at the equator, dividing the parent cell into two new daughter cells. All these changes in the cytoplasm are known as *cytokinesis*. In plant cells, a cell plate structure derived from the coalescence of Golgi vesicles forms across the midline of the parent cell. This becomes the plasma membrane upon which the cell wall forms.

Cell Division in Prokaryotes

A prokaryotic cell does not have an organized nucleus. Bacteria and blue-green algae are examples of prokaryotes. The DNA is contained in a circular chromosome that is attached to a site on the cell membrane known as the *mesosome*. Prior to cell division, the chromosome replicates, producing another chromosome identical to itself. The replicated chromosome is attached to its own mesosome on the cell membrane. The cell now elongates, and as it does so, the chromosomes are pushed further apart. At the time when the cell doubles in length, the cell membrane pinches inward. New cell wall material surrounds the pinched-in membrane and separates the two new cells (Figure 5.10).

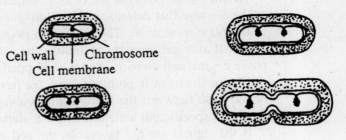

Figure 5.10   Division of a bacterium

Meiosis

Sexual reproduction begins with the process of meiosis. Meiosis is a reduction division that takes place in the primary sex cells (*oocytes* in the female and *spermatocytes* in the male), which have the *diploid* number of chromosomes. Through the process of *meiosis*, primary sex cells produce haploid gametes. In *spermatogenesis*, four haploid sperm cells are produced from one primary sex cell in the testes. In *oogenesis*, one haploid egg cell is produced from one primary sex cell in the ovary. Thus the chromosome number in the egg and sperm cell is one-half that of the primary sex cell or the body cells, known as *somatic cells*. During the process of fertilization, an egg cell nucleus fuses with a sperm cell, and thus the diploid chromosome number is restored in the *zygote*.

Reduction of chromosomes is accomplished by two meiotic divisions appropriately called Meiosis I and Meiosis II. Meiosis I results in the reduction in the number of chromosomes. Meiosis II produces four haploid cells. Figure 5.11 shows the events that take place during both stages of meiosis.

One major event of meiosis is *genetic recombination*. Meiosis makes new combinations of genes possible in two ways. First, new assortments of chromosomes result from the separation of homologous chromosomes during anaphase I. Either chromosome of a pair can end up in a daughter cell with either member of any other pair of chromosomes. Second, during meiosis, homologous chromosomes wrap around each other in an event known as *synapsis*. During synapsis, *crossing over* may take place in which chromosomes exchange like parts. This, in effect, breaks the linkage of genes that occupy sites on the same chromosome. Genes that were previously on one chromosome are now on different chromosomes and can be passed on independently of each other. Genetic recombination explains, in part, the variability of inherited characteristics made possible by sexual reproduction.

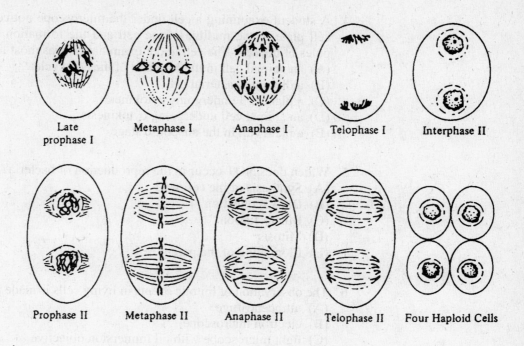

Figure 5.11   Stages of meiosis

Model Questions Related to Cell Division

1. Which of the following is NOT associated with nuclear and cell division in animals?
   (A) Formation of spindles
   (B) Duplication of chromosomes
   (C) Formation of cell plate
   (D) Constriction of the dividing cells
   (E) Separation of chromosomes

2. All of the following statements are true of mitosis EXCEPT
   (A) the cells arising from the process are genetically alike
   (B) the process may not be followed by cytokinesis
   (C) spindle fibers are involved in the movement of chromosomes
   (D) both sexually and asexually reproducing organisms utilize the process
   (E) the cells arising from the process are monoploid

3. The process that ensures equal and identical distribution of genetic material to each cell is known as
   (A) meiosis
   (B) mitosis
   (C) cleavage
   (D) parthenogenesis
   (E) epigenesis

4. A student examining a cell under the microscope noticed the formation of a cell plate in the midline of the cell and the formation of nuclei at opposite poles of the cell. The cell under examination was most likely
   (A) an animal cell in the M phase of the cell cycle
   (B) a dividing bacterial cell
   (C) a plant cell undergoing cytokinesis
   (D) an animal cell undergoing cytokinesis
   (E) a plant cell in the anaphase stage

5. Which does NOT occur in the reproduction of bacteria?
   (A) Synthesis of the cell wall
   (B) DNA replication
   (C) Binary fission
   (D) Mitosis
   (E) Elongation of the parent cell

6. The observation of mitotic events in living cells is made possible by use of the
   (A) ultracentrifuge
   (B) electron microscope
   (C) light microscope with oil immersion objective
   (D) phase-contrast microscope
   (E) dissecting binocular microscope

7. During mitosis, the chromosome is attached to the spindle fiber by means of the
   (A) centrosome
   (B) centriole
   (C) centromere
   (D) kinetosome
   (E) chromatid

8. Which statement concerning the nucleolus is NOT true? The nucleolus
   (A) is produced anew after each cell division
   (B) is capable of self-replication during mitosis
   (C) is produced by a chromosome nucleolar organizer
   (D) is packed with ribonucleoprotein
   (E) and its bounding membrane disappear during mitosis

9. DNA and histones form beadlike globules known as
   (A) nucleosides
   (B) the chromatin network
   (C) centrosomes
   (D) nucleosomes
   (E) mesosomes

10. Which process takes place during the G_1 phase of the cell cycle?
    (A) Mitotic spindle formation
    (B) DNA replication
    (C) Formation of chromatids
    (D) Replication of centromeres
    (E) Protein synthesis

11. The replication of cytoplasmic organelles takes place during
    (A) karyokinesis
    (B) cytokinesis
    (C) the leptotene phase
    (D) diakinesis
    (E) the zygotene stage

12. Chromosome movement during mitosis is explained by the hypotheses of
    (A) sliding filament and subunit disassembly
    (B) one gene–one enzyme
    (C) supercoiling and condensation
    (D) all of these
    (E) none of these

13. Which of the following statements about the immediate products of meiotic division is CORRECT?
    (A) They are genetically identical.
    (B) They have one-half the number of chromosomes of the parent cells.
    (C) They are gametes in higher plants.
    (D) They are spores in animals.
    (E) They are found in archegonia.

Answers and Explanations to Model Questions: Cell Division

1. C The cell plate is found in plants. Since plant cells have a rigid cell wall, the furrowing of the cell membrane to divide the cytoplasm in half is not possible. Various vesicles coalesce with portions of spindle fibers in the center of the cell to form the focal point upon which the endoplasmic reticulum's secretions accumulate. This constitutes the cell plate that forms the new cell wall.

   Incorrect Choices

   A, B, E The general process and features of mitosis (nuclear division) are the same in all cells, with minor variations.

   D Constriction of a cell depends on the activity of microfilaments interlaced with the pliable membrane. Since the cell membrane of a plant cell is surrounded by an inflexible wall, microfilament contractions cannot produce constriction or furrowing.

2. E The process of mitosis provides each cell with the same number of chromosomes as the parent cell. This could be either monoploid (haploid) or diploid.

   Incorrect Choices

   A Since the genetic material is supplied to each new cell in an unaltered form, all cells are genetically alike.

   B Cytokinesis is cytoplasmic division and is a process separate from nuclear division (mitosis).

   C Like contractile muscle protein, spindle fibers provide the movement of chromosomes to opposite poles in the cell.

   D In asexually reproducing organisms, one cell gives rise to a daughter cell that is genetically like the parent cell. This is accomplished through mitosis. In sexually reproducing organisms, the replacement of worn-out or injured body cells occurs as a result of mitosis.

3. B Chromosome duplicates are separated into two identical nuclei as a result of mitosis.

   Incorrect Choices
   A In meiosis, each cell receives one-half of the genetic material from the parent cell.
   C Cleavage is cell division without growth and occurs in embryos. Each embryonic cell contains the same genetic material. This is accomplished through mitosis.
   D The development of an organism from an unfertilized egg is called parthenogenesis.
   E The gradual development of an embryo is epigenesis.

4. C Cytokinesis is the division of the cytoplasm following mitosis that produces two daughter cells. A cell plate is characteristic of plant cell cytokinesis.

   Incorrect Choices
   A The M phase of the cell cycle includes mitosis and cytokinesis. Animal cell cytokinesis is characterized by a cleavage furrow.
   B A bacterial cell lacks a nucleus.
   D Animal cytokinesis is characterized by the formation of a cleavage furrow. There is no cell plate formation.
   E Chromosomes are moved to the opposite pole of the cell during the anaphase stage of mitosis.

5. D Mitosis is the process of cell division that occurs in eukaryotic cells. Bacterial cells are prokaryotes.

   Incorrect Choices
   A Bacterial cells synthesize their own cell walls.
   B A bacterial chromosome replicates before fission.
   C Bacterial cells reproduce by binary fission, in which each daughter cell receives a copy of a single chromosome.
   E The shape of the parent cell changes before fission.

6. D The stages of mitosis can be followed by observing the cells of aquatic embryos through the phase-contrast microscope.

   Incorrect Choices
   A The ultracentrifuge is used to break tissue cells apart by spinning in lysing solutions.
   B Living cells cannot be observed through an electron microscope.
   C The standard light microscope, even if the oil immersion objective is used, cannot offer the light contrast needed to observe the mitotic figures in living cells.
   E A dissecting binocular is used for microsurgery. It gives no light contrast.

7. C During mitosis, the chromosomes are anchored to the spindle fibers by the centromeres of each chromosome. The centromere is the only part of the chromosome that can bind with the spindle fibers.

Incorrect Choices

A The centrosome is a cytoplasmic organelle that lies outside the nucleus and functions during cell division.

B The centriole is the center core of the centrosome.

D The kinetosome is an organelle from which the flagella of protozoans extend.

E A chromatid is a chromosome strand.

8. B The nucleolus is not capable of self-replication. During cell division, the nucleolus and its bounding membrane disappear. After cell division, the nucleolus reappears, having been produced by a nucleolar organizer that is in a region of a particular chromosome. Each nucleolus is packed with ribonucleoprotein.

Incorrect Choices
A, C, D, E See answer 8.

9. D Nucleosomes are the beadlike globules formed by DNA and its attached histones.

Incorrect Choices

A A nucleoside is a nitrogenous base linked to the 1' carbon of a ribose or deoxyribose sugar.

B The chromatin network refers to the thinly stretched chromosomes in the nondividing nucleus.

C Centrosomes lie outside of the nucleus and function during cell division.

E A mesosome is the site on the cell membrane of a bacterium where the circular nucleus is attached.

10. E Protein synthesis takes place during the G_1 phase of the cell cycle.

Incorrect Choices

A Mitotic spindle formation takes place during mitosis, the M phase.

B DNA replication takes place during the synthesis phase (S).

C Chromatid formation takes place during the S phase.

D Replication of centromeres takes place during the S phase.

11. B The replication of cytoplasmic organelles takes place during cytokinesis.

Incorrect Choices

A Karyokinesis is another term for mitosis.

C, D, E Leptotene, diakinesis, and zygotene are stages of the first prophase in meiosis.

12. A Chromosome movement during mitosis is explained by the hypotheses of sliding filament and subunit disassembly.

Incorrect Choices

B One gene–one enzyme refers to the work done by Beadle and Tatum in identifying mutant genes in the mold *Neurospora crassa*.

C Supercoiling and condensation refers to the shortening and thickening of chromosomes at the onset of mitosis.

D, E are incorrect without explanation.

13. B The *immediate products* of meiosis are cells with half the number of chromosomes (haploid number) as the parent cell.

Incorrect Choices

A Meiosis produces haploid cells that are genetically different from the parent cell and genetically different from each other.

C Gametes in higher plants are not the first products of meiosis. For example, when microsprocytes undergo meiosis, microspores are formed. Each microspore undergoes a mitotic division, becoming an immature male gametophyte or pollen grain.

D Animal cells do not have spores.

E Archegonia are structures in Bryophytes where egg cells are produced.

CHAPTER SIX
Energy Transformations

Capsule Concept

All living organisms require energy to carry out the many functions necessary for life. The source of all energy is the sun's light energy, which is trapped by chlorophyll, the green pigment in plant cells. Energy transformations in the cell are the source of all food and energy for all living things. Organisms that make their own food are called *autotrophs*.

Vocabulary

Anaerobic respiration   a biochemical pathway that does not require oxygen to produce ATP

Cellular respiration   biochemical pathways through which chemical bond energy is released from food and changed into ATP

Cytochrome   iron-bearing protein that functions as an electron carrier

Electron transport system   a series of oxidation-reduction reactions in aerobic cellular respiration

FAD   flavin adenine dinucleotide—a hydrogen carrier

Glycolysis   anaerobic respiration

Krebs cycle   a series of biochemical reactions in the cell that result in the production of ATP molecules

NAD   nicotinamide adenine dinucleotide—an electron acceptor and hydrogen carrier used in respiration

NADP   nicotinamide adenine dinucleotide phosphate—an electron acceptor and hydrogen carrier used in photosynthesis

Cellular Respiration

There are two phases of cellular respiration: anaerobic and aerobic. During the anaerobic phase (*glycolysis*), glucose is phosphorylated. Phosphoric acid is added to the glucose molecule through the action of enzymes and the phosphoric acid-contributing compound, adenosine triphosphate (ATP). The double phosphorylation of glucose causes an electronic rearrangement of elements, making it more unstable. Glucose diphosphate breaks into two three-carbon compounds of phosphoglyceraldehyde (PGAL). Further phosphorylation by inorganic phosphoric acid and enol-shifting produces electron rearrangement and high-energy phosphate bonds, which are formed within pyruvic acid. Removal of the high-energy phosphate from PGAL produces ATP. Thus two molecules of ATP are utilized in glycolysis, while four are produced. There is a net gain of two ATP molecules.

It is to be noted that in the conversion of PGAL to PGA (phosphoglyceric acid), hydrogen atoms and electrons are carried by an electron transport system to a final hydrogen acceptor, oxygen. The *electron transport system* is a series of molecules arranged in a graded electron affinity from lowest to highest reduction potentials. Such an arrangement allows an electron to pass down a chain of acceptor molecules spontaneously until the highly stable compound water is formed. In cellular respiration, the electron-carrying molecules are nicotinamide adenine dinucleotide (NAD), flavin adenine dinucleotide (FAD), and the cytochromes. Both NAD and FAD are vitamin B complex derivatives.

Cytochromes (Figure 6.1) contain the pigment heme, which is much like hemoglobin and chlorophyll. Cytochromes are proteins with a prosthetic group of iron in a tetrapyrrole ring. There are several kinds: b, c_1, c, a, a_3 (cytochrome oxidase).

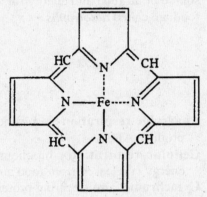

Figure 6.1   Structure of cytochrome

NAD^+ is the first molecule to be reduced in the chain of carrier molecules (Figure 6.2). It passes the electrons on to FAD, which, in turn, passes them on to the cytochromes. The final electron acceptor in the chain is oxygen.

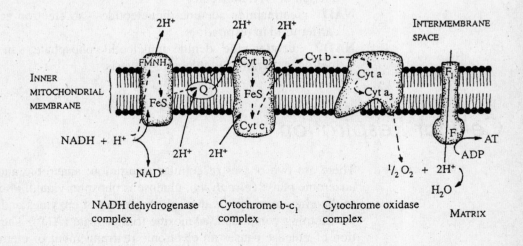

Figure 6.2   Electron transport chain—chemiosmotic ATP synthesis

Although one molecule each of NAD^+ and FAD can handle two electrons at a time, one molecule of cytochrome can handle only one electron at a time. Therefore, for every one dehydrogenase reaction, two molecules of cytochrome are involved.

As the electron moves through the series of redox reactions, its energy is reduced. If the energy released totals 8,000 cal/mole, phosphoric acid is coupled to ADP to form ATP. The resulting high-energy bond can later be broken to furnish energy to power endogenous cellular reactions.

The electron transport system is embedded in the inner membranes of the mitochondria. According to Mitchell's chemiosmotic hypothesis, the coupling of ADP to phosphoric acid to produce ATP in the mitochondria depends on both an electron and a proton gradient proton-motive force. As the electrons flow along the respiratory chain, the protons (H^+) are shunted to the outside of the inner membrane. This establishes a potential difference across the membrane. The only way the protons can cross the membrane is through channels or stalks of the F_1-F_0 complex.

Stalks (F_0) protruding into the matrix from the inner membrane terminate in spheres. The spheres (F_1) are composed of ATP synthase. The passage of the protons drives the reaction which couples ADP to phosphoric acid.

When oxygen is present, pyruvic acid is further degraded to carbon dioxide, water, and energy through cyclic reactions. The reactions are part of the Krebs (tricarboxylic) or citric acid cycle (Figure 6.3). Of the 36 molecules of ATP produced during cellular respiration, 32 are a result of the operation of the electron transport system. The production of ATP during *electron transport* is called *oxidative phosphorylation* or chemiosmotic phosphorylation.

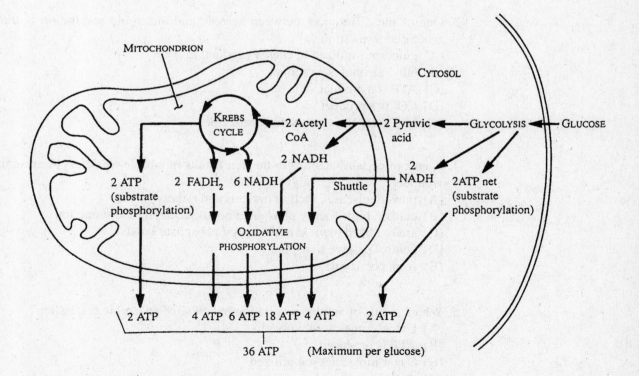

Figure 6.3   ATP yield from citric acid cycle

Organisms that cannot utilize oxygen lack a cytochrome system of the electron transport chain. The hydrogens of $NADH_2$ are carried to pyruvic acid, which is converted to alcohol and/or lactic acid. The term *fermentation* is used to describe anaerobic respiration. Although fermentation is characteristic of biologically simpler organisms, an alternative metabolic pathway similar to fermentation occurs during muscle contraction under conditions of oxygen deficiency. Glucose derived from glycogen is converted into lactic acid, which causes muscle fatigue.

Besides glucose, other organic compounds may enter the respiratory pathway at different points. Various compounds in the pathway, such as pyruvic acid, may also serve as a carbon backbone for the synthesis of organic compounds (amino acids, fatty acids, fats, and so on).

Model Questions Related to Cellular Respiration

1. Which of the following is an outcome of the Krebs cycle?
   (A) The synthesis of coenzymes
   (B) The reduction of glucose to glucose phosphate
   (C) The production of pyruvate and oxygen
   (D) The production of energy stored in NADH and $FADH_2$
   (E) The synthesis of carbohydrates

2. The driving force in the production of ATP in the mitochondria comes from
   (A) ATPase
   (B) electrochemical proton gradients
   (C) osmotic gradient
   (D) diffusion of electrons
   (E) ionic gradient

3. One of the differences between aerobic and anaerobic respiration is that in anaerobic respiration
   (A) glucose is utilized in energy production
   (B) ethyl alcohol is a product
   (C) ATP is a product
   (D) CO_2 is a product
   (E) NAD (DPN) is one of the coenzymes

4. A lack of certain vitamins in the diet results in muscle weakness because these vitamins
   (A) prevent diseases such as beriberi and rickets
   (B) contain high-energy phosphate bonds required for contraction
   (C) are oxidized to yield high-energy phosphate bonds
   (D) control passive transport
   (E) form coenzymes

5. Which is true of aerobic respiration but NOT of anaerobic respiration?
   (A) Cytochromes are involved.
   (B) ATP is produced.
   (C) Sugar molecules are utilized.
   (D) ADP is produced.
   (E) CO_2 is a by-product.

6. Which statement BEST describes an important event in the process of oxidative phosphorylation?
(A) ATP is changed to creatine phosphate.
(B) Reduced cytochrome b is oxidized directly to cytochrome a^3.
(C) Phosphorus is added to glucose molecules.
(D) Lactic acid is produced.
(E) ADP is converted into ATP.

7. Muscle fatigue results from the accumulation of lactic acid in muscles because of
(A) enzyme deficiency
(B) oxygen deficiency
(C) excess number of mitochondria in the cell
(D) too much oxygen in the cells
(E) inability to metabolize glucose

Questions 8–11 are based on the results of the experiment that follows and on your knowledge of biology. The purpose of the experiment is to measure the capacity of pigeon breast tissue cell fractions to carry out oxidation of glucose and pyruvate.

Pigeon breast tissue and lysing fluid are put into a homogenizer where the tissue cells are separated from each other and cell membranes are broken open. The homogenate is transferred to a centrifuge and spun at 70,000 rotations per minute. The nuclei fall to the bottom. The supernatant fraction above the nuclei is spun at a higher speed to sediment the mitochondria. The supernatant fraction above the mitochondrial sediment contains ribosomes. This is the supernatant referred to in the table. Each of the separated fractions is put into a manometer to measure changes in the volume of oxygen. The data chart below shows the volume of oxygen consumed expessed as percent of oxygen used up by the whole homogenate.

Fractions Measured for O_2	Glucose Substrate	Pyruvate Substrate
Entire homogenate	100	100
Cell nuclei	12	6
Supernatant	6	6
Mitochondria	3	75
Mitochondria + nuclei	5	50
Mitochondria + supernatant	150	130
Mitochondria + nuclei + supernatant	95	100

8. From the data given, the ability to oxidize pyruvate exists most extensively in the fraction of
(A) nuclei
(B) supernatant
(C) mitochondria
(D) supernatant + ribosomes
(E) ribosomes

9. The data presented in the table indicates that nuclei
   (A) are sites for transcription of tRNA
   (B) and ribosomes have the same oxidizing capacity
   (C) lower the oxidation level of mitochondria
   (D) are absent from the entire homogenate
   (E) and the supernatant oxidize glucose most efficiently

10. Of the following, which is a CORRECT statement about the supernatant?
    (A) It oxidizes pyruvate more effectively when combined with mitochondria.
    (B) It effectively converts pyruvate to glucose.
    (C) When combined with nuclei, the supernatant has a greater capacity for oxidizing glucose than has the entire homogenate.
    (D) When combined with mitochondria and nuclei, it has a greater capacity to oxidize glucose than pyruvate.
    (E) The supernatant fraction exhibits a greater capacity to oxidize pyruvate than glucose.

11. During cellular respiration, glucose is converted to pyruvate in the cytosol and pyruvate is converted to CO_2 and H_2O in the mitochondria. Where are the GREATEST number of ATP molecules produced?
    (A) In the cytoplasm during phosphorylation of glucose
    (B) In the outer membranes of the mitochondria
    (C) In the outer membranes of the ribosomes
    (D) In the inner membranes of the mitrochodria during electron transport
    (E) In cystol surrounding the mitochondria

Answers and Explanations to Model Questions: Cellular Respiration

1. D Most of the energy produced by the oxidative steps of the Krebs cycle is stored in NADH. For each molecule of acetate that enters the cycle, three molecules of NAD are reduced to NADH. In one oxidative step, electrons are transferred to FAD. Through electron transfer, FAD is reduced to $FADH_2$.

   Incorrect Choices
   A Coenzymes NAD and FAD are not synthesized during the Krebs cycle but are present in the system functioning as electron carriers.
   B Glucose is phosphorylated during the anaerobic phase known as glycolysis.
   C Pyruvate is the end product of the glycolytic pathway.
   E Carbohydrates are not synthesized during respiration. The outcome of cellular respiration is energy and carbon dioxide and water.

2. B The driving force in the production of ATP in the mitochondria comes from electrochemical proton gradients. The movement of electrons along the transport system results in the establishment of a proton gradient across the membranes. The movement of the protons across the membranes at specific points provides the force of phosphorylation of ADP to ATP.

   Incorrect Choices
   A ATPase is the enzyme that catalyzes the coupling of ADP to P.
   C Osmotic gradient is the difference in the concentration of water molecules across a membrane.

D Electrons do not diffuse across the mitochondrial membrane. Electrons are carried by special molecules.

E Ionic gradients are differences in concentrations of ions across a membrane.

3. B Alcohol fermentation is a means by which glycolysis is terminated when oxygen is not available. Pyruvic acid is decarboxylated to acetaldehyde and CO_2. The acetaldehyde becomes a hydrogen acceptor and is converted into ethyl alcohol and oxidized NAD. Alcohol fermentation takes place in yeast cells.

Incorrect Choices

A Chemical energy in the form of high-energy phosphate bonds is released from glucose during anaerobic and aerobic respiration. The difference between these two respiratory phases is the number of ATP molecules produced.

C During anaerobic and aerobic respiration, ADP molecules are upgraded to ATP as illustrated: ADP + P $\rightarrow$ ATP.

D Carbon dioxide is produced as a result of anaerobic and aerobic respiration. The enzyme carboxylase effects the removal of CO_2 from pyruvic acid.

E NAD functions as a hydrogen carrier. As a consequence of anaerobic respiration, the hydrogen from NADH + H (reduced NAD) is transferred to pyruvic acid, which becomes a final hydrogen acceptor. When respiration occurs aerobically, hydrogen moves from NADH through a transport chain consisting of FAD and cytochromes. At the end of this chain, oxygen is the final hydrogen acceptor.

4. E In order for some enzymes to function, a portion of the enzyme molecule must be able to accept the atoms (or electrons) of the substrate that is being removed. Coenzymes are nonprotein, organic portions of an enzyme that accept atoms (or electrons) from substrate molecules. Without coenzymes, the reaction cannot be completed.

Incorrect Choices

A Rickets is caused by a deficiency of vitamin D, a substance needed to promote the retention of calcium and phosphorus in the bones. Beriberi is caused by a lack of thiamine. This disease is characterized by an accumulation of pyruvic and lactic acids in the blood and in the brain, impairing the cardiovascular, nervous, and gastrointestinal systems.

B ATP provides the high-energy bonds necessary for muscle contraction.

C Glucose is the primary energy source.

D Passive transport is the movement of molecules from an area of greater concentration to an area of lesser concentration. Energy is not utilized for diffusion.

5. A Oxygen is the final hydrogen (electron) acceptor at the end of the chain of cytochrome carriers.

Incorrect Choices

B, C, D, E All of these activities take place during anaerobic respiration.

6. E As hydrogen is passed down through the cytochrome transport system, energy is released to convert ADP and inorganic phosphate into ATP.

Incorrect Choices

A Creatine phosphate is produced from ATP, but it is not part of the ordinary cellular respiratory pathway. It occurs primarily in muscle cells, where it is the major supplier of energy for contraction.

B Cytochrome b is oxidized by cytochrome c.

C When phosphorus is added to ADP, ATP is formed.

D Lactic acid is a product of anaerobic respiration, which does not employ an electron system.

7. B If oxygen is not available, the hydrogens from the anaerobic phase of cellular respiration are passed on to pyruvic acid, forming lactic acid. Lactic acid may act as a respiratory enzyme inhibitor.

Incorrect Choices

A The accumulation of lactic acid results from a lack of oxygen and is not related to the amount of enzymes present. The respiratory enzymes are jammed up since there is no oxygen to act as an acceptor of hydrogen.

C It is not the number of mitochondria present, but rather the amount of oxygen available, that results in lactic acid production.

D Excess of O_2 would be advantageous in converting pyruvic acid into CO_2 and H_2O; lactic acid would not be formed or would not accumulate in such quantities.

E Inability to metabolize glucose would result in a diabetic state. Other organic compounds would be utilized, resulting in accumulation of products other than lactic acid.

8. C The mitochondrial fraction exhibits the greatest ability to oxidize pyruvate as shown in the figures on the chart.

Incorrect Choices

A The data indicates that oxidation of pyruvate by nuclei is only 6% as compared with 75% by the mitochondria.

B Oxidation by the supernatant is only 6%.

D and E There is no chart listing of supernatant + ribosomes or ribosomes. Although information given is that ribosomes are present in the supernatant, data shows that a very small percentage of the oxidation of pyruvate occurs in the supernatant fraction.

9. C Examination of the data shows that the percentage of oxidation in the mitochondrial fraction is 74% while that in the fraction of mitochondria + nuclei drops to 50%. Mitochondria + supernatant is 130%, but mitochondria − nuclei + supernatant drops to 100%. These figures indicate that nuclei lower the oxidation level of the mitochondria.

Incorrect Choices

A The data is unrelated to DNA transcription.

B This data is not given.

D Nuclei are present in the entire homogenate.

E This answer is not supported by the data.

10. A The data given indicates that when mitochondria are combined with the supernatant, the oxidizing power is 130% as compared with 75% of the uncombined mitochondrial fraction.

Incorrect Choices
B This choice presents incorrect information and is not related to the problem.
C This choice is not supported by the data.
D Data show that nuclei lower the rate of oxidation of pyruvate.
E The figures on the chart show that the supernatant oxidizes pyruvate and glucose equally at 6%.

11. D Your knowledge of the biology of cellular respiration will enable you to select the correct answer. Electron transport chains are built into the inner mitochondrial membranes that are expanded by infoldings called cristae. The electron transport chains are energy converters that are coupled to ATP synthesis. During cellular respiration, the greatest number of ATP molecules (34) are produced in the inner membranes of the mitochondria.

Incorrect Choices
A An ATP molecule is used up in the phosphorylation of glucose.
B Pyruvate enters the outer membrane of the mitochondria, but the major work of ATP synthesis takes place in the inner membrane.
C Ribosomes do not function in cellular respiration.
E Anaerobic respiration occurs in the cytoplasm.

Photosynthesis

Photosynthesis is the process by which green plants manufacture carbohydrates out of carbon dioxide and water in the presence of light. Chlorophyll is essential to the process.

Chlorophyll (Figure 6.4), a green pigment, is related in molecular structure to hemoglobin and cytochrome. Each of these pigment molecules consists of a metallic porphyrin ring and protein. Unlike other pigments, chlorophyll is necessary for the process of photosynthesis.

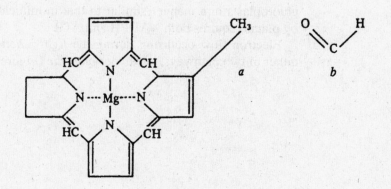

Figure 6.4   Chemical formulas for chlorophyll *a*
and *b*–CH$_3$ in chlorophyll *a*;–CHO in chlorophyll *b*

There are several kinds of chlorophyll molecules: *chlorophyll* a, *chlorophyll* b, *chlorophyll* c, *chlorophyll* d, and *chlorophyll* e. Among photosynthetic organisms, chlorophyll *a* is the most common type of molecule. This blue-green chlorophyll differs from yellow-green chlorophyll *b* in that the latter has a formyl group

instead of a methyl (–CH$_3$) group. This slight difference causes the two pigments to differ in their ability to absorb light.

In addition to the chlorophylls, other pigments known as accessory pigments are present in plant cells. These include the carotenoids (carotenes and xanthophylls) and the phycobilins (phycoerythrin and phycocyanin).

The process of photosynthesis depends on the action of light on chlorophyll. Remember that chlorophyll is a metallic pigment. "White" light is composed of seven color components: violet, indigo, blue, green, yellow, orange, and red. Since light is transmitted by wave motion, each component of the visible spectrum can be distinguished by a specific wavelength.

When light strikes a metallic surface, an electric current is produced. This is known as the *photoelectric effect*. Because of this phenomenon, the modern biologist has found the Newtonian theory of the particulate nature of light useful in the explanation of photosynthesis.

Each wavelength of light consists of packets of energy called *photons*. The amount of energy for each proton is qualified by this formula:

$$\text{Energy} = \text{Constant } (h) \times \frac{1}{\text{Wavelength } (v)}$$

Photons from the blue and red ends of the spectrum contain enough energy to dislodge electrons from chlorophyll molecules, elevating them from the ground state to the excited state. Chlorophyll *a* molecules are arranged in two distinct ways called photosystems. These are used to collect light. The arrangements are designated as photosystem I (PSI) and photosystem II (PSII). PSI is also known as P700 and PSII as P680. These numbers refer to their absorption spectra. The movement of chlorophyll electrons through an electron transport chain and the establishment of a proton-motive force are responsible for the production of NADPH and ATP (*photosphophorlyation*). ATP is generated by chemiosmosis in chloroplasts in a manner similar to that in mitochondria. Carbohydrate synthesis by plants requires both NADPH and ATP.

Electron flow occurring during the *light reaction* of photosynthesis may take either of two pathways, *cyclic* or *noncyclic* (Figure 6.5).

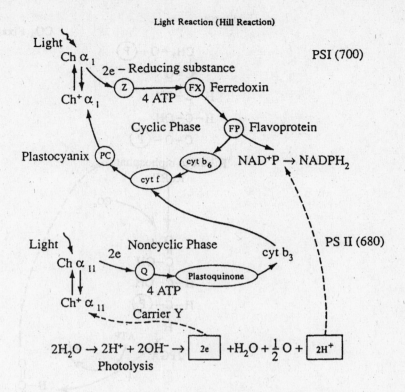

Light Reaction (Hill Reaction)

Figure 6.5   Cyclic and noncyclic pathways of electron flow

During the cyclic phase, chlorophyll *a* electrons in the reaction center of PSI are dislodged by wavelengths of light in the 700nm range. The dislodged electrons return the chlorophyll *a* molecules by way of the electron transport chain. The result is the production of ATP. Not all electrons that escape from chlorophyll *a* are returned to the parent molecule. Some of the electrons follow a noncyclic course from chlorophyll *a*. Meanwhile, the photolysis of water makes hydrogen protons available. These join with the electrons that have traveled through the flavoprotein. However, chlorophyll *a* in the reaction center of PSII absorbs light in the 680nm range. The electrons dislodged by the light are transferred to the electron-deficient chlorophyll *a* in PSI by way of an electron transport chain.

To date, the mechanism of *photolysis* of water is not completely understood. By experimentation, which employed the oxygen isotope ^{18}O, it has been established that oxygen is evolved during photolysis. Through the technique of chlorophyll spectral analysis, it has been shown that PSI generates ATP but produces no oxygen.

The second major stage of photosynthesis involves reductive carbon dioxide fixation. Because the cyclic reactions that function at this time do not require light as a source of energy, the term *dark reaction* has been used to designate this phase of photosynthesis (Figure 6.6). Calvin and his associates determined the path of carbon in the carbon dioxide by the use of ^{14}C.

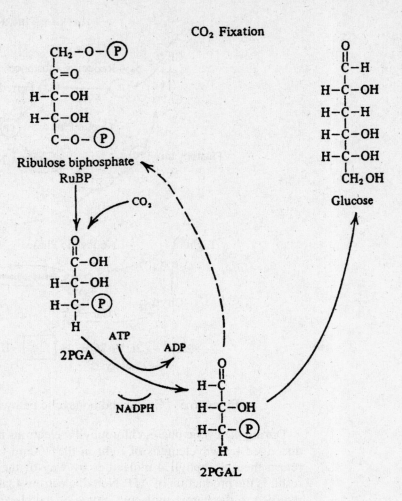

CO_2 Fixation

Figure 6.6   Calvin cycle or dark reaction

In the dark reaction, CO_2 combines with ribulose biphosphate (RuBP). Through a series of enzymatically controlled reactions, PGAL is formed.

PGAL is an important intermediate product of photosynthesis. Note the many metabolic pathways that lead to and away from the molecule (Figure 6.7).

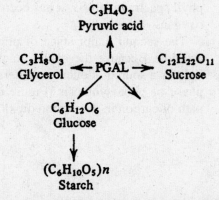

Figure 6.7   Synthetic pathways from PGAL

PGAL is a 3-carbon compound. Therefore the Calvin cycle is also known as C_3 photosynthesis.

The C_3 pathway operates efficiently under high concentrations of CO_2. When the CO_2 concentration is reduced, O_2 combines with RuBP and glycolic acid is formed. Both O_2 and CO_2 compete for the RuBP enzyme. The destruction of RuBP and the loss of CO_2 is known as *photorespiration*. Photorespiration occurs in all temperate region C_3 plants.

In the late 1960s, a second photosynthetic pathway was discovered. The pathway is known as the C_4, or Hatch-Slack photosynthetic pathway. In the C_4 pathway, CO_2 combines with a compound known as PEP. The result is a 4-carbon compound, malate, that is transferred to special cells in the leaf. The 4-C compound is decarboxylated and CO_2 enters the Calvin cycle in the bundle sheath cells of the leaf.

The C_4 pathway is found in plants that inhabit the desert and tropical environments.

$$CO_2 + PEP \rightarrow \text{oxaloacetic acid} \rightarrow \text{malate} \rightarrow CO_2 \rightarrow \text{Calvin cycle}$$

C_4 plants have a special leaf anatomy known as Kranz anatomy (Figure 6.8). The mesophyll cells produce the 4-C compound. The mesophyll cells surround the bundle sheath cells. Bundle sheath cells also contain chloroplasts. The cells decarboxylate the 4-C compound and contain the mechanism for the Calvin cycle.

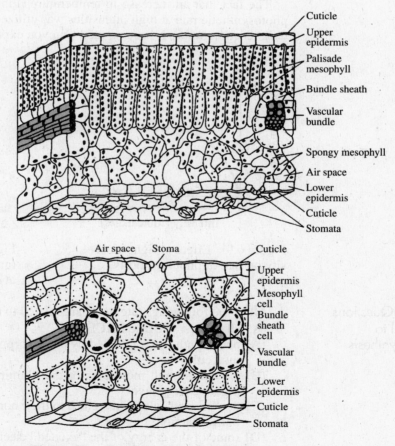

Figure 6.8 C_3 leaf cross section (above),
Kranz anatomy C_4 leaf cross section (below)

Because the stomates of some water-storing plants (succulents and cacti) are closed during the day to prevent water loss, carbon dioxide intake is prevented. These plants take in carbon dioxide during the night and convert it to organic acids. This method of carbon fixation is called *crassulacean acid metabolism*, CAM. Carbon dioxide for photosynthesis is released from the acids during daylight.

In most green plants, chlorophyll is contained in membranous structures, the *chloroplasts*. Chloroplasts (like mitochondria) contain their own DNA and ribosomes, making self-replication possible. Small, thin-membraned sacules called *thylakoids,* stacked on top of each other like coins, form the *grana*. The ground substance between each pair of interconnected grana is the *stroma*.

Embedded in the thylakoid membranes are chlorophyll molecules and the electron transport chain. As electrons move through the electron transport chain, protons (H^+) are pumped into the thylakoid compartment. Protons can recross thylakoid membranes only through ATP synthase channels where the heads of these enzyme molecules face into the stroma. As protons cross thylakoid membranes, ATP is produced in the stroma. The process of ATP production is known as chemiosmosis.

The enzymes that function during the light reaction of photosynthesis are located in the grana. Enzymes necessary for the dark reaction are located outside the grana in the stroma.

The several factors that influence the rate of photosynthesis are light intensity, temperature, CO_2 concentration, and availability of minerals. Light intensity refers to the number of photons per unit of time. According to the graph in Figure 6.9, the rate of photosynthesis increases up to a point with the increase in light intensity. Why?

The fact that an increase in temperature (Figure 6.10) has a greater effect on photosynthetic rate at high intensities was utilized to suggest the possibility of the dark reaction before it became fact. Can you explain this?

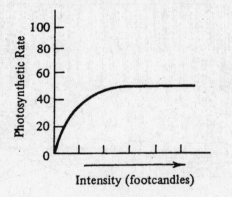

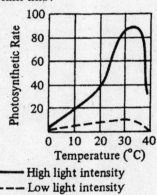

Figure 6.9   Effect of light intensity on rate of photosynthesis

Figure 6.10   Effect of temperature on rate of photosynthesis at different light intensities

Model Questions Related to Photosynthesis

1. All of the following statements are related to the light reaction of photosynthesis. All of these are true EXCEPT
   (A) photosynthesis is initiated by the absorption of light energy by chlorophyll molecules
   (B) the chlorophyll molecules of the illuminated chloroplasts are raised to a higher energy level
   (C) the "excited" electrons return to their normal energy levels if not captured in the reaction
   (D) some of the energy of the "excited" electrons is used to split carbon dioxide into carbon and oxygen
   (E) the biochemical activities of the light reaction occur in the grana

2. An activity performed by plants but not by animals is
   (A) digestion of carbohydrates
   (B) fixation of carbon dioxide
   (C) adaptation
   (D) cellular respiration
   (E) resistance to disease

3. In the process of photosynthesis, the oxygen released comes from
   (A) PGAL
   (B) starch
   (C) carbon dioxide
   (D) carbon dioxide and water
   (E) water only

For Questions 4–10, choose the letter of the item that BEST matches each numbered statement

   (A) Hydrolysis   (D) Amination
   (B) Phosphorylation   (E) Decarboxylation
   (C) Reduction with hydrogen

 4. Reaction that converts ADP into ATP

 5. Reaction that converts fats into fatty acids and glycerol

 6. Reaction that converts pyruvic acid into acetaldehyde

 7. Reaction that converts proteins into amino acids

 8. Reaction that converts unsaturated fats into saturated fats

 9. Reaction that converts pyruvic acid into alanine

10. Reaction that converts pyruvic acid into lactic acid

11. Which is NOT a by-product of the light reaction of photosynthesis?
    (A) $NADPH_2$
    (B) Oxygen
    (C) PGAL
    (D) ATP
    (E) Hydroxyl ions

12. Which statement concerning chloroplasts is NOT true?
    (A) They have a DNA separate from nucleic DNA.
    (B) They can function outside a cell.
    (C) They have their own ribosomes.
    (D) They are divided into grana.
    (E) They manufacture insufficient amounts of ATP for complete cellular metabolism.

13. Below are four statements. Which statements are true for BOTH aerobic respiration and photosynthesis?
   - (1) Synthesizes ATP
   - (2) Makes use of molecular oxygen
   - (3) Makes use of an electron transport system
   - (4) Is controlled by enzymes

   - (A) 1, 2, 3, 4
   - (B) 1, 2, 3
   - (C) 1, 3, 4
   - (D) 2, 3, 4
   - (E) 2 only

Base your answers to Questions 14–16 on the paragraph below.

Englemann placed a strand of the filamentous alga *Cladophora* onto a slide, to which he added aerobic bacteria. Under conditions of white light, microscopic observation revealed that the bacteria were evenly distributed along the length of the alga filament. The beam of white light was broken down into its component wavelengths, which extended across the spectrum from the blue end to the red end. It was then focused on the alga. Englemann noticed that the bacteria migrated to and accumulated around the portions of the alga filament that were positioned within the zones of the blue and red wavelengths of light.

14. A probable reason for the clustering of the bacteria in the red and blue portions of the spectrum is
   - (A) the bacteria prefer to feed in red and blue light
   - (B) the bacteria prefer red and blue colors to green and yellow colors
   - (C) the bacteria clustered at the red and blue ends of the spectrum by chance only
   - (D) the bacteria exhibit a positive red and blue phototropism
   - (E) the bacteria require some substance found in greater abundance in the red and blue areas of light

15. The difference in conditions of the alga in the blue and red wavelengths of light might be due to the presence of
   - (A) carbon dioxide
   - (B) water
   - (C) oxygen
   - (D) carbohydrates
   - (E) protein

16. The process that was investigated by the experiment was
   - (A) fermentation
   - (B) photosynthesis
   - (C) cellular respiration
   - (D) plasmolysis
   - (E) bacteria adaptation

Answers and Explanations to Model Questions: Photosynthesis

1. D Some of the energy of photosynthesis is used to the process of photolysis of water. Oxygen is released from the water molecule.

Incorrect Choices

A The photosynthetic process is initiated by the excitation of electrons in the prophyrin portion of the chlorophyll molecules.

B The excited electrons are raised to a higher energy level in the process of photosynthesis.

C If the electrons do not enter an electron transport chain, they fall back down to their original energy levels. The excess energy appears as a fluorescence.

E Photosynthesis occurs in the chloroplasts. Grana are organized units within the chloroplasts that contain the enzymes and the chlorophyll molecules needed for light reactions.

2. B Reductive carbon fixation is part of the photosynthetic reaction that occurs during the dark phase.

Incorrect Choices

A Both plants and animals require a source of energy for metabolic activities. Carbohydrates furnish the energy source.

C Adaptation allows organisms to survive in their environment. The adaptative mechanism is related to the genetic mechanism that exists in both plants and animals.

D Respiration produces energy, which is necessary to both plants and animals.

E Survival depends on the disease resistance in both plants and animals.

3. E The oxygen released during photosynthesis comes from water. The chloroplast splits water into hydrogen and oxygen. The use of ^{18}O, a heavy isotope, to trace the pathway of oxygen atoms during photosynthesis proved that O_2 comes from the splitting of water.

Incorrect Choices

A PGAL, phosphoglyceraldehyde, is formed during carbon dioxide fixation known as the Calvin cycle. Oxygen is not an end product of PGAL formation.

B Starch is a by-product of photosynthesis in plants. It is formed in a series of dehydration syntheses or condensations of carbohydrate molecules. During each condensation, a molecule of water is formed from the hydrogen and oxygen is released.

C During the Calvin cycle in photosynthesis, CO_2 is joined to ribulose biphosphate, forming 2 molecules of PGA (phosphoglycerate). Oxygen is not an end product of this reaction.

D Carbon dioxide and water form carbonic acid.

4. B ADP is converted into ATP by the addition of a phosphoric acid molecule. Phosphorylation is the addition of phosphorous, usually from phosphoric acid, to a molecule.

5. A The addition of water splits fats into their component parts. Hydrolysis is the separation of substances into component parts by the addition of water.

6. E $CH_3COCOOH$ (pyruvic acid) is converted into CH_3COH (acetaldehyde) by decarboxylation. Decarboxylation is the removal of carbon dioxide from a molecule.

7. A The addition of water to proteins splits them into their component parts, amino acids. Hydrolysis is the separation of substances into component parts by the addition of water.

8. C The term "unsaturated" implies that double bonds are available. The addition of hydrogens saturates the bonds. Reduction with hydrogen is the addition of hydrogen to a molecule.

9. D An amino group must be added to convert pyruvic acid into alanine. Amination is the addition of ammonia or amine to a molecule.

10. C Pyruvic acid accepts hydrogen and forms lactic acid. Reduction with hydrogen is the addition of hydrogen to a molecule.

11. C PGAL is produced in the dark reaction.

Incorrect Choices
A Electrons from chlorophyll molecules and hydrogen from water are carried to NADP. The electrons become available from the excitation of chlorophyll by light.
B Oxygen results from the photolysis of water, which is light dependent.
D Energy is produced as the chlorophyll electrons are transported through an electron chain.
E Hydroxyl ions result from the photolysis of water.

12. E Although enough ATP is produced during the light reaction to drive the dark reaction, there are many other cellular metabolic activities that require energy, which is furnished by the process of cellular oxidation.

Incorrect Choices
A Chloroplasts are self-duplicating structures; therefore they must contain the genetic material (DNA) necessary for duplication.
B Because chloroplasts can function outside a cell, the chemical mechanisms involved in photosynthesis could be studied.
C Since the chloroplasts can function outside cells, they must contain the mechanism necessary to produce enzymes, ribosomes.
D Stacks of lamellae comprise the areas known as grana.

13. C Since the process of photosynthesis produces oxygen as a by-product, (2) is incorrect. Statements (1), (3), and (4) are main features in both process-

es. Any choice of answers that excludes statement (2) is correct.

14. **E** Since the bacteria are evenly distributed along the filament prior to the use of a spectrum, there must be something special in the regions of blue and red light. The highest rates of photosynthesis are known to occur in plants exposed to both red and blue wavelengths of light. More oxygen was liberated at these regions, attracting greater numbers of the aerobic bacteria.

Incorrect Choices

A, B Organisms such as bacteria are not capable of having preferences.

C Since the rate of photosynthesis is influenced by the wavelength of light, the bacteria are not clustered by chance.

D There are no photoreceptors or wavelength-detecting apparatuses in bacteria cells.

15. **C** Since photosynthesis occurs most efficiently in blue and red light, there must have been a substance produced in great abundance which attracted the organisms. There are two possibilities—nutritive material and oxygen. Since the experiment emphasized the use of aerobic bacteria, oxygen must have been the special substance.

Incorrect Choices

A Carbon dioxide is utilized in the process of photosynthesis.

B Water is evenly distributed along the filament.

D, E Both substances are present in the filament.

16. **B** Light wavelengths influence the rate of photosynthesis.

Incorrect Choices

A Process by which energy is produced anaerobically.

C Process by which energy is produced aerobically.

D Process in which the cell shrinks because of water loss.

E Prior to formation of the spectrum, bacteria are capable of surviving; therefore adaptation is not a factor.

References for Area I: Molecules and Cells

All of the following are from *Scientific American*:

Bayley, H. "Building Doors Into Cells." September 1997

Bazzazz, F. A., and E. D. Fajer. "Plant Life In a CO_2-Rich World." January 1992

Ben-Jacob, E., and H. Levine. "The Artistry of Microorganisms." October 1998

Bernstein, M. P., S. Sandford, and L. Allamandola. "Life's Far-Flung Raw Materials." July 1999

Bisceglie, A. M., and B. Bacon. "The Unmet Challenges of Hepatitis C." October 1999

Bollinger, J. J., and D. J. Wineland. "Microplasmas." January 1990

Coleman, W. J., and G. Coleman. "How Plants Make Oxygen." February 1990

Deduve, C. "The Birth of Complex Cells." April 1996

Deyo, R. A. "Low Back Pain." August 1998

Fischetti, V. A. "Streptococcal M Protein." June 1991

Foster, K. R., M. F. Jenkins, and A. C. Toogood. "The Philadelphia Yellow Fever Epidemic of 1793." August 1998

Hinkle, P. C., and R. E. McCarty. "How Cells Make ATP." March 1978

Hogle, J. M., M. Chow, and D. J. Filman. "The Structure of Poliovirus." March 1987

Horwith, A. F. "Integins and Health." May 1997

Jordon, V. C. "Designer Estrogens." October 1998

Kalil, R. E. "Synapse Formation in the Developing Brain." December 1989

Kartner, N., and V. Ling. "Multidrug Resistance in Cancer." March 1989

Kimelberg, H. K., and M. D. Norenberg. "Astrocytes." April 1989

Laver, W. G., N. Bischofberger, and R. Webster. "Disarming Flu Viruses." January 1999

Levy, S. "The Challenge of Antibiotic Resistance." March 1998

Lysaght, M. J., and P. Aebischer. "Encapsulated Cells as Therapy." April 1999

Mazia, D. "The Cell Cycle." January 1974

McDonald, J. W. "Repairing the Damaged Spinal Cord." September 1999

McIntosh, J. R., and K. L. McDonald. "The Mitotic Spindle." October 1989

Mooney, D. J., and A. Mikos. "Growing New Organs." April 1999

Oldstone, M. B. A. "Viral Alteration of Cell Function." August 1989

Orci, L., J. Vassalli, and A. Perrerlet. "The Insulin Factory." September 1988

Renner, R. "Asbestos in the Air." February 2000

Richards, F. W. "The Protein Folding Problem." January 1991

Rothman, J., and L. Orci. "Budding Vesicles in Living Cells." March 1996

Smith, K. A. "Interleukin-2." March 1990

Stoeckenius, W. "The Purple Membrane of Salt-Loving Bacteria." June 1976

Storey, K. B., and J. Storey. "Frozen and Alive." December 1990

Stossel, T. "The Machinery of Cell Crawling." September 1994

Tiollais, P., and M. Annick-Buendia. "Hepatitis B Virus." April 1991

Todorov, I. N. "How Cells Maintain Stability." December 1990

Wickramasinghe, H. K. "Scanned-Probe Microscopes." October 1989

Wolpert, L. "Pattern Formation in Biological Development." March 1978

Zewail, A. H. "The Birth of Molecules." December 1990

Zivin, J. A., and D. W. Choi. "Stroke Therapy." July 1991

PART FOUR

Area II of Advanced Placement Biology: Heredity and Evolution

CHAPTER SEVEN
Molecular Genetics

Capsule Concept

Modern genetics has made great strides beyond the theories of classical Mendelian inheritance, especially in biochemistry. The modern investigator attempts to answer "how" and "why" questions using all of the tools of modern technology. The secrets of cell function are contained in the biochemical events that take place in the cell nucleus and cytoplasm. Much scientific knowledge and technological skill has been applied to learning about the functions of the cell's organelles and its fine structure. In addition, modern geneticists devote investigative effort to the correction of gene flaws by using techniques known as genetic engineering.

Vocabulary

Autosomal dominant   a dominant gene that is carried on a body cell or somatic cell chromosome

Autosomal recessive   a recessive gene that is carried on a nonsex chromosome

Biotechnology   the broad field of genetic engineering

Capsid   the protein coat of a virus

Chimeras   plasmids produced in test tubes from separate genetic elements

Clone   a group of genetically identical individuals; an identical copy of a gene from one organism that is replicated from another organism

Double helix   the shape of DNA

Exon   the coding region of DNA

Gene   a blueprint for a protein

Gene splicing   the set of techniques used to incorporate a group of DNA molecules from one organism into a DNA strand of another organism

Genetic engineering   techniques used to remove faulty genes from chromosomes and replace them with functional genes

Genetic therapy   processes involved in correcting genetic disorders by means of genetic engineering

Genome   the entire set of genes for an organism

Infect   to enter a cell (used with reference to a virus)

Intron   the noncoding region of DNA

Karyotype   a photograph of chromosomes that have been arranged in homologous pairs and that have been assigned numbers

Nondisjunction   failure of a chromosome pair to separate at meiosis

Plasmid   a ring of DNA present in a bacterial cell

Pleiotropy   the many effects caused by one defective gene

Polygenic disease   a disease caused by the interaction of several genes

Recombinant DNA   spliced DNA

Retrovirus   an RNA virus containing reverse transcriptase

Transduction   the transfer of genes from one bacterium to another, using a virus that acts as the carrier of the genes

Historical Search for Genetic Material

Seventy-five years ago, the molecular structure and function of DNA were unknown. Toward the end of the 1920s, research to determine the composition and function of genetic material began. This section contains a summary of some of the important research that provided the groundwork for later discoveries in modern molecular genetics.

In 1928, Frederick Griffith discovered that the genetic characteristics of bacterial cells killed by heat could be transferred to living bacterial cells of a different strain. The principle was called *transformation*. Griffith did not know the nature of the transforming agent. However, he assumed that it could not be a protein because heat denatures proteins. Sixteen years later in 1944, Oswald Avery, Maclyn McCarty, and Colin MacLeod discovered that the transforming agent was deoxyribonucleic acid (DNA).

Further evidence confirming the role of DNA in heredity was provided by Alfred Hershey and Martha Chase in 1950. By using radioactive isotopes, they were able to demonstrate that the DNA from a bacteriophage was capable of infecting the bacterium *Escherichia coli*. The presence of viral DNA within a bacterial cell altered its characteristics.

Edwain Chargaff analyzed the base composition of DNA and found that the number of thymine residues equaled the number of adenine residues. He also found that the number of guanine residues equaled the number of cytosine residues. This information was crucial to the discovery of the double-helix model of DNA.

DNA Structure and Replication

Structure of DNA

The genetic material, located in the chromosomes, is a combination of nucleic acid and histones (short-chained proteins). Deoxyribonucleic acid (DNA) is made up of units called *nucleotides*. Each nucleotide contains three components: the five-carbon sugar dexoyribose, a phosphate group, and a nitrogenous base. The nitrogenous base may be a double-ringed purine—adenine (A) or guanine (G), or a single-ringed pyrimidine—thymine (T) or cytosine (C). These four bases permit the formation of four different nucleotides.

Each DNA molecule is composed of two strands of nucleotides. Study Figure 7.1. Notice that a phosphate group is joined to a sugar and the sugar to a nitrogenous base, forming a nucleotide. A nucleotide is linked with its complement through the nitrogenous bases. Figure 7.2 shows that the two strands of DNA resemble a ladder. The sugar-phosphate backbones form the sides of the ladder, while the protein base pairs form the rungs. The diagram shows that, when nucleotides are joined together in a strand of DNA, the phosphate group attached to the 5' (five prime) carbon of the sugar of one nucleotide links up with the 3' (three prime)

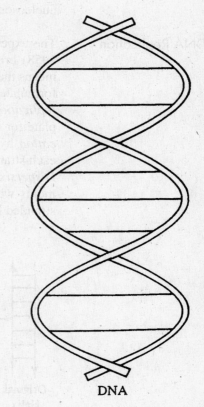

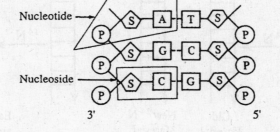

Figure 7.1   Structure of DNA   Figure 7.2   DNA double helix

carbon of the sugar in the adjacent nucleotide. This type of linkage in the sugar-phosphate backbone causes the sides of the DNA "ladder" to be uneven. The nitrogenous bases stick out to one side of the sugar-phosphate backbone.

The nitrogenous bases of the nucleotides are joined together in a specific way. Adenine links only with thymine, requiring two hydrogen bonds. Guanine will join only with cytosine. This linkage requires three hydrogen bonds. In each DNA molecule, the number of adenine molecules always equals the number of thymines and the number of guanine molecules equals the number of cytosines.

In 1962, James Watson and Francis Crick were awarded the Nobel Prize for working out the model of DNA structure. The DNA molecule is referred to as a *double helix*, two strands wound around each other. The two strands are *antiparallel*, extending in opposite directions. One has a 5' phosphate group attachment at one end; the other has a 3' attachment. The asymmetrical backbones cause the twisting with ten nucleotide pairs per turn. Figure 7.2 shows the Watson-Crick model of DNA structure. In 1952, Rosalind Franklin showed through X-ray crystallography that DNA is a double helix. By using her work, Watson and Crick worked out the model that is now commonly accepted.

Chromosomes   In eukaryotic cells, a double-stranded helix of DNA is located in a *chromosome*. A chromosome consists of DNA and proteins. Together they make up *chromatin*, which is folded and coiled. Chromatin is diffuse and invisible in the cell until it condenses in preparation for nuclear division.

Individual units of DNA wound around histones (proteins) are known as nucleosomes. Nucleosomes give chromosomes a beaded appearance.

DNA Replication

The experimental work of Matthew Meselson and Franklin Stahl (published in 1958) brought to light the sequence of events in DNA *replication*. Replication means that a DNA molecule can make an exact copy of itself. Meselson and Stahl formulated three hypotheses that might account for the replication. *Conservative replication* suggested that the double-stranded DNA molecule serves as a template for a new two-stranded DNA moleule (Figure 7.3). *Semiconservative replication* hypothesized that the two strands of a DNA molecule "unzip" and that each strand serves as a template for the formation of one new strand (Figure 7.4). *Dispersive replication* proposed that the DNA molecule broke up into short segments which then serve as templates for the restructuring of two new double-stranded DNA molecules.

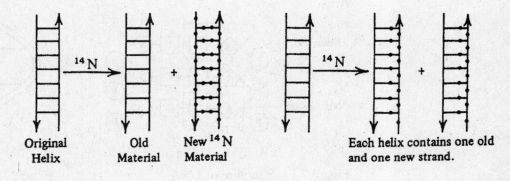

Figure 7.3   Conservative replication

Figure 7.4   Semiconservative replication

A summary of the Meselson and Stahl experiment that gave strong support for semiconservative replication as the logical method follows:

Bacteria were grown in a medium of heavy nitrogen, ^{15}N. The DNA of these organisms revealed a density factor greater than the DNA density of bacteria cells grown in a normal medium. At the end of the first generation, a DNA of intermediate density was obtained. The bacteria grown in ^{15}N medium were transferred to a normal medium. At the end of the second generation, one half of the DNA was of intermediate density and the other half was of normal density. Density-gradient centrifugation indicated that the two nucleotide strands of DNA unzip. Each becomes a template for the formation of a new nucleotide strand. Every new mitotically produced cell contains DNA molecules consisting of an old and a new nucleotide strand joined together.

Just before cell division begins, DNA makes an exact copy of itself. As stated before, this process of DNA duplication is known as *replication*. Replication is a complex process involving dozens of enzymes and many proteins.

The hydrogen bonds between the base pairs of the two nucleotide strands weaken, and the two strands separate. The separating of the nucleotide strands is accomplished by the action of the enzyme, *helicase*. Each strand then acts as a template for the formation of a new nucleotide chain. Single-stranded binding proteins keep the template straight until complementary strands are synthesized. The site where the helicase works and where replication takes place is known as the *DNA fork*.

Replication begins at specific sites along the chromosome. Special proteins (primers) recognize these sites. DNA polymerase interacts with these proteins and helps the nitrogenous base in a given nucleotide to pick up and attach to a complementary base of the DNA template. DNA *polymerase* is a complex of proteins that adds nucleotides to the 3'–5' or leading strand. Thus the new strand produced is in the 5'–3' direction.

The 5'–3' strand or *lagging strand* is duplicated in short segments called *Okazaki fragments*. Ligase is the enzyme that joins these fragments together. In addition, there are 50 different enzymes that repair errors in the replication process.

The nucleotides contain pyrophosphate groups. Cleavage of the pyrophosphate groups releases the energy required for polymerization of the nucleotides.

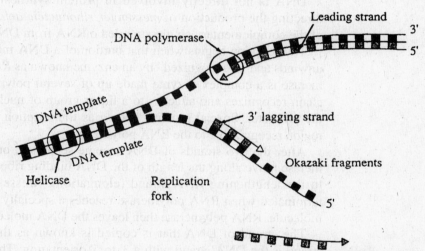

Figure 7.5   DNA replication

About RNA Molecules

Ribonucleic acid (RNA) molecules are made of nucleotide subunits similar to those present in DNA. Like the nucleotides in DNA, each RNA nucleotide consists of a phosphate group, a sugar, and a nitrogenous base. However, the sugar in RNA is ribose, which contains one more oxygen atom than the deoxyribose sugar contained in DNA. RNA differs from DNA in other ways also. RNA usually consists of a single strand of nucleotides, although it can form double-stranded sections. DNA is commonly doublestranded. The nitrogenous bases that compose RNA nucleotides are adenine, uracil, guanine, and cytosine. RNA does not contain the amino acid thymine; adenine pairs with uracil.

There are three kinds of RNA. *Messenger RNA* (mRNA) carries the code that specifies the sequence of amino acids in a polypeptide chain. The code carried by mRNA is copied from DNA. mRNA carries the genetic code for a protein from DNA to the ribosomes, where protein synthesis takes place. *Transfer RNA* (tRNA) ferries amino acids to the ribosomes and fits them into the correct place in the polypeptide chain. Each kind of amino acid is serviced by its own tRNA. *Ribosomal RNA* (rRNA) is present in large quantities in the ribosomes, but its exact function is not known.

DNA and Transcription

As you have just read, specific pairing of the nitrogenous bases is the rule in double-stranded DNA. Adenine binds to thymine, and guanine binds to cytosine. Each set of three nucleotides in linear sequence represents a code for an amino acid. This three-nucleotide sequence is known as a *triplet codon* or simply a *codon*. The entire amino acid sequence in a protein is coded in a DNA molecule. The codon is the part of the genetic code that specifies the particular amino acid that is to be put into a polypeptide chain.

DNA is not directly involved in protein synthesis. It serves as a template directing the production of *messenger ribonucleic acid* (mRNA). The production of the complementary single-stranded mRNA from DNA is called *transcription*.

Transcription begins when that portion of a DNA molecule containing the code unwinds and is "recognized" by an enzyme known as *RNA polymerase*. RNA polymerase is a complex enzyme made up of several polypeptide chains. At least one chain recognizes and attaches to a linear group of nucleotides on the DNA called the promoter. Special proteins known as transcription factors make the promoter region recognizable to the RNA polymerase.

After the two strands of DNA become separated, other sections of RNA polymerase move along the length of the DNA binding ribonucleotides (A, C, G, or U) to the lengthening RNA strand (elongation). Messenger RNA transcription is terminated when RNA polymerase reaches a specially coded portion of the DNA molecule. RNA polymerase then leaves the DNA molecule.

The strand of DNA that is copied is known as the *sense* strand. The sense strand is the DNA strand with a 3' to 5' orientation. Therefore, mRNA elongation occurs in the 5' to 3' direction. In eukaryotic cells, the mRNA produced contains nucleotide sequences, cap and tail, that do not code for a protein. The intron—cap and tail nucleotides—are later removed. The introns are thought to facilitate the movement of mRNA from the nucleus into the cytoplasm.

The messenger RNA, which now contains the complementary code of its template DNA, moves out of the nucleus. The mRNA attaches to a ribosome in the cytoplasm.

Protein Synthesis

Protein synthesis begins when mRNA becomes attached to a small subunit of the ribosome. The positioning of the mRNA on this ribosomal subunit is important. The first codon (AUG) on mRNA must be positioned correctly in order for protein synthesis to begin. Meanwhile, individual amino acid molecules are activated by enzymes using energy from ATP. An AMP-amino acid-enzyme complex is formed:

Activating Enzyme + Amino Acids + ATP $\longrightarrow$ Amino Acid-AMP-Enzyme Complex

Enzymes known as aminoacyl tRNA synthases bind amino acids to specific tRNA molecules. This complex is activated by transfer RNA (tRNA).

Transfer RNA consists of a short chain of about 75 nucleotides arranged in a four-leaf clover configuration (Figure 7.6). Two segments of the four-leaf clover are active. A third segment is composed of three protein bases that form an *anticodon*. The anticodon is attracted to its complementary codon on mRNA. The fourth segment of tRNA recognizes the amino acid complex and binds with the amino acid molecule. Now tRNA is ready to carry its specific amino acid molecule to the activated ribosome. The anticodon of tRNA binds to the AUG codon on mRNA. The amino acid carried by tRNA is methionine, the beginning amino acid in every protein sequence translation.

At this point, the small ribosomal unit is a complex consisting of mRNA and its AUG codon, the anticodon of tRNA, and the amino acid methionine. Next, a large ribosomal unit becomes attached to the complex. This large ribosomal unit has two binding sites for tRNA, know as a peptidyl (P) site and an aminoacyl (A) site. Now the ribosome is in a condition to effect the building of a polypeptide chain.

Then another tRNA with a complementary anticodon binds to the next codon in mRNA. Through the action of the enzyme peptidyl transferase, the first amino acid and the second amino acid are joined by a peptide bond, leaving the first tRNA empty. Both amino acids are held by the second tRNA. The empty tRNA detaches from the ribosome and is now free to pick up more methionine.

Translocation

Next the ribosome moves along the mRNA in a process known as *translocation* (Figure 7.7). The second tRNA and mRNA are moved along the ribosome from the A site to the P site. By means of translocation, the third codon in mRNA is brought to the A site. The second codon is now at the P site. (The cycle of tRNA anticodon binding to its complementary codon on mRNA starts over again.) The complementary anticodon from an amino-acid bearing tRNA binds to the third mRNA codon. Peptidyl transferase caused peptide bond formation between the second and third amino acids. The second tRNA is now free, detaching from the ribosome. The energy for this process is provided by the molecule GTP (guanine triphosphate).

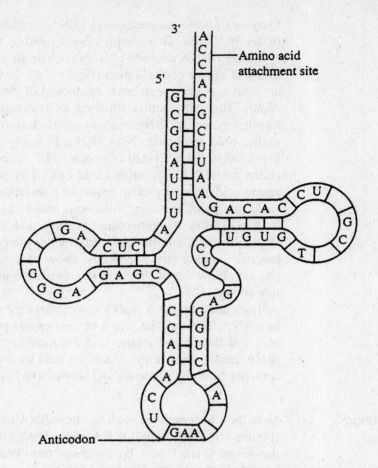

Figure 7.6   Four-leaf clover configuration of tRNA

This process repeats until the ribosome reaches a *stop* codon. A special protein molecule known as a *releasing factor* binds to the stop codon. This results in the detaching of mRNA from the ribosome. Then the ribosomal subunits separate, and the ribosome releases the newly formed polypeptide.

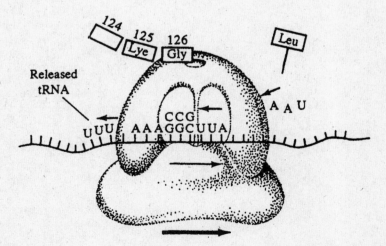

Figure 7.7   Ribosome moves one codon to the right along the mRNA and is now ready for the next charged tRNA.

About Ribosomes

The mRNA is much longer than a single ribosome. Several ribosomes, therefore, can move along the mRNA at one time. This group of ribosomes is known as a *polysome* or *polyribosome*. Polyribosomes have been observed in cells actively producing enzymes.

Ribosomes are organelles that consist of 50% ribosomal nucleic acid, or rRNA, and 50% proteins. They are assembled outside the nucleus from two separate nucleotide chains, 40S and 60S. The 40S chain provides the site for the attachment of mRNA, while the 60S chain functions as a site for the binding of amino acids. Both chains are constructed from a DNA template. The genes for the production of rRNA are located on a chromosome that is usually surrounded by nucleotides and by rRNA molecules in various stages of completeness. This nuclear site is known as the *nucleolus*.

When the mRNA is no longer required, it is decomposed into nucleotides by enzymes. The fate of mRNA depends on whether its transcription is continued or stopped. If mRNA is produced faster than it is decomposed, the enzyme synthesis continues. Obviously, there must be a mechanism for activating and deactivating genes.

Gene Expression

Two modern theories attempt to explain how genes become activated. One is the result of studies of gene action in the bacterium *Escherichia coli*. This theory proposes that a gene consists of a regulator gene capable of producing repressor protein, two areas of a few nucleotides known as a *promoter* and an *operator*, and one or more structural genes. The entire genetic unit is called an *operon* (Figure 7.8). If the operator area is bound, the polymerase from the promoter cannot proceed down the DNA strand and no mRNA is produced. The operator may be bound by a repressor substance. The repressor may be the product of a metabolic activity or may be synthesized by another gene, the inhibitor. If the repressor is prevented from binding the operator by the action of an inducer, mRNA transcription occurs. The inducer, a substance that initiates enzyme synthesis, may be a molecule from the internal or external environment of the cell.

Another theory proposes that there are regulatory genes that constantly produce inactive repressor molecules. The inactive repressor is activated by the presence of a compound (corepressor) from the internal or external environment. Activation of the repressor results in the binding of the operator.

It is possible that both control systems work. However, the experimental models for these theories were prokaryotic cells, which contain a single chromosome and in which genes are arranged in single file (operon). The actual operation of gene control in eukaryotic cells, where there are many chromosomes and several genes for the production of one enzyme, appears to be very complex.

In eukaryotic cells, control of gene expression may occur at different levels: the gene level, transcription, translation, and posttranscription. The complex DNA packaging in chromosomes exerts control over which genes are expressed. Processes similar to those in prokaryotic transcription operate in eukaryotes as well. Initiating factors are also required at the translation level to express a gene. RNA splicing, the removal of introns (noncoding mRNA segments) from exons (coding mRNA segments), is a posttranscriptional control of gene expression.

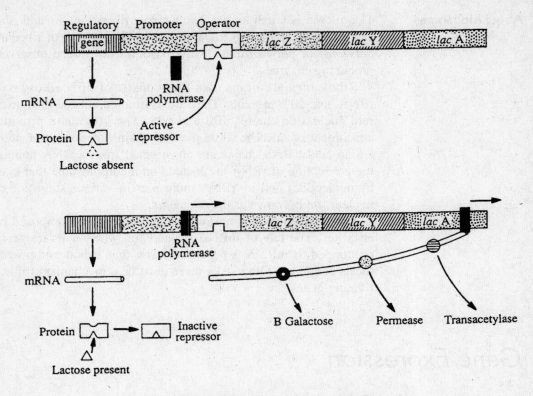

Figure 7.8   Lac operon regulation

The one gene–one enzyme theory was formulated from experimental data obtained from the mold *Neurospora*. However, there are many enzymes that are complex proteins composed of several polypeptide chains. Each chain is produced by a separate gene. Thus the new theory suggests a one gene–one enzyme polypeptide chain.

Model Questions Related to Nucleic Acids and Protein Synthesis

1. Regardless of species, the basic structure of the DNA molecules of all frogs will demonstrate an unchanging pattern in the ratio of
   (A) thymine:uracil
   (B) adenine:cytosine
   (C) purine:pyrimidine
   (D) pyrimidine:ribose
   (E) adenine:guanine

2. Of the following, which is an INCORRECT statement about tRNA?
   (A) It is transcribed from special genes in mRNA in the cytosol.
   (B) There is a specific tRNA molecule for each amino acid.
   (C) All tRNA molecules have the same general shape.
   (D) Clusters of tRNA genes are in the same part of a DNA molecule.
   (E) The aminoacyl attachment site and the anticodon are parts of the tRNA molecule.

Answers to Questions 3–5 are based on the experimental data presented in the table below and on your knowledge of biology.

In a research laboratory, DNA samples were taken from the tissue cells of some multicelluar organisms and from some protists. The data presented in the table below compares the percentages of the nitrogenous bases in the nucleotide components of each of the experimental organisms:

| Cells of | *Percentages of Nitrogenous Bases* | | | |
	Adenine	Thymine	Guanine	Cytosine
Hog Liver	30.8	31.0	23.0	23.1
Goat Liver	32.3	32.3	21.5	21.9
Human Liver	32.9	31.4	21.9	21.8
Mackerel sperm	29.7	29.5	24.1	24.6
Amoeba	33.7	34.6	20.3	19.4
Euglena	34.5	34.6	20.2	18.7

3. The data in the table indicates that
   (A) purine molecules are present in DNA in a greater percentage than pyrimidine molecules
   (B) purine and pyrimidine molecules are present in DNA in equal percentages
   (C) the percentage of adenine in euglena indicates that it is larger in size than the amoeba
   (D) in mackerel sperm the pyrimidines outnumber the purines
   (E) the nitrogenous bases in cells of human liver are different from the nitrogenous bases in goat liver cells

4. The figures in the data table support the fact that
   (A) the purine molecule is a five-sided ring fused to a pyrimidine ring
   (B) the guanine base extracted from hog liver cells is smaller in size than the guanine in mackerel
   (C) in all samples shown the percentage of guanine in DNA is less than the percentage of adenine
   (D) the DNA in human liver cells has a greater amount of cytosine than other samples
   (E) in all samples the percentage of thymine equals the percentage of cytosine

5. Which statement is substantiated by the data?
   (A) The DNA in liver cells is more reactive than the DNA in sperm cells.
   (B) The ratio of purine to pyrimidine molecules is the same in protists as in multicellular organisms.
   (C) The presence of chloroplasts decreases the number of nucleotides in DNA.
   (D) In goat liver cells, the ratio of cytosine to adenine is 1:2.
   (E) There is more DNA in the tissue cells of humans than in the tissue cells of hogs.

6. The nucleolus
   (A) regulates the reproduction of the cell
   (B) functions in the synthesis of RNA
   (C) secretes the nucleotides of DNA
   (D) synthesizes and stores lipid molecules
   (E) acts as a reservoir for excess DNA molecules

7. The "one gene–one enzyme" theory can be expanded to explain how genes exert control in cells by
   (A) determining polypeptide chains
   (B) interacting with proteins in chromosomes
   (C) assorting independently during meiosis
   (D) inhibiting enzyme formation in some cell reactions
   (E) replicating before the beginning of mitosis

8. Working independently, Federick Griffith and Oswald Avery provided the first evidence that DNA carries genetic information by showing that
   (A) DNA is present in all tissue cells
   (B) DNA brings about heritable transformation of bacterial cells
   (C) DNA is present in chromosomes
   (D) DNA is composed of a double helix
   (E) DNA is contained in chloroplasts

9. Modern theories of protein synthesis indicate that
   (A) transfer RNA molecules specific for particular amino acids are activated by peptidyl transferase to link up with the complementary codon in mRNA
   (B) amino acid molecules bind with their complementary codons on mRNA and thus cause the linkage of tRNA molecules
   (C) messenger RNA, transcribed from a DNA template in the nucleus, provides information that governs the sequence of amino acids in a polypeptide chain
   (D) messenger RNA molecules are moved from the nucleus to the ribosomes by tRNA
   (E) enzymes that catalyze protein-synthesizing activities at the ribosomes are transcribed from activator genes

10. Protein synthesis begins when
    (A) the anticodon of tRNA binds with its mRNA complement when mRNA is attached to the small ribosomal unit
    (B) transfer RNA picks up its specific amino acid from the cytoplasmic fluid and transports it to the ribosome
    (C) messenger RNA, along with activating enzymes and amino acids, forms an AMP-mRNA-activating enzyme complex
    (D) messenger RNA is moved by translocation along the ribosome so that the A site is free of the mRNA codon
    (E) messenger RNA binds to the small ribosomal unit so that its first codon is at the appropriate site

11. The active production of protein in the cell occurs
    (A) along a DNA sense strand
    (B) along a tRNA strand
    (C) along an mRNA strand
    (D) along an rRNA strand
    (E) along the length of a chromosome

12. Messenger RNA is always read from which direction?
    (A) From start codon to stop codon
    (B) The 5' to the 3' direction
    (C) The 3' to the 5' direction
    (D) The third codon on the strand
    (E) From any direction on the strand

13. Which of the following must occur in order for the elongation of a polypep-
    tide chain to continue?
    (A) Translocation of the last tRNA to the P site
    (B) Formation of a peptide bond between the last two amino acids
    (C) Appropriate base pairing of the next tRNA
    (D) Movement of the ribosome three places to the right
    (E) All of these events must take place.

14. The AUG codon with which all mRNA begins is known as the
    (A) inhibitor
    (B) accelerator
    (C) initiator
    (D) sensor
    (E) none of the above

15. During translation, polypeptide chain elongation continues until
    (A) all amino acid molecules are used up
    (B) the tRNA molecules become inactive
    (C) the polypeptide chain breaks in half
    (D) the stop codon is reached
    (E) the ribosomal units become undone

16. Meselson and Stahl were able to support the hypothesis that DNA replication
    is semiconservative by the evidence of
    (A) the two equal sedimentary bands that appeared in the cesium chloride
        solution
    (B) the single band of DNA that appeared in the centrifuge tube
    (C) the 14_N band that contained three times as much DNA as the 15_N band
    (D) no DNA sedimentation bands appearing in the culture tube
    (E) uniformity of the DNA bands formed in cesium chloride

Answers and Explanations to Model Questions: Nucleic Acids and Protein Synthesis

1. C Regardless of species, the ratio of purine to pyrmidine molecules is always
      the same in DNA. There is a 1:1 relationship between adenine and thymine
      as well as between guanine and cytosine.

Incorrect Choices
A DNA does not contain uracil.
B Adenine is not a complement of cytosine.
D Ribose is the 5-carbon sugar that forms the skeleton of RNA.
E Adenine and guanine are both purines and are not complements in DNA nucleotides.

2. A Transfer RNA is transcribed from tRNA genes (arranged in clusters of seven or eight tRNA genes) in the same part of a DNA molecule. Transfer RNA is not transcribed from mRNA.

Choices B, C, D, E are all correct statements about tRNA.

3. B In DNA molecules, adenine always pairs with thymine and guanine with cytosine. The A-T and G-C base pairs are always in equal ratios.

Incorrect Choices
A The data does not show this. See explanation for B above.
C The data provides no information about the size of organisms. Both euglena and amoeba are microscopic.
D Note that the percentages of purine and pyrimidine in mackerel are almost equal. Your knowledge of biology should tell you that this statement is impossible.
E The nitrogenous bases are the same in all DNA molecules as indicated on the table.

4. C The data in the table shows that in all samples the percentage of guanine in DNA is less than that of adenine.

Incorrect Choices
A The data provides no information about the structure of molecules.
B No data is given about the size of the molecule. There is a slightly greater percentage of guanine in mackerel sperm than in hog liver.
D and E The data does not support these answers.

5. B Look at the table. Notice that the ratio of purine to pyrimidine molecules is the same in protists (amoeba and euglena) as is the multicelluar organisms.

Incorrect Choices
A, C, D, E are all incorrect and not supported by the information given.

6. B The nucleolus functions in the synthesis of RNA.

Incorrect Choices
A The nucleus regulates cell reproduction.
C, D, E The nucleolus does none of these things.

7. A When a protein is composed of more than one chemically different polypeptide chain, each chain is determined by its own gene. Therefore you can see how one gene can determine one polypeptide.

Incorrect Choices
B, C, D, E These incorrect answers have no logical explanation.

8. B The experimental work that was done utilized the pneumococcus organism, in which the rough and smooth colony characteristics were shown to be determined genetically and could be controlled by DNA.

Incorrect Choices
A, C, D, E None of these choices is germane to the question of genetic information in DNA.

9. C Messenger RNA is transcribed in the nucleus. A "sense strand" of DNA is the template for the construction of mRNA. Messenger RNA contains the instructions for the sequence of amino acids that are necessary to form a given polypeptide chain.

Incorrect Choices
A Peptidyl transferase is the enzyme that catalyzes the peptide bond between two amino acids held by tRNAs on the ribosome.
B The anticodon of a tRNA binds with the complementary codon on mRNA. There is no binding between codons and amino acids.
D Transfer RNA does not carry mRNA.
E Activator genes function in transcription, not in protein synthesis.

10. E Protein synthesis begins when mRNA attaches to the small ribosomal unit so that its first codon (AUG) occupies the correct site on the ribosome for the initiation of protein synthesis.

Incorrect Choices
A, B, D are events that occur after mRNA attaches to the small ribosomal unit.
C This choice is entirely incorrect. Messenger RNA does not form an AMP complex.

11. C The actual production of protein occurs along an mRNA strand that is attached to a ribosome. Choices A, B, D, and E are incorrect.

12. B mRNA is read from the 5' to the 3' direction. The initiator codon is 5' AUG 3' and binds with a tRNA anticodon that is organized from the 5'–3' direction.

Incorrect Choices
A, C, D, E are incorrect because mRNA can work only as explained above.

13. E All of the events listed in A through D are necessary for the elongation of a polypeptide chain.

14. C The "initiator" is the proper terminology for the AUG codon. Choices A, B, D, and E are incorrect.

15. D Polypeptide chain elongation continues until the stop codon is reached. When this happens, no other tRNAs become attached to the mRNA at the anticodon/codon sites. The ribosome disassembles and the mRNA becomes detached. Choices A, B, C, and E are incorrect.

16. **A** Evidence of semiconservative replication of DNA was provided by the appearance of two equal sedimentary bands in the cesium chloride solution. Two strands of the DNA double helix separate during replication, but each remains intact. Each strand serves as a template for the assembly of a new strand.

Incorrect Choices
B, C, D, E are incorrect without explanation.

Gene Functions and Mutations

A gene is a blueprint for a protein. A defect in a single gene causes the production of a faulty protein. Faulty proteins cannot function appropriately in the biochemical processes of cells nor can they serve adequately as structural molecules. Scientists estimate that the human body contains 100 trillion cells, each carrying the same set of genes. A single defective gene can promote severe defects in the structure of body organs and in the function of tissue cells.

A change in a gene (DNA molecule) is a *mutation*. Mutations are inheritable, passing from one generation to the next. A mutation in a body cell gene is known as a *somatic mutation*. Somatic mutations cause changes in characteristics of cells and of tissues and organs comprised of cells derived from the original mutation-bearing cell. Mutations in sex cells are called *germ cell mutations*. These are passed on to an organism's offspring.

There are two types of mutations: point mutations and chromosome aberrations. *Point mutations* occur at the nucleotide level. Errors in duplication of the DNA nucleotide sequence by base-pair substitutions or by insertions or deletions of a nucleotide produce genetic alterations. *Frame-shift mutations* force the genetic code to be read from the wrong place. It also may happen that a nucleotide sequence is inverted, so that the DNA is backwards.

Lesch-Nyhan disorder is an example of a genetic disorder caused by a change in the nucleotide sequence. In this disorder, the mutant gene is one that codes for hypoxanthine guanine phosphoribosy transferase, or HPRT, an enzyme that controls the production of uric acid. Since the enzyme does not function, Lesch-Nyhan victims suffer a buildup of uric acid in the blood (uremia), causing painful gout, severe kidney damage, and mental retardation. Albinism, phenylketonuria, sickle-cell anemia, Cooley's anemia, hemophilia, color blindness, Huntington's chorea, and Tay-Sachs disease are other genetic disorders caused by altering of the nucleotide sequences.

Some allele mutations have remained in the population because they are recessive. The heterozygous individuals may be completely normal, or the heterozygous condition may have been favored by some environmental factor. Persons heterozygous for the sickle-cell anemia trait are resistant to malaria. In the parts of the world where malaria is a problem, therefore, the heterozygous individual has survived. In other circumstances, as in the case of Huntington's chorea, the disorder does not manifest itself until the individual has reached the reproductive age and has passed on the allele to the offspring.

Deletions, duplications, translocations, and inversions of chromosomes are *chromosome aberrations*. Pieces of chromosomes play an important role in the rearrangement of genetic material that alters the size, color, behavior, and fecundity of organisms. Translocation Down's syndrome results from the fusion of chromosomes 14 and 21. Another form of Down's syndrome is caused by trisomy of chromosome 21. During meiosis, sometimes a homologous chromosome pair fails to separate. This may result in a fertilized egg's having one chromosome less or one chromosome more. In either case, the result is harmful to the organism. Recently, DNA segments that move in and out of a DNA site (*transposons*) and alter the genetic material have been discovered.

Viruses and Bacteria

Historically, viruses and bacteria occupy important positions in the elucidation of the structure and function of heredity material. A virus is a noncellular particle consisting of a nucleic acid, either DNA or RNA, enveloped by a protein coat, the *capsid*. A virus can exist in three states: as an infectious particle, as a vegetative virus directing the production of viral particles, and as a provirus.

Viruses are obligate intracellular parasites. The genetic material of the virus, once it enters the host cell, converts the cell into a viral factory. In general, viral DNA in a host cell serves as a template for the production of viral mRNA and new viral DNA. These viruses may demonstrate either a lytic or a lysogenic cycle. In a lytic cycle, new viral particles are produced and released from a host cell. In a lysogenic cycle, viral DNA is duplicated along with the host's DNA (provirus) and remains dormant.

RNA viruses operate in different ways in a host. The RNA may serve as a template for the synthesis of viral RNA and proteins, or the RNA may contain a code for the synthesis of *reverse transcriptase*. The reverse transcriptase uses RNA as a template for DNA synthesis. The DNA directs the production of new viral particles. This type of virus, known as a retrovirus, is important in genetic engineering techniques.

Bacteria are prokaryotic cells. Their DNA is located in one circular chromosome. The bacterial chromosome has been used in determining the biochemistry of gene expression. In addition, bacterial cells contain *plasmids*, smaller rings of DNA carrying accessory genes. Plasmids are vectors in the transfer of genes from one organism to another in the genetic engineering process.

Genetic Engineering

Genetic engineering is part of the broad field of *biotechnology*, in which technical skills are used to solve biological problems. Genetic engineering is the technology that enables scientists to change the genetic material (DNA) in the cells of an organism. Specifically, a single gene (DNA segment) within the total score of an organism's DNA can be isolated, removed, and replaced by a gene from another organism.

Steps in Changing Genes

Changing genes is not easy. It involves a series of precise steps under exacting laboratory conditions. To understand these steps, let us follow the genetic engineering procedures used to reproduce the human gene that codes for insulin.

As you recall, insulin is a hormone produced by the beta cells of the Isles of Langerhans in the pancreas. Insulin controls the metabolism of sugar in cells by making plasma membranes permeable to sugar. Without insulin, sugar cannot enter body cells and accumulates in the blood, damaging organs. The disorder of faulty sugar metabolism is known as diabetes mellitus. Traditionally, diabetics have been treated with insulin taken from the pancreas glands of cattle and swine. However, bovine and swine insulins are slightly different from human insulin and stimulate serious allergic reactions in some people. Genetic engineering has solved the problem: human insulin can now be mass produced relatively inexpensively.

Gene Splicing

The first step in the genetic engineering process is *gene splicing*. In our example, the gene to be spliced is a human gene from the pancreas that controls the production of insulin. This gene is removed from the parent DNA molecule by cutting the DNA at the sequence that codes for the insulin gene. (Scientists know the genetic sequence that codes for insulin and for many other proteins.)

Cutting a DNA strand at the right places is accomplished by using a *restriction enzyme* to serve as a molecular knife. Restriction enzymes belong to a class of enzymes known as endonucleases; each restriction enzyme is specific for a particular nucleotide sequence contained in a DNA strand. In our example, the restriction enzyme used recognizes the sequence that codes for the insulin gene and attaches itself to the appropriate sites on the DNA molecule. By chemical action not clearly understood, the restriction enzyme is able to cleave the DNA, releasing the desired gene from the parent molecule.

Transduction

The gene that is snipped away from the DNA molecule is called the *passenger*, because it is going to be carried by DNA from another organism (called the *vehicle*) into a foreign cell. A ring of DNA known as a *plasmid* is removed from the bacterial cell *Escherichia coli*, the common bacterium found in the human intestine. A restriction enzyme is used to snip away a piece of the plasmid at a site compatible with the insertion of the human insulin gene. The human gene is inserted into the plasmid matching the sticky ends and is made to stick by another type of enzyme known as a *DNA ligase*. The bacterial plasmid together with its inserted gene is known as *recombinant DNA*. The plasmid acts as a vehicle for transporting a human gene into a bacterial cell. This process is known as *transduction*.

The recombinant DNA is then introduced into a receptive *E. coli* cell. Once inside the bacterial cell, the plasmid will replicate, as will the new gene it contains. The replication of a foreign gene inside a host cell is known as *cloning*. The clones of the human gene in the *E. coli* cell are identical to those residing in the DNA of human pancreatic cells. Through the amazing techniques of genetic engineering, *E. coli* now becomes a source of human insulin.

Source of Genes

Genes on a chromosome may be located by *in situ hybridization*. Radioactively tagged pieces of laboratory-synthesized nucleic acid (probes) are added to metaphase cells on a slide. The radioactive pieces bind to the cell's DNA, forming a hybrid that is located by autoradiography.

Once genes are located on the chromosome, they may be removed and directly inserted into a vector, or the genes may be artifically synthesized as cDNA (complementary DNA). Frederich Sanger developed a method of DNA sequencing using gel electrophoresis and computers to determine the nucleotide sequence of a gene. This information is then used to produce cDNA.

Gene Therapy in Human Cells

Theoretically, current knowledge of genetic engineering should permit genetic engineers to remove a faulty gene and replace it with a normal gene. However, one hundred trillion human body cells each carry the same set of genes, although all genes do not function in all cells. In human gene therapy, the genetic engineer must solve a number of problems. In which cells shall the new gene be placed? How can engineered genes be made to enter human cells?

Retroviruses as Vehicles

When a retrovirus enters a cell, it discards the protein coat and injects its RNA strand into the host cell. Unlike other viruses, retroviruses contain reverse transcriptase, an enzyme that changes the viral RNA into a DNA form, reversing the flow of genetic information from RNA to DNA. As you know, it is DNA that forms the coded template against which nucleotide sequences are formed. (Retroviruses are considered the causes of immunological disorders.)

In research laboratories, retroviruses are proving to be useful as vehicles (vectors) for transporting human engineered genes into human cells. Before a new gene can be inserted into a cell, a method of moving this gene into the cell must be devised. Since viruses enter human cells, they can be used as vehicles to carry human genes into human cells. Not only must a selected gene be carried into a human cell, but it must also be spliced into the host cell's DNA. The DNA must then be able to replicate, making more of the new and desirable gene. However, when a virus enters a cell, it captures the cell's genetic material and uses the cell's DNA to make viral DNA. To make retroviruses useful ferries of human genes, special techniques are required to make the experimental viral DNA produce more genes without the virus. Viruses are specific for the types of human cells they enter. Thus not all viruses can enter any cell indiscriminately.

Making a Vehicle

To make a vehicle, a specific type of retrovirus is introduced into a culture of mouse cells. As the retrovirus enters a cell, it sheds its protein coat and releases its RNA strand. Special enzymes cause the formation of DNA genes that then enter the nucleus of the host cell. The DNA made by the virus is removed from the mouse cell nucleus. From this DNA, viral genes are removed, including those needed by the virus to make its protein coat. However, the regulator genes which control the activity of structural genes are kept intact.

The desirable gene is spliced into the remaining part of the viral gene sequence at a site near the regulator genes that stimulate gene action. This engineered gene sequence is then made to enter bacterial cells, where many copies of the recombinant DNA are cloned. This cloned DNA is now introduced into a fresh culture of mouse cells so that they can enter these new host cells. The cloned DNA moves into the nucleus of each host mouse cell. The recombinant DNA in the nucleus of the mouse cell must be packaged in such a way that it can be removed from the mouse cell and made to enter human cells.

To do this, the new DNA with the desirable gene must be enclosed in a virus that then can be used to infect human cells. A "helper" virus is added to the culture of mouse cells. This virus contributes the viral gene that codes for the synthesis of a viral protein coat. Another "helper" virus contains the packaging gene that enables the protein coat to be wrapped around the recombinant DNA. The finished vector viruses can now infect cells but cannot reproduce themselves. It is necessary to structure virus vehicles that cannot reproduce themselves. The genetic engineer determines what human cells get the virus with the recombinant DNA. Uncontrolled viral reproduction is thus prohibited.

Ideally, the genetic engineer cultures these vector cells (vehicles) with human bone marrow cells. In vitro, the vector cells enter the human bone marrow cells and make copies of the DNA that carries the desirable gene. Treated bone marrow cells are then injected back into the patient's marrow, where, it is hoped, the enzyme will be produced. However, to date, this has not been possible.

Accomplishments of Genetic Engineering

Proteins that inhibit the replication of RNA viruses are called *interferons*. Interferons are produced by certain cells after they have been infected by viruses. Viruses such as those that cause influenza or poliomyelitis stimulate cells to produce interferons. These "immune" proteins have been medically useful in preventing certain viral diseases.

During the 1970s, all interferon used for medical treatment was extracted from human cells with great effort and, of course, at great cost. Since then, techniques of genetic engineering have markedly reduced the cost and effort involved in obtaining interferon. To date, researchers have introduced 15 human interferon genes into bacterial and yeast cells, making it possible to produce abundant and inexpensive interferons useful in treating several kinds of viral infections. (However, it should be noted that interferon has not been successful in the treatment of cancer.)

So far the technology of producing recombinant DNA has been very useful in the

> treatment of hepatitis
> prenatal and postnatal diagnosis of genetic disorders
> development of weaker versions of vaccines (which are safer)
> development of processes for safer and more efficient cleanup of organic wastes generated in food-processing industries
> production of human insulin using bacterial cells
> production of enzymes for the cheese industry
> production of human growth hormone
> human gene therapy

The Genome Search

In 1910, Thomas Hunt Morgan changed the entire direction of inquiry from the search for hereditary factors into the science of genetics. Morgan chose the fruit fly, *Drosophila melanogaster,* as the animal for research on which he and his students would work out many of the basic principles of genetics. *Drosophila* has four large chromosomes (two pairs). These can be easily located in and isolated from the cells in the fly's salivary glands. The fruit fly has become so entrenched as the organism of choice for genetics study that it is the focus of research for at least 5,000 scientists, familiarly called "fly people."

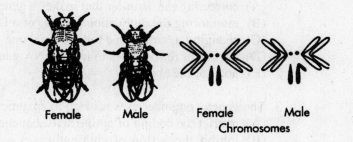

Female   Male   Female   Male
                               Chromosomes

Figure 7.9   *Drosophila melanogaster*—fruit fly

In March of 2000, J. Craig Venter of the private research company, Celera Genomics, and Gerald l. Rubin of the University of California announced that they, assisted by 195 scientists, had decoded *Drosophila's* genome. A *genome* is the total genetic makeup of an organism. The chromosomes contain DNA, which contains sites or codes for genes.

The Venter-Rubin teams have identified 13,601 of the fruit fly's genes. The next step is to find out what each gene does. Interestingly enough, of the 289 human disease genes, 177 have analogs (counterparts) in the fly. The genes and proteins of *Drosophila* are quite similar to those of humans.

The monumental task of decoding the human genome has begun in earnest. It is believed that human cells carry up to 100,000 genes. Teams of scientists in the privately owned research companies and in government-funded laboratories are both vying and cooperating with each other in the race to decode the human genome. That the Human Genome Project is about two-thirds of the way completed is evidenced by the fact that scientists now know 2 billion of the 3 billion "letters" that make up human DNA. You know that each human cell contains DNA structured into 23 pairs of chromosomes. Each chromosome has hundreds to thousands of codes for building proteins. Proteins build structural parts of cells and body organs. In addition, proteins constitute chemical substances such as hormones and enzymes that regulate chemical processes in the body. The human genome contains all the biochemical instructions in the form of combinations of the DNA bases adenine (A), thymine (T), cytosine (C), and guanine (G) for making and maintaining a human being.

Researchers have identified more than 8,000 genes on human chromosomes, including those linked to diseases such as prostate cancer, breast cancer, Parkinson's disease, and Alzheimer's disease. Decoding the human genome may provide answers about how to prevent or cure diseases.

Model Questions Related to Genetic Engineering

1. Genetic engineering is the modern technology that allows a scientist to
   (A) change the order of nucleotides in a given gene
   (B) substitute one DNA segment for another in the genetic material of a cell
   (C) double the mRNA that is stored in the cell nucleolus
   (D) replace all the genes that are in the cells of a multicellular organism
   (E) manufacture synthetically most of the enzymes used in the biochemical events of the cell

2. Gene therapy refers to
    (A) correcting the disorder that makes a gene faulty
    (B) monitoring the replication process of a DNA segment
    (C) changing a faulty gene for a normal one
    (D) the use of retroviruses to make DNA genes
    (E) none of the above

3. The genetic engineer uses restriction enzymes to
    (A) restrict the coding of amino acid sequences along a given strand of mRNA
    (B) inhibit the action of endonucleases as nucleotide sequences are being organized
    (C) function in the isolation of plasmids in bacteria
    (D) serve as a vehicle for mobile genes
    (E) serve as a molecular knife to cleave a specific DNA segment from a DNA strand

4. A passenger gene is used to
    (A) replace viral DNA in a retrovirus
    (B) unzip nucleotides as it moves from cell to cell
    (C) replace a mutant gene
    (D) carry genes from one cell to another
    (E) remove a plasmid from a bacterium

5. A plasmid is
    (A) a ring of DNA in a bacterial cell
    (B) a fragmentary particle in a eukaryotic cell
    (C) a vehicle that carries a human gene into a mouse cell
    (D) none of the above
    (E) all of the above

6. DNA ligase is an enzyme that
    (A) strips away antagonistic nucleotide sequences from a DNA strand
    (B) makes an inserted gene stick to the plasmid vehicle
    (C) causes the loss of a specific nucleotide from a DNA segment
    (D) does all of the above
    (E) does none of the above

7. Transduction refers to
    (A) the production of complementary strands of mRNA from DNA
    (B) the movement of tRNA from one attachment site to another on a ribosome
    (C) the conversion of the genetic information carried by a tRNA molecule into the amino acid sequence of a polypeptide
    (D) the transporting of a human gene into a bacterial cell
    (E) movement of a ribosome along mRNA

8. Which of the following is a true statement about mutations?
    (A) Mutations are heritable.
    (B) Mutations are changes in genes.
    (C) Mutations may be caused by mutagenic agents.
    (D) All of the above
    (E) None of the above

9. Point mutations
    (A) remain fixed in the germplasm and cannot be passed on to progeny
    (B) occur at the 5' end of a DNA strand
    (C) force the genetic code to be read from the wrong place
    (D) are caused by a break in a chromosome
    (E) result when one nucleotide is changed for another

10. Of the following, which is the most accurate statement about mutations?
    (A) Mutations are usually harmful.
    (B) Mutations are usually dominant over the normal gene.
    (C) Mutations are mostly the result of faulty genes.
    (D) Mutations in somatic cells have little effect.
    (E) Mutations in sperm-cell genes are more powerful than those in egg-cell genes.

11. Retroviruses differ from other viruses in that
    (A) retroviruses are the only viruses capable of reproducing outside of the cell
    (B) retroviruses direct the flow of genetic information from RNA to DNA
    (C) retroviruses contain the enzyme reverse transcriptase, which inhibits the formation of the protein coat
    (D) retroviruses change the flow of information from DNA to RNA
    (E) retroviruses produce genes but cannot reproduce viral particles

12. Genetic engineers use vector cells to
    (A) reproduce a selected gene in its recombinant DNA
    (B) splice mutant genes from DNA strands
    (C) produce enzymes at a faster rate than is possible with other cells
    (D) inhibit the action of virus vehicles
    (E) prevent penetration of cell nuclei by infective viruses

Answers and Explanations to Model Questions: Genetic Engineering

1. **B** Genetic engineering technology permits the substitution of one gene (DNA segment) for another.

    Incorrect Choices
    A To date nucleotide rearrangement in a gene has not been accomplished.
    C, D, E have no logical explanations.

2. **C** Gene therapy is the correcting of a genetic disorder by changing a mutant gene for a normal one. Thus far, gene therapy has not been accomplished in humans.

    Incorrect Choices
    A Faulty or mutant genes cause disorders. The disorder is not the causative agent of gene mutations.
    B, D, E are incorrect without logical explanation.

3. **E** Restriction enzymes belong to a group of enzymes called endonucleases. Each endonuclease is specific in its ability to cleave a given DNA segment. Restriction enzymes are popularly referred to as molecular knives because they function on the molecular level.

   Incorrect Choices
   A Restriction enzymes do not function in protein synthesis.
   B, C have no logical explanation.
   D Viruses usually serve as vehicles on which human genes are carried into bacterial cells.

4. **C** As the name implies, a passenger gene or DNA segment replaces a mutant gene in a DNA strand. It is called a "passenger" because it is taken from one cell and placed into another.

   Incorrect Choices
   A, B, D, E are incorrect without logical explanation.

5. **A** A plasmid is a ring of DNA in a bacterial cell bearing additional genes from those of the larger DNA strand in bacteria. Plasmids are used in genetic engineering.

   Incorrect Choices
   B, D, E are incorrect without explanation.
   C A plasmid could not penetrate the nucleus of a mouse cell.

6. **B** DNA ligase is an enzyme that matches up the sticky ends of the passenger gene and its plasmid vehicle.

   Incorrect Choices
   A, C, D, E are incorrect without logical explanation.

7. **D** Transduction is a term used to describe the transporting of a human gene into a bacterial cell.

   Incorrect Choices
   A The production of complementary strands of mRNA from its DNA template is called *transcription*.
   B The movement of tRNA from one attachment site on the ribosome to another is known as *translocation*.
   C The conversion of genetic information carried by a tRNA molecule into the amino acid sequence of a polypeptide is known as *translation*.
   E The movement of a ribosome along a strand of mRNA is *translocation*.

8. **D** All of the statements presented about mutations are true.

   Incorrect Choices
   E is obviously incorrect since A, B, and C are correct statements.

9. E A point mutation occurs when one nucleotide is changed for another. The defect in the gene occurs at a particular location.

Incorrect Choices

A Genes are never fixed in the germplasm. Genes in the germplasm are heritable. This, however, is not a point mutation.

B Answer is incorrect without explanation.

C A *frame-shift* mutation forces the genetic code to be read from the wrong place.

D A break in a chromosome will cause deletion of genes, but this is not a point of mutation.

10. A Mutations are usually harmful to the individual.

Incorrect Choices

B Mutations are usually recessive to the normal. Usually two recessive mutated genes are required for the characteristic to be expressed in the progeny.

C Some mutations are caused by faulty genes. Others may be caused by breaks in the chromosomes that cause deletions of genes, or by nondisjunction, in which chromosomes fail to separate during meiosis, or by translocation, in which a broken piece of chromosome is added to a whole chromosome.

D Mutations in genes that are in body (somatic) cells do have harmful effects. The tissues in which the cells have a mutated gene may fail to function properly.

E Not true. Mutations in sperm and egg cells are equally disadvantageous.

11. B Retroviruses direct the flow of genetic information from RNA to DNA. Under usual circumstances, the flow of genetic information is directed from the DNA template to mRNA.

Incorrect Choices

A Like other viruses, retroviruses reproduce inside cells.

C The enzyme *reverse transcriptase* directs the flow of information in retroviruses from RNA to DNA.

D See explanations A and C.

E Retroviruses reproduce viral particles.

12. A Genetic engineers use vector cells to reproduce a selected gene in its recombinant DNA. For example, the insulin-producing gene inserted into an *E. coli* bacterium enables the bacterium to produce human insulin.

Incorrect Choices

B, C, D, E have no logical explanation.

CHAPTER EIGHT
Heredity

Capsule Concept

The first organized mathematical study of how traits are passed from one generation to another was accomplished by an Austrian monk, Gregor Mendel. After beginning his work in the year 1856, Mendel designed his experiments in heredity around inheritance in the garden pea. He identified seven distinct, easily recognizable traits in this self-pollinating plant and called each of these traits a *unit character*. For each unit character, he identified one contrasting or opposite trait. If the unit character was height, for example, the opposite traits were tall and short. From observations based on his experiments with the garden pea, Mendel formulated three principles of heredity.

Vocabulary

Alleles   alternative forms of the same genes
Crossing over   the exchange of two homologous chromosomes of like parts; crossing over breaks linkage groups
First filial generation (F_1)   a heterozygous generation
Genotype   the genetic makeup
Heterozygous   having two different alleles for the same trait
Homologous chromosomes   a pair of like chromosomes having the same length and the same genetic loci, and with their centromeres in the same position
Homozygous   having two of the same alleles
Homozygous recessive   the genotype of a recessive trait
Hybrid   mixed genes
Intermediate inheritance   incomplete dominance
Linkage group   genes that are linked on the same chromosome; linked loci
Locus (plural: loci)   a site on a chromosome where a gene for a particular trait is located
Parental generation (P_1)   pure breeding stock
Phenotype   the expression of genes
Pleiotropy   the ability of a single gene to affect many characteristics
Second filial generation (F_2)   the result of crossing hybrids
Testcross   a cross between an organism with a dominant phenotype and one with a homozygous recessive

Mendelian Principles

The theory of evolution proposed by Charles Darwin was based on unexplainable events and observations. Darwin noted that variation occurs among organisms and that the resulting variants are able to transmit their characteristics to their offspring. He did not attempt to explain these facts. However, the principles of genetics, as set forth by Gregor Mendel, give meaning to evolution.

Mendelian genetics is based on three general principles that were formulated after extensive experimentation on garden peas. Mendel observed that a cross between two pure organisms of contrasting traits results in offspring (hybrids) that exhibit only one of the parental traits. This trait is dominant over the masked, recessive one (*Law of Dominance*). Since each parent contributes one half of the genetic material to the offspring, the genes (factors that control traits) occur in pairs. The hybrid is heterozygous, that is, it receives two different genes for a given trait. Different forms of the same gene are called *alleles*. Each parent is pure, or homozygous, for the trait and contains one allele for it.

When two hybrids are crossed, both grandparental phenotypes are observed. However, the dominant trait appears in a 3:1 ratio over the recessive trait. During the formation of eggs and sperm, the two alleles for the trait separate (*Law of Segregation*). The crossing of dihybrids yields two different characteristics which are inherited independently from each other (*Law of Independent Assortment*). In garden peas, for example, characteristics of shape and seed color are inherited independently of each other as a result of meiosis. A dihybrid garden pea plant can produce four genotypically different gametes. When these gametes are crossed, the offspring exhibit a 9:3:3:1 phenotype ratio.

Whether Mendel worked with organisms that did not show true dominance is not clear. The phenomenon of incomplete dominance is observed in many plants and animals, for example, snapdragons, four-o'clocks, cattle, and Andalusian fowl.

Mendelian genetics can be clarified by current chromosome theory. Genes are nucleotide units within a chromosome. Chromosomes are arranged in pairs. Each member of a pair may have identical or opposite alleles. During egg and sperm formation, the chromosome pairs separate. Each gamete receives one member of the pair of chromosomes (Law of Segregation). Upon recombination, the zygote may have a combination of alleles different from that of either parent. If two different traits are controlled by two genes located on two separate chromosome pairs, each gene separates independently of the other (Law of Independent Assortment).

Intermediate Inheritance

In Mendel's experiments with garden peas, the phenotypes of the offspring were either dominant or recessive. For example, green pods (*GG*) are dominant over yellow pods (*gg*). An offspring heterozygous for pod color (*Gg*) is green because of the complete dominance of *G* over *g*.

Some species exhibit *intermediate inheritance*. When red four o'clock flowers are crossed with white, the offspring are pink. This type of inheritance is also known as *incomplete dominance*. When hybrid four o'clocks are crossed, the red and white phenotypes segregate out.

Parent:	*RR* red	×		*WW* white
F_1	*RW*	× pink		*RW*
F_2	*RR* red	*RW* *RW* pink		*WW* white

The fact that the red and white phenotypes appear when hybrids are crossed indicates that genes for flower color maintain their identity. These genes are separate entities remaining unchanged. The pink color of the hybrids means that they have only one half the red pigment of the homozygous red parent.

Multiple Alleles

A characteristic may be determined by two alleles or by multiple alleles. Such a multiple-allele phenomenon was observed in fruit flies. Eye color in fruit flies may be expressed as red, white, eosin, wine, apricot, ivory, or cherry. In human beings, multiple alleles, designated as A, B, and i determine blood type. The chart below indicates the possible allele combinations for each blood type.

Type	Genotype
A	$I^a I^a$ or $I^a i$
B	$I^b I^b$ or $I^b i$
AB	$I^a I^b$
O	ii

The development of certain phenotypic traits depends on the interaction of two or more genes. The shape of the comb on Leghorn roosters is determined by the interaction of two opposite alleles: r produces a single comb, and R produces a rose comb. Another gene, P, produces a pea comb. The R gene and the p gene interact (collaboration) to produce a totally different phenotype (walnut comb). As long as any one of the two dominant genes is present, R or P, the rooster will reflect that characteristic. The chart below summarizes the genotype and phenotype relationships.

$$P \qquad \underset{\text{rose}}{RRpp} \qquad \times \qquad \underset{\text{pea}}{rrPP}$$

$$F_1 \qquad \underset{}{RrPp} \qquad \times \qquad RrPp$$
$$\text{walnut}$$

$$F_2 \qquad \underset{\text{walnut}}{R\_P\_} \qquad \underset{\text{rose}}{R\_pp} \qquad \underset{\text{pea}}{rrP\_} \qquad \underset{\text{single}}{rrpp}$$

Many characteristics in organisms cannot be explained except on the basis of gene interaction. Flower color, coat color of mammals, and eye or hair color of human beings are a few of the traits that result from the interaction of several genes. Interacting genes may be located on the same or on different chromosomes.

Pleiotropy

Pleiotropy is the ability of a single gene to affect many phenotypic characteristics. The human disease sickle-cell anemia provides a ready example of a pleiotropic gene. Pleiotropy and the sickling of cells is described in the section titled Human Genetic Diseases.

Epistasis

In the garden pea, purple allele P is dominant. However, another gene, C, which is inherited independently of the P gene, determines whether the flowers will have the purple pigment. Flowers that have the genotypes PP or Pp will be white if the homozygous recessive gene, cc, is present. For the purple pigment to be synthesized, C genes must be present. Therefore, the P gene "stands on" or is epistatic to the C gene.

Polygenic Phenotypes Polygenic inheritance results from the effect of several genes on a single trait. Human skin color is determined by three independently inherited genes. The three genes are incompletely dominant over each other and in various combinations of dominant and recessive genes, determine if the skin color is dark, intermediate, or light.

Linkage Genes located on the same chromosome are inherited together and are called a *linkage group*. Linkage groups are altered by crossing over, an event that occurs during prophase 1 of meiosis (see pages 114–115 for a detailed description of meiosis). This is the time when homologous chromosomes intertwine and may exchange homologous parts. A mixing of the mother's and father's genes occurs during crossing over.

Effects of Sex Chromosomes According to the chromosome theory of inheritance, all chromosomes occur in identical pairs. However, it was originally observed in *Drosophila* (fruit flies) that the chromosomes in one pair were not identical. It was determined that this pair was always associated with the male of the species.

The cells of females contain two identical sex chromosomes designated as the XX chromosomes, while the male sex chromosomes are designated as XY. In human beings and perhaps in other mammals, the presence of the Y chromosome confers maleness on the organism.

Sex-Linkage In 1910, Thomas Hunt Morgan observed white-eyed fruit flies in a pure culture of red-eyed flies. When red-eyed females were mated with mutant white-eyed males, all the offspring had red eyes. The red eye trait was dominant over the white-eye trait. When white-eyed females were mated with red-eyed males, however, all the male offspring were white-eyed and all the females were red-eyed. If heterozygous females were mated to red-eyed males, 50% of the male offspring were red-eyed and 50% were white-eyed. The female offspring were all red-eyed. These results indicate that the genes for eye color are located on the X chromosome. Since the genes on the X chromosome have no alleles on the Y chromosome, the male inherits the traits controlled by the genes on the X chromosome. This is known as *sex-linkage*.

Color blindness and hemophilia are common sex-linked traits in human beings. Since chromosomes are inherited as a unit, genes move together during meiosis (linkage).

According to Mendelian genetics, a dihybrid testcross produces four phenotypes in a 1:1:1:1 ratio. This ratio is expected because the two genes for the traits are located in separate chromosomes. However, many test results do not show the typical ratio. This is interpreted to mean that the two different genes are linked on the same chromosome. If the genes were located in the same chromosome, a 1:1 testcross ratio would be expected. Since crossing-over occurs the 1:1 ratio is altered and all four phenotypes are observed.

There are several possible linkage arrangements for a dihybrid. The actual linkage is determined by observing the percentages of phenotypes that result from the testcross.

In the following problem, a 4:4:1:1 ratio is obtained from a *Drosophila* dihybrid testcross. Because a 1:1:1:1 ratio does not occur, linkage can be assumed. The offspring that occur in greatest numbers reflect the outcome of linkage. The offspring appearing in fewest numbers are the products of crossing-over. Therefore the two dominant genes are located on the same chromosome, while the two recessives are on the other chromosome.

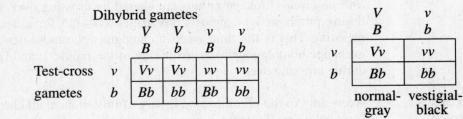

V = normal wings B = gray body
v = vestigial wings b = black body

$VvBb$ × $vvbb$
dihybrid   test-cross

Expected results if genes are on separate chromosomes:

Dihybrid gametes

		V B	V b	v B	v b
Test-cross	v	Vv	Vv	vv	vv
gametes	b	Bb	bb	Bb	bb

Expected results if genes are on same chromosome:

	V B	v b
v	Vv	vv
b	Bb	bb
	normal-gray	vestigial-black

$VvBb$ = normal wings, gray body
$Vvbb$ = normal wings, black body
$vvBb$ = vestigial wings, gray body
$vvbb$ = vestigial wings, black body

OBSERVED RESULTS

	40%	40%	10%	10%
Test-cross gametes	Normal-gray	Vestigial-black	Vestigial-gray	Normal black
v	Vv	vv	vv	Vv
b	Bb	bb	Bb	bb

Crossover products
V v
b B

The lower the frequency of recombination (crossing-over), the closer the genes are on the chromosome. Genes that are farther apart are more likely to be involved in crossing-over than those that are closer together. As long as there are marker genes (genes that produce detectable effects) on a chromosome, the relationship of other genes to the marker can be determined.

The study of linkage and recombination has led to determination of the order of genes on chromosomes of such organisms as bacteria, molds, and *Drosophila*. The actual physical position of a gene on a chromosome is much more difficult to determine, and much depends on the study of deficiency mutations within a chromosome.

Human Pedigrees   Geneticists study human inheritance by tracing a particular trait through a family's history for as many generations as possible. To organize the information in a meaningful way, geneticists make use of a chart-like device called a *pedigree*. Figure 8.1 is a human pedigree that traces the inheritance of sickle cell anemia in a family. Mendel's laws are useful in deducing the genotypes of the parent generation by studying the phenotypes of their offspring. The pedigree not only presents the past genetic history, but also serves the purpose of predicting the future.

For example: what are the chances of a person with the sickling trait in his family producing offspring with sickle cell anemia if the sickling trait also appears in his wife's family?

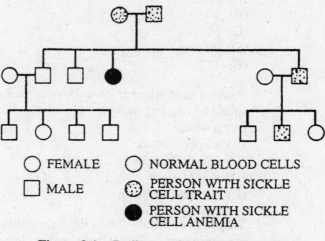

FEMALE   NORMAL BLOOD CELLS

MALE   PERSON WITH SICKLE CELL TRAIT

PERSON WITH SICKLE CELL ANEMIA

Figure 8.1   Pedigree of Sickle Cell Anemia

Genetic counselors work on problems such as the one posed above by assembling the family history of a given trait into a pedigree. On the basis of the inheritance pattern of a trait in the family, informed reproductive decisions can be made.

Testcross

To determine the genetic makeup of progeny, animal breeders make use of a mating device called a *testcross*. When it is uncertain whether a phenotypic characteristic is pure or hybrid in a particular animal, the breeder will mate the animal with an animal that is recessive for that trait. For example, in guinea pigs black coat is dominant over white coat. Therefore, a guinea pig with a black coat may be either pure (*BB*) for coat color or hybrid (*Bb*). By mating the black guinea pig in question with a guinea pig recessive for coat color (*bb*, a white guinea pig) the genotype of the dominant parent will be revealed. If guinea pigs with white coats appear in the litter, then the parent in question is hybrid. Plant breeders also use testcrosses to determine genotype.

Model Questions Related to Mendelian Principles

1. An experiment in probability required that two like coins be flipped simultaneously. The results for 100 flips were recorded. Which one of the following choices indicates the results to be expected?
   (A) two heads—50%; two tails—50%
   (B) two heads—25%; one tail, one head—50%; two tails—25%
   (C) two heads—25%; two tails—75%
   (D) two heads—50%; one head, one tail—50%
   (E) two tails—50%; one head, one tail—50%

2. In guinea pigs, black is dominant. One half of a particular litter is white. If it is assumed that the laws of chance operate, the parent cross was
   (A) *BB* × *Bb*
   (B) *Bb* × *Bb*
   (C) *Bb* × *bb*
   (D) *bb* × *bb*
   (E) *BB* × *bb*

3. As a result of crossing two hybrid yellow garden peas, 120 offspring are produced. According to the laws of chance, the most probable number of yellow offspring is
   (A) 0
   (B) 30
   (C) 60
   (D) 90
   (E) 120

4. To determine whether an unknown black guinea pig is pure or hybrid black, it shoud be crossed with
   (A) a white
   (B) a hybrid black
   (C) a hybrid white
   (D) a pure black
   (E) another unknown

5. The diagram below shows three generations of deafness in a family. A dark circle represents a deaf female, and a clear circle represents a normal female. A dark square represents a deaf male, and a clear square represents a normal male.

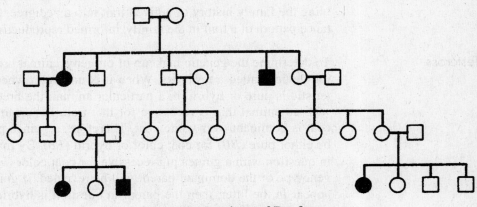

Three Generations of Deafness

The condition of deafness shown in the pedigree is inherited as
(A) a dominant autosome characteristic
(B) a recessive autosome characteristic
(C) a sex-linked dominant characteristic
(D) a sex-linked recessive characteristic
(E) a holandric characteristic

6. In human beings, brown eyes are usually dominant over blue eyes. Suppose that a blue-eyed man marries a brown-eyed female whose father was blue-eyed. The percentage of their children with blue eyes would be closest to
   (A) 0%
   (B) 25%
   (C) 50%
   (D) 75%
   (E) 100%

7. An infant whose sister is normal has thalassemia major. What are the probable genotypes of the parents?
   (A) $TT \times TT$
   (B) $Tt \times Tt$
   (C) $tt \times tt$
   (D) $Tt \times TT$
   (E) None of the above

8. One hundred experimental matings of brown birds with white birds produced speckled offspring. The mating of two speckled birds will probably result in offspring that are
   (A) 75% white, 25% brown
   (B) 25% brown, 75% white
   (C) 25% brown, 50% speckled, 25% white
   (D) 100% speckled
   (E) 50% brown, 25% white, 25% speckled

9. In tomatoes, cut leaf is dominant over potato leaf and purple stem is dominant over green stem. In the table below, the result of a mating producing an F_1 is given

Purple-cut	Purple-potato	Green-cut	Green-potato
1790	620	623	222

   The most probable parent genotypes are
   (A) $PpCc \times PpCc$
   (B) $PpCc \times Ppcc$
   (C) $ppcc \times ppcc$
   (D) $PPCc \times ppCc$
   (E) $Ppcc \times ppCc$

10. Which set of parents could NOT be the parents of a child with type O blood?
    (A) Father type A, mother type O
    (B) Father type A, mother type B
    (C) Father type B, mother type O
    (D) Father type AB, mother type O
    (E) Father type O, mother type O

11. A dominant gene, A, causes yellow coat color in rats. The dominant allele of another independent gene, R, produces black coat color. When the two dominants occur together, they interact to produce gray. When the two recessives interact, they produce cream color. A mating between a gray male and a cream female produced a litter in which $3/8$ of the offspring were yellow, $3/8$ were gray, $1/8$ were black, and $1/8$ were cream. If the genotype of the female was $aarr$, what was the genotype of the male?
    (A) $AARR$
    (B) $AaRr$
    (C) $AaRR$
    (D) $AARr$
    (E) $aarr$

12. There is evidence that a certain color in cats is sex-linked. Yellow is recessive to black. A heterozygous condition results in tortoise shell or calico color. A calico cat has a litter of 8 kittens: 1 yellow male, 2 black males, 2 yellow females, and 3 calico females. What was the male parent's probable color?
    (A) Yellow
    (B) Black
    (C) Calico
    (D) Yellow and black
    (E) Albino

Questions 13–16 are based on the following information:

> In *Drosophila*, there is a dominant gene for gray body color and another dominant gene for normal wings. The recessive alleles of these two genes result in black body color and vestigial wings, respectively. Flies homozygous for gray color and normal wings were crossed with flies that had black bodies and vestigial wings. The F_1 progeny were then testcrossed with the following F_2 results:

Gray body, normal wings	236
Black body, vestigial wings	253
Gray body, vestigial wings	50
Black body, normal wings	61

13. The genes for these characteristics are
    (A) not linked
    (B) linked without crossing-over
    (C) linked with 18.6% crossing-over
    (D) linked with 37.2% crossing-over
    (E) linked with 81.4% crossing-over

14. Which statement is TRUE about the individuals in the F_1 generation?
    (A) There are no linked alleles.
    (B) The linked alleles are GG and VV.
    (C) The linked alleles are Gv and gV.
    (D) The linked alleles are GV and gv.
    (E) The linked alleles are in a different combination from the ones described above.

15. Which of the following is TRUE about the 61 black, normal flies in the F_2 generation?
    (A) There are no linked alleles.
    (B) The linked alleles are gV and gv.
    (C) The linked alleles are the same as in the F_1 generation.
    (D) The linked alleles cannot be determined without another testcross.
    (E) None of the above

16. If some of the black bodied, normal-winged flies were crossed with black-bodied, vestigial-winged individuals, the offspring would probably be
    (A) like those obtained in the F_2 generation
    (B) unpredictable
    (C) about 50% black-vestigial, 50% black-normal
    (D) about 50% gray-vestigial, 50% black-normal
    (E) about 25% black-normal, 25% black-vestigial, 25% gray-normal, 25% gray-vestigial

17. The sites of four gene loci on a chromosome are based on the following crossover frequencies:

A and B	40%
B and C	20%
C and D	10%
C and A	20%
D and B	10%

Which of the following best represents the relative positions of the four genes?

(A) [A   C B D]

(B) [A   C D B]

(C) [A   D   C   B]

(D) [A   C   B   D]

(E) [A   B   C   D]

18. In a diploid organism with the genotype *AaBBCCDDEE*, how many genetically distinct kinds of gametes would be produced?
    (A) 2
    (B) 4
    (C) 8
    (D) 16
    (E) 32

19. Which of the following is true of a gene that is dominant?
    (A) It is usually detrimental.
    (B) It will occur more frequently than its recessive allele.
    (C) It will occur less frequently than its recessive allele.
    (D) It is more likely to be passed on to the next generation than its recessive allele.
    (E) It will have the same phenotypic effect whether it appears in a homozygous or heterozygous condition.

Answers and Explanations to Model Questions: Mendelian Principles

1. B When a coin is flipped, there is a 50-50 chance that it will turn up heads. In statistics, the chance that two independent events will occur simultaneously is equal to the product of the chances of the two events occurring separately. Thus the chance that two heads will occur on two different coins is $1/2 \times 1/2$ or 25%. (Since there are two possible ways of obtaining a heads and tails pattern, this pattern will occur 25% + 25% = 50% of the times.)

$$HH \quad 1/2 \times 1/2 = 1/4$$
$$\left. \begin{array}{l} HT \quad 1/2 \times 1/2 = 1/4 \\ TH \quad 1/2 \times 1/2 = 1/4 \end{array} \right\} 50\%$$
$$TT \quad 1/2 \times 1/2 = 1/4$$

2. C The only way that white individuals can occur is if the recessive genes combine. Since each parent must contribute one recessive gene, each parent has one recessive gene. However, 50% of the offspring are black. To produce black, only one black gene need be present. This means that at least one parent must possess a gene for blackness. If both parents had a gene for blackness, a 3:1 ratio, not a 1:1 ratio, would be expected.

3. D The offspring from a union of hybrids are expected to exhibit a 3:1 phenotype ratio. It is expected that 75% will reflect the dominant trait, yellow: 75% of 120 is 90.

4. A A testcross involves a mating between a homozygous recessive individual and a genotypically unknown mate. In this way there is a 50–50 chance that a recessive trait will appear if it is contained within the genotypically unknown individual.

   Incorrect Choices
   B A hybrid black mating would decrease the odds that the recessive would appear (3:1 chance instead of a 1:1 chance).
   C Since white is the recessive trait, it appears only in a homozygous condition.
   D Since black is a dominant trait, it would mask the presence of a recessive.
   E Another unknown would only complicate matters and contribute nothing to the solution of the problem.

5. B Since the trait appears in the F_1 generation from unaffected parents, it must be recessive in nature.

   Incorrect Choices
   A The trait would appear in every individual in the F_1 generation if it were dominant and should not appear in the offspring of unaffected individuals.
   C The inheritance of the trait does not appear to be related to the sex of the individual. If it were an X sex-linked dominant, then the unaffected female of the P_1 generation would not exist.
   D If the trait is sex-linked recessive, only the males would be affected.
   E Holandric traits are those whose genes are present only on the Y chromosome (that is, they are inherited exclusively through male descent). Since normal females do not have Y chromosomes, the situation is impossible in the pedigree.

6. C Since the wife of the man had a blue-eyed father, she is carrying the gene for blue eyes. A 1:1 ratio is expected from the marriage of a person who is hybrid for eye color to a person exhibiting the recessive trait.

7. B The infant received one allele for the defective trait from each parent. The sister, who is normal, received one allele for the normal trait from each parent. The parents are heterozygous for the alleles.

8. C Since the mating of two genotypically different individuals produced offspring with characteristics distinctly different from either parent, incomplete dominance is suspected. The results from the mating of the heterozygotes is indicated in the Punnett square:

	B	*W*
B	*BB*	*BW*
W	*BW*	*WW*

9. **A** Since a 9:3:3:1 ratio of the offspring is observed, we are dealing with a dihybrid cross.

Incorrect Choices

B A 3:3:1:1 ratio would occur.
C All green-potato individuals would be produced.
D A 3:1 purple-cut to purple-potato ratio would be observed.
E A 1:1:1:1 ratio would be observed.

10. **D** Type O blood results from the combination of two recessive alleles. Since a person with AB blood can contribute only the dominant alleles, *A* and *B*, to the offspring, there is no possible way for the recessives to combine.

Incorrect Choices

A The father may be heterozygous (*Ai*) for the trait.
B Both parents may be heterozygous (*Ai* or *Bi*) for the trait.
C The father may be heterozygous (*Bi*) for the trait.
E Both parents are homozygous for the *i* allele.

11. **B** Since the cream-colored individual can contribute only *ar* alleles, the other alleles must be supplied by the gray parent. The gray parent must contribute a set of *ar* alleles to produce cream-colored offspring; therefore the gray parent is heterozygous for both traits.

Incorrect Choices

A This mating would result in 100% gray offspring.
C This mating would result in 50% gray, 50% black offspring.
D This mating would result in 50% gray, 50% yellow offspring.
E This mating would result in 100% cream-colored offspring.

12. **A** A yellow female kitten must inherit two X chromosomes, each containing the recessive allele for yellow. Since she receives one allele from her female parent, she must receive the other allele from her male parent. Because the male has only one X chromosome, he must have had yellow coat color.

Incorrect Choices

B Only black females would have resulted.
C A male cannot be calico because the characteristic is produced in the heterozygous condition. The male has only one X chromosome.
D A male will be either black or yellow because of the presence of the allele on his X chromosome.
E Albinism is not a sex-linked characteristic.

13. **C** In a simple dihybrid testcross, a 1:1:1:1 ratio is expected since the alleles for each gene will be located in different chromosomes. Independent assortment will occur. Because this ratio is not observed, linkage is suspected. If the genes are linked, the organisms in the minority are produced because of crossing-over. Of the offspring, 18.6% are the products of crossing-over.

14. **D** The testcross individual can contribute only two recessive alleles for each gene. To produce the majority of the offspring, which are gray-normal and black-vestigial, the two dominant genes must be present in the same chromosome.

15. **B** If the genes for the two characteristics are linked in the same chromosome, there must be crossing-over to produce the gray-vestigial and black-normal individuals.

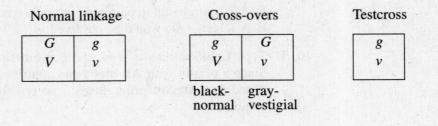

16. **C** black-normal   black-vestigial
    ggVv   *ggvv*

17. **B** The greater the distance between linked genes in a chromosome, the greater is the percentage or frequency of crossing-over. *A* and *B* are farthest apart. Since *C* is equidistant from both *A* and *B*, it occupies a middle position. Because *D* is equidistant from *C* and *B*, it occupies a middle position between them.

18. **A** With the exception of the allele pair *Aa*, all the genes are homologous; therefore there are two possible gametes that can be formed, *ABCDE* and *aBCDE*.

19. **E** A dominant gene is one that expresses itself regardless of its allele partner.

    Incorrect Choices
    A There is no correlation between dominance and beneficial or detrimental characteristics. Actually, if the dominant allele is detrimental, it will eventually disappear from the population.
    B, C, D According to the Laws of Segregation, the alleles separate during meiosis. There is a 50-50 chance that the gamete with either allele will be involved in reproduction.

Genetic Versus Infectious Diseases

Differences Between Two Types of Diseases

Infectious diseases are caused when microorganisms invade the body and do harm to it. Genetic diseases are caused by defective genes or by chromosomes that are abnormal in either structure or number. Genetic diseases are determined at the time of conception when genetic material in the egg and sperm fuse to form a zygote (fertilized egg). Faulty hereditary material in the zygote becomes a blueprint for impaired physical structure of the developing organism or for underactivity of one or more biochemical processes in the tissue cells. If the defect in the nuclear material of the cleaving zygote is not lethal, the embryo develops into an individual that is usually mentally underdeveloped and often physically handicapped.

In contrast with infectious diseases, genetic diseases are not curable and may or may not be treatable. Treatment has been devised for certain genetic diseases, including phenylketonuria, if diagnosed early enough. However, the genetic disorder remains in the cells of the individual and can be passed on to progeny. Genetic disorders remain with those afflicted for life.

Mode of Inheritance of Genetic Diseases

Genetic diseases are transmitted in any of three ways: single-gene defects, chromosomal abnormalities, or multiple-gene defects. Sometimes multiple-gene defects are modified by environment. As said previously, genes are DNA segments that occupy specific sites on chromosomes, and a change in a gene is a mutation. Most mutations are harmful to the individual and recessive to the normal gene.

Radiation and certain chemicals are *mutagenic agents*. Mutagens, or mutagenic agents, cause genes to mutate by having adverse effects on the nucleotides that compose DNA molecules. Some chemicals become incorporated into the nucleotides and prevent them from functioning normally. Other chemicals react with the nucleotide bases and prevent them from forming proper base pairs. Some mutagens interfere with the hydrogen bonding that links protein bases in nucleotide pairs. Defective genes are passed on from parent to offspring.

Chromosomal abnormalities are passed on to progeny in the same manner as gene disorders by way of the fertilized egg. Changes in chromosomes change heredity. Sometimes a piece of a chromosome is broken off and is lost as are the genes that are on the chromosome fragment. Loss of a piece of chromosome is called a *deletion*. If the missing genes do not control vital traits, the organism probably will survive with its heredity being altered somewhat.

Human Genetic Disorders

Single-Gene Defects

Sickle-cell anemia is a genetic disease caused by a defect (mutation) in a single gene. A base substitution in a nucleotide is responsible for the abnormal hemoglobin that causes the sickling of red blood cells. In the DNA molecule, a substitution of adenine (A) for thymine (T) changes the codon in mRNA from GAA to GUA. Therefore valine replaces glutamic acid in the polypeptide chain that forms hemoglobin. This change in a single amino acid drastically alters the nature of the hemoglobin, causing it to have reduced oxygen-carrying capacity and changing the shape of red blood cells from round to crescent.

The gene responsible for sickle-cell anemia is best described as *peiotropic*— one defective gene causing many adverse effects. Hemoglobin-S, the hemoglobin of sickle cells, changes its shape when oxygen is released from it. This causes thickening of the membranes of the red blood cells and also induces the half-moon shape. The misshapen cells cannot flow through the capillaries easily. The capillaries become clogged, causing pain, muscle weakness, and debilitating fatigue in the afflicted.

The expression of sickle-cell anemia is determined by the presence of two autosomal recessive genes. A person who inherits one normal hemoglobin gene and one abnormal gene for hemoglobin-S is clinically normal and does not suffer the painful and fatiguing effects of sickle-cell anemia.

Phenylketonuria (PKU) is another disease caused by a single-gene defect. Like sickle-cell anemia, PKU is an autosomal recessive disorder. The individual inheriting two faulty genes for PKU cannot produce the liver enzyme phenylalanine hydroxylase, which is necessary to catalyze the change of phenylalanine to tyrosine. Without the enyzme, phenylalanine and its metabolites accumulate in the body (especially in the brain tissue), producing severe mental retardation in infants. Brain damage can be prevented if PKU babies are maintained on diets low in phenylalanine. However, the genetic condition, two defective genes that prevent the building of the enzyme phenylalanine hydroxylase, does not change.

Other autosomal recessive disorders are:

cystic fibrosis	persistant lung infections occuring at birth
Tay-Sachs disease	fatal muscle weakening disease
thalassemia	severe anemia
Hartnup disease	faulty metabolism of tryptophan, necessary for the synthesis of vitamin B_3

A few mutant genes are dominant over the normal. The following are genetic disorders caused by an autosomal dominant gene:

galactosemia	faulty enzyme prevents digestion of galactose sugar
achondroplastic dwarfism	short legs and arms, big head and face; condition not an endocrine disorder
congenital cataracts	opacity of the lens of the eye due to an abnormal biochemical condition
syndactyly	webbed fingers and toes
Huntington's chorea	progressive deterioration of the nervous system beginning in middle life
osteogenesis imperfecta	brittle bones

Sex-Linked Disorders

Sex is determined by a pair of chromosomes known as *sex chromosomes*. The XX chromosome pair specifies femaleness; the XY chromosome configuration determines maleness. Genes carried on the X chromosome are said to be sex-linked, and mutant X chromosome genes cause sex-linked disorders. The Y chromosome carries very few genes.

Sex-linked disorders are determined by defective genes carried on the X chromosome. In females, two defective gene pairs are necessary in order for the disease to show. If one gene of the pair is normal and the other mutant, the female will show the normal condition. However, she becomes a carrier for the mutant gene. In males, only one mutant gene on the X chromosome determines the presence of a sex-linked genetic disorder because there is no counteracting gene on the Y chromosome. Sex-linked diseases are usually passed from a mother carrier to the son. Some examples of sex-linked diseases follow:

hemophilia	faulty blood-clotting mechanism
Duchenne muscular dystrophy	muscle weakness that becomes progressively worse

| color blindness | inability to distinguish certain colors, usually red, blue, and green |
| hydrocephalus | water in the brain, causing undue pressure and resulting in mental retardation |

Polygenic Disorder

A polygenic disorder is caused by the interaction of several genes. The interacting genes may occupy different loci on the same homologous chromosome pair, or they may be positioned on nonhomologous chromosomes. Nevertheless, one phenotypic trait is determined by the interaction of several genes. Polygenic disorders are difficult to study because it is hard to sort out the roles of the interacting genes. Some genetic disorders caused by multiple gene interaction are as follows:

allergy
atherosclerosis
cancer (certain types)
cleft palate and lip
schizophrenia
peptic ulcer
congenital dislocation
  of the hip

clubfoot
hypertension
kidney stone disease
manic depressive illness
pyloric stenosis (narrowing of
  the stomach opening into the
  small intestine)
spina bifida (open spine)

Disorders Involving Whole Chromosomes

As you have learned, sometimes a homologous chromosome pair fail to separate during meiosis. If a sex cell with the extra chromosome unites with a normal sex cell, the zygote will have an extra chromosome—a condition called *trisomy*. Trisomy of chromosome 21 results in the condition known as *Down's syndrome*. A person so afflicted is mentally retarded, has poor muscle tone, has a large tongue, and has abnormal palm and foot prints.

There are other common trisomies of homologous autosomes. These occur in the 17–18 chromosome group and in the 13–15 chromosome group. In both cases, the individuals may have severe malformations which are usually fatal within the first year of life.

Nondisjunction can lead to trisomy of the sex chromosomes. There are individuals who inherit XXX sex chromosomes. These females have a high incidence of mental retardation and exhibit a wide range of physical abnormalities. In Klinefelter's syndrome, the individual's sex chromosomes are XXY. These persons are males, sterile, and often mentally defective.

Through nondisjunction, it is also possible that a zygote receives one less chromosome, making its full chromosome number 45. The absence of a single chromosome, known as *monosomy*, is a condition that is almost always lethal to the embryo. However, monosomy of the sex chromosomes may not be lethal. A condition known as *Turner's syndrome* is caused by the presence of only one X chromosome without the accompanying X or Y chromosome. This pattern is shown as XO. The individual with the XO sex chromosome configuration is essentially female but does not develop secondary sex characteristics. These females are abnormally short and remain sterile throughout life. Some are mentally handicapped.

Disorders Involving Other Chromosome Abnormalities

There are several types of chromosome abnormalities. Some are caused by deletions, broken-off pieces of chromosome. During synapsis, when homologous chromosomes come together and wrap around each other, the chromosome with the deleted fragment has missing genes. The normal chromosome therefore has a segment against which no genes are matched. This may interfere with the normal production of cells.

If the broken bit of chromosome attaches itself to the companion chromosome, an abnormal situation results. One chromosome has a missing piece; the other has additional genes attached to it. It now has two genes for certain traits when it should have only one. This condition is known as *duplication*. If this chromosome is passed on to the zygote stage, the resulting individual will have in all probability some rather severe genetic disorder. If the chromosome fragment becomes attached to a nonhomologous chromosome, the result is known as *translocation*. This affects the progeny adversely.

At times a piece of chromosome breaks off and then becomes reattached backward. Although the same genes are present on the chromosome, they are not in proper sequence. This condition is known as an *inversion*. A chromosome with inverted genes cannot match up its genes completely with those on the homologous chromosome. In some way, inversions do affect heredity and the expression of certain traits adversely.

Sometimes errors in chromosome numbers develop during cleavage when the embryo is going through a series of rapid cell divisions. It happens that through nondisjunction in cleaving cells, an individual's body has cells of normal chromosome number, trisomic cells, and monosomic cells. Such a condition is known as a *mosaic*. Mosaicism results in physical abnormalities.

Detection of Abnormal Chromosomes by Amniocentesis and Karyotyping

Abnormal chromosomes can be detected in developing embryos by a clinical procedure known as amniocentesis. During the 16th or 17th week of pregnancy, a bit of the fluid that surrounds the developing embryo is withdrawn from the amniotic sac for study. The amniotic fluid contains cells that are shed from the developing embryo. These cells are cultured, removed from the culture, and photographed. A karyotype (Figure 8.2) is made. The investigator can pick out from the karyotype abnormal chromosome types or patterns.

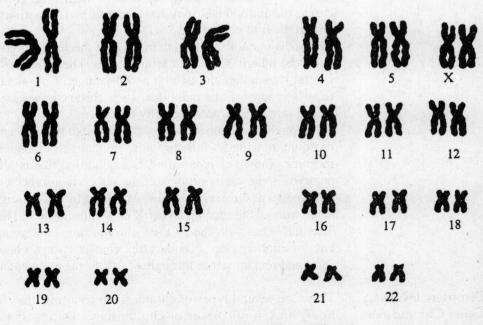

Figure 8.2   Karyotype of normal female

1. When a specific lipoprotein in the blood becomes attached to the plasma membrane of the fibroblast cells, the synthesis of cholesterol is inhibited. A single mutant gene prohibits the attachment of the lipoprotein. Most probably the mutant gene causes
   (A) abnormal activity of lipoprotein molecules
   (B) decreased reproduction of fibroblast cells
   (C) defective plasma membrane receptor sites
   (D) destruction of the lipoprotein molecule
   (E) erratic activity in the abnormal cytoplasm

2. Biochemically each human gene can be recognized by
   (A) special fluorescent dyes
   (B) the protein the gene produces
   (C) the speed of the gene in an electric field
   (D) the rate of replication
   (E) the gene's position on a chromosome

3. Plasmids and viruses are useful in genetic engineering because
   (A) they contain nucleic acids
   (B) they are very small
   (C) foreign DNA can be spliced into them
   (D) they enter the host cells easily carrying extra genes into the host cell
   (E) all of the above

4. The procedure used by molecular biologists to study a cell's store of chromosomes is
   (A) tissue culture
   (B) binding
   (C) karyotyping
   (D) fusion
   (E) cloning

5. Hybrid cells are used in genetics research laboratories to
   (A) map genes on chromosomes
   (B) locate erratic chromosome bands
   (C) determine the mode of replication
   (D) induce rapid mitosis
   (E) measure enzyme insufficiency

6. Each cell of the human body contains a collection of genes that number approximately
   (A) 10,000
   (B) 20,000
   (C) 30,000
   (D) 40,000
   (E) 100,000

7. Epoxides, triazene, carbamates, and formaldehyde have the ability to
   (A) serve as antimutagens
   (B) increase the reproduction of fruit flies
   (C) prevent the irradiation of germ cells
   (D) increase the frequency of mutations
   (E) react as purines in *Escherichia coli*

8. The Philadelphia chromosome is really chromosome 22 in a much shortened state. This autosomal abnormality is present in all cases of chronic myelogenous leukemia. Of the following, which is a true statement?
   (A) The Philadelphia chromosome is passed from a carrier mother to a son via the X chromosome.
   (B) The Philadelphia chromosome is passed from one generation to the next in accordance with the Mendelian rules of inheritance.
   (C) Chromosome 22 goes through transformation in the blood-forming cells before the onset of the disease.
   (D) The abnormality of chromosome 22 can be detected by applying the techniques of amniocentesis and karyotyping.
   (E) Chromosome 22 can be stretched to normal length through procedures used in genetic engineering.

9. Genetic diseases are not curable because
   (A) gene interactions prevent the identification of faulty genes and the sorting out of their various roles
   (B) defective genes remain in cells and are passed on to progeny
   (C) defective genes cannot readily be isolated and removed from body cells
   (D) all of the above
   (E) none of the above

10. A deletion will most certainly result in
    (A) destruction of gametes before fertilization
    (B) spontaneous abortion of the embryo
    (C) faulty activity of homologous chromosomes
    (D) loss of certain genes
    (E) breaks in H bonds between complementary bases

11. Sickle-cell anemia is caused by
    (A) lack of iron in the diet
    (B) faulty base-pairing in complementary nucleotides
    (C) the coding of valine in place of glutamic acid
    (D) trisomy of chromosome 13
    (E) translocation between nonhomologous chromosomes

12. Tryptophan is an essential amino acid necessary for the synthesis of vitamin B_3. Persons afflicted with Hartnup's disease have critically low levels of tryptophan in the blood and very high levels of it in the urine, indicating that
    (A) a pair of faulty genes prevent the synthesis of tryptophan by the somatic cells
    (B) autosomal recessive mutant genes code for the faulty metabolism of vitamin B_3
    (C) Hartnup's disease is a sex-linked disorder caused by a defective gene on the X chromosome
    (D) a pair of mutant genes have altered the reabsorptive ability of the kidney tubules
    (E) a chromosome deletion has rendered the glomerulus inoperative

13. Phenylketonuria is a disease caused by a single-gene defect. Unlike sickle-cell anemia, it can be treated by
    (A) controlling the diets of PKU infants
    (B) using vaccines to counteract the effect of the mutant gene
    (C) correcting the mutant gene by genetic engineering
    (D) applying low-level radiation doses to affected cells
    (E) using the "wet diaper" test at stated intervals

14. Trisomy of chromosomes 17–18 results in
    (A) Down's syndrome, in which the individual is physically impaired and mentally defective
    (B) many severe malformations which cause death of infants in the first year of life
    (C) Kleinfelter's syndrome, a condition producing sterility and underdevelopment of the sex organs
    (D) *Cri de chat,* a series of physical malformations accompanied by mental retardation of the afflicted
    (E) the development of very tall, thin individuals who are overaggressive

15. Centric fusion is a type of translocation in which two nonhomologous chromosomes join (fuse) at their centers. It is not unusual for chromosome 15 to fuse with chromosome 21. The result of this fusion produces a person who has 46 chromosomes but has Down's syndrome. The reason for the appearance of this disorder is
    (A) chromosome 15 undergoes mutation and its genes resemble those of chromosome 21
    (B) the individual has sufficient chromosome material for 47 chromosomes
    (C) Down's syndrome is a catch-all term for several genetic disorders
    (D) translocated chromosomes stimulate the rapid replication of adjacent chromosomes
    (E) centric fusion of chromosome 15 is the usual cause of Down's syndrome

Answers and Explanations to Model Questions: Human Genetic Disorders

1. C The lipoprotein can be attached only at a specific receptor site on the cell membrane. If these receptor sites are defective, the lipoprotein cannot fit into the receptor site.

   Incorrect Choices
   A, B, D, E are incorrect without logical explanation.

2. B Each DNA segment or gene serves as a template for a particular mRNA which receives the genetic code for a specific protein.

   Incorrect Choices
   A, C, D, E are incorrect without logical explanation.

3. E Plasmids and viruses are very small, nucleic acid-containing particles into which foreign DNA can be spliced. Plasmids and viruses enter host cells easily into which they carry extra genes.

Incorrect Choices
A, B, C, D are all reasons for using plasmids and viruses in genetic engineering.

4. C Chromosomes are removed from human cells and photographed. The photograph is cut apart, matching each chromosome with its homologue. A standardized numbering system is used to identify 23 pairs of human chromosomes. This technique allows the investigator to identify physical abnormalities in chromosomes.

Incorrect Choices
A, B, D, E are incorrect without logical explanation.

5. A Hybrid cells are used to isolate human chromosomes for better study. Once isolated, the investigator can follow the chromosome through several cell division generations to identify and map genes. To date maps have been prepared for 100 human chromosomes.

Incorrect Choices
B, C, D, E are incorrect without logical explanation.

6. E Current information supports the count of 100,000 genes in each human cell.

Incorrect Choices
A, B, C, D are incorrect without logical explanation.

7. D All of the chemicals listed are mutagens, agents of mutation.

Incorrect Choices
A, B, C, E are incorrect without logical explanation.

8. C A much shortened chromosome 22 is known as the Philadelphia chromosome. Transformation (deletion) takes place in the somatic blood-making fibroblast cells. The transformation of chromosome 22 is the cause of chronic myelogenous leukemia.

Incorrect Choices
A The disorder is not sex-linked and is unrelated to the X chromosome.
B, D The transformed chromosome 22 is not present in the cells of the embryo and therefore does not follow the principles of Mendelian inheritance and cannot be detected by karyotyping.
E is incorrect without logical explanation.

9. D All of the statements presented are correct.

10. D When chromosomes lose pieces, they lose the genes on the broken fragments.

Incorrect Choices
A, B, C, E are incorrect without logical explanation.

11. **C** Sickle-cell anemia is caused by a faulty genetic code in which adenine (A) is substituted for thymine (T). Transcription of the faulty code results in valine being substituted for glutamic acid in the polypeptide chain of hemoglobin.

Incorrect Choices
A Sickle-cell anemia is not caused by nutritional deficiency.
B, D, E are incorrect without logical explanation.

12. **D** You have been given information that the blood level of tryptophan is low but the level in the urine is high. This means that kidney tubules are not reabsorbing this amino acid, allowing it to diffuse out. Mutant genes are not coding for the enzymes that control the reabsorbtion of tryptophan.

Incorrect Choices
A Tryptophan is not synthesized in the body. It is one of the essential amino acids that must be taken in by way of food.
B Vitamin B_2 is synthesized from tryptophan.
C, E are incorrect without explanation.

13. **A** By preventing the intake of the amino acid phenylalanine, a pile-up of this amino acid in the brain and body tissues is avoided. Controlled diet prevents the mental retardation that accompanies excessive phenylalanine in the tissues. However, the genetics of the person remains unchanged.

Incorrect Choices
B, C, D, E are incorrect without explanation.

14. **B** Trisomy of chromosomes 13–15 and 17–18 result in severe malformations that cause the death of infants in the first year of life.

Incorrect Choices
A Down's syndrome is caused by trisomy of chromosome 21.
C Kleinfelter's syndrome is caused by the trisomy of sex chromosomes XXY.
D *Cri de chat* ("cry of the cat") is a severe condition of malformation and mental retardation caused by chromosome deletion.
E is incorrect without explanation.

15. **B** Chromosomes 15 and 21 have fused. Although two chromosomes have fused, their genetic material remains the same. Essentially, the individual in this question has a chromosome number of 46, genetic material that is equal to 47 chromosomes.

Incorrect Choices
A, C, D, E are incorrect without logical explanation.

CHAPTER NINE
Evolution

Capsule Concept

Evolution is concerned with change. It is the biological discipline that links present-day living organisms to their ancestral prototypes. To find answers in retrospect, a scientist has to utilize various kinds of information to elucidate past biological phenomena. Therefore the evolutionist must lean heavily on the biological and physical sciences in order to establish relationships between living phyla and the past.

As you read this chapter, think of evolution as a synthesis of many sciences. Principles of genetics, biochemical concepts, morphological studies, and embryological facts are pertinent to the sorting, classifying, and assembling of many pieces of information into reasonable hypotheses. The sciences of physics, chemistry, geology, archeology, and paleontology provide supporting data that give credence to the basic theories upon which the concept of evolution is built.

Vocabulary

Abiotic   without life

Adaptive radiation   an evolutionary process by which an ancestral species gives rise to many different species, in a branchlike manner, with special adaptations for various environments

Autotrophic   having the ability to convert inorganic substances to organic molecules

Barrier   a geographic factor that prevents the intermingling of species

Chemical selection   the process through which nonliving aggregates in the early seas became stable

Coacervates   combinations of polymers in water believed to be the forerunners of living cells

Convergent evolution   an evolutionary process in which organisms that are not related become more similar in one or more characteristics because of independent adaptation to similar environmental situations

Fossil   the preserved remains or imprints of plant and animal bodies

Gene pool   all of the dominant and recessive genes present in a population

Hardy-Weinberg Law   a mathematical statement relating to sexually reproducing population genetics; it proposes that, under certain conditions of stability, both gene frequencies and genotype ratios remain constant from generation to generation

Heterotrophic   nutritionally dependent on preformed organic molecules

Microspheres   proteinoid droplets of equal size that are more stable than coacervates

Natural selection   Darwin's concept that only the fittest will survive in a natural environment

Organic evolution   the process by which new organisms develop over long periods of time from previous organisms

Proteinoids   proteinlike molecules produced by U.S. biochemist Sidney Fox in his laboratory experiments attempting to simulate the origin of living forms

Protobionts   the forerunners of living cells; early aggregates

Reducing atmosphere   the early atmosphere, which contained a high volume of hydrogen

Spontaneous generation   a belief of the ancients that nonliving matter can give rise to living organisms

"Use and disuse"   a phrase used to summarize Lamarck's theory of the inheritance of acquired characteristics

Vestigial structure   a structure that has lost its function but persists in some form in modern animals

The Origin of Life

Today's living organisms are very different from earlier life forms. The fossil record reveals not only the kinds of living things of the past but also their relationships to modern species. The Darwinian concept suggests that all life descended, with modification, from preexisting life. Genetics supports this theory. The idea that life originates from life was expounded by Spallanzani, Redi, and Pasteur when they refuted the theory of spontaneous generation. However, if one traces the history of life on earth as far back as its origin, one must inevitably come to a point in time when there was no preexisting life. The theory of organic evaluation attempts to explain the origin of life.

At one time, the oldest available fossils were remnants of organisms that lived during Cambrian times. Newer, more modern techniques have brought to light the presence of organic compounds that could have been synthesized only by living organisms. It is believed that these fossil compounds are over 3 billion years old, indicating that life was present on our planet when it was a little more than 1 billion years old. The organic composition of the fossils led investigators to believe that they were produced by organisms similar to our present-day bacteria and blue-green algae. Although these prokaryotic cells are anatomically simple, they are, no doubt, more complex than the systems that represented the first living organisms. Scientists believe that the earth came into being about $4\frac{1}{2}$ billion years ago. Although the earliest known fossils are about $3\frac{1}{2}$ billion years old, life must have originated before this time.

The Early Atmosphere

Today all living organisms have one metabolic process in common: glycolysis. However, bacteria exhibit a diversity of substrates utilized for anaerobic respiration. The many modes of bacterial metabolism offer clues concerning the early atmosphere. The primitive atmosphere contained hydrogen, methane, ammonia, and water vapor, but lacked oxygen. It was a reducing atmosphere.

Development of Complex Molecules— Prebionts

Stanley Miller has demonstrated that basic organic compounds, such as amino acids, purines, pyrimidines, sugars, and complete nucleotides, can be synthesized experimentally in a specially prepared chamber. He showed how electric discharges from lightning (heat from radioactivity, and ultraviolet light) may have been the sources of energy for protein synthesis. The adenine nucleotide is synthesized with ease because it is a polymer of hydrogen cyanide, a compound formed in the primitive atmosphere. Therefore is it not reasonable to assume that ATP, ADP, and AMP—compounds utilized to power metabolic reactions—were the forerunners of life? One must picture a warm sea teeming with molecules essential for life (a hot, thin soup).

Melvin Calvin demonstrated how complex molecules can be formed by polymerization. Sidney Fox produced microspheres, proteinoids (long-chain peptides) surrounded by what resembles a membrane. A. I. Oparin found that he could produce coacervate droplets that had the ability to incorporate simple enzymes within their structures. Perhaps, with all of the aforementioned conditions, a stable, organized system of metabolizing molecules could have arisen (prebionts). However, without the ability to reproduce, life could not have continued beyond one generation.

From Prebiotic to Living

Nonliving systems arrived at the threshold of life through the process of *chemical selection.* As nonliving droplets broke up and reformed, those that acquired stable components lasted much longer. A stage was reached in prebiotic systems in which there was an accumulation of stable aggregates that were able to carry out chemical reactions. These chemical reactions preserved the aggregates and resulted in their continuance.

To boost a prebiotic system into a living state, two major requirements had to be met:
- the development of means of storing and using energy
- the ability to pass genetic information along to progeny

Let us now consider the origin of metabolic energy.

Origin of Energy Metabolism

An initial requirement of any metabolic system is the ability to acquire energy, to store it temporarily, and to use it as needed. The use of energy is the basis of life. Scientists now theorize that a membrane-bound, light-driven hydrogen-ion pump provided the necessary energy for coacervates or microspheres to gather certain small molecules and ions from the primordial seas. As the H-ion pumps became more complex, electron transport systems of respiration and photosynthesis were formed.

Some scientists further theorize that energy metabolism was originally based on ATP. It has been shown experimentally that ATP can power certain chemical reactions in prebiotic aggregates, just as might have happened before the appearance of living cells. A combination of a H-ion pump and ATP may have been sufficient to provide energy for anaerobic respiration.

Origin of Biological Information

A living cell differs from a coacervate and microsphere in that the cell is able to pass on genetic information. DNA forms the template for RNA which, in turn, governs the synthesis of proteins. These proteins regulate the exchange of materials between the cell and its external environment, catalyze chemical activities within the cell, and also govern the synthesis of more protein.

DNA needs enzymes to do its work, and enzymes need DNA to provide instructions. The question arises: Over evolutionary time from coacervates to living cells, which came first—DNA or enzymes? Modern research indicates that both must have evolved at the same time through some type of complex feedback cycle involving interaction among many primitive "genes" and "enzymes."

According to modern theory, RNA was the first nucleic acid to evolve. The 2' hydroxyl group of the ribose sugar can bind to amino acids and hold them in a certain sequence. DNA does not have the 2' hydroxyl group and therefore cannot bind amino acids. Modern retroviruses are able to assemble DNA against an RNA template. Thus it is quite probable that primitive RNA preceded DNA as an informational macromolecule. It is quite conceivable that the nucleic acids (the storage molecules of genetic information) and proteins (the molecules of enzymes) evolved together, making possible the evolution of the living cell.

First Organisms—Heterotrophs

The first organisms were undoubtedly heterotrophs utilizing the organic pool for nutrition. As time went on, the early heterotrophs must have faced a serious crisis because oxygen from photodissociation (atmospheric reactions) began to increase in concentration. This changed the atmosphere. Oxygen and hydrogen peroxide (H_2O_2) destroyed many heterotrophs. Eventually, photosynthetic organisms—autotrophs such as blue-green algae—evolved, increasing the oxygen content of the atmosphere and thereby threatening the continued existence of the heterotrophs. An ozone layer in the atmosphere developed from the high concentration of oxygen, which further diminished the organic compounds available to the beleaguered heterotrophs. Some heterotrophs developed pathways for utilizing oxygen in energy production (aerobic respiration). The carbohydrates produced by autotrophs and the oxygen of the atmosphere supplied the new heterotrophs with the nutritive materials necessary for survival (heterotroph-autotroph hypothesis). Thus the stage was set for the development of life on the scale that is known today.

Model Questions Related to the Origin of Life

1. According to the heterotroph-autotroph hypothesis, which is the proper sequence?
   (A) Autotroph—heterotroph—organic molecules—aggregate molecules
   (B) Organic molecules—aggregate molecules—autotroph—heterotroph
   (C) Heterotroph—aggregate molecules—autotroph—organic molecules
   (D) Aggregate molecules—organic molecules—heterotroph—autotroph
   (E) Organic molecules—aggregate molecules—heterotroph—autotroph

2. The primitive reducing atmosphere of the earth contained all of the following gases EXCEPT
   (A) ammonia
   (B) oxygen
   (C) methane
   (D) water vapor
   (E) hydrogen

3. Although the earth is over 4 ½ billion years old, the earliest fossils date from about
   (A) 1 billion years ago
   (B) 3 billion years ago
   (C) 500 million years ago
   (D) 3 million years ago
   (E) 5,000 years ago

4. Which of the following statements is TRUE about the origin of life on earth?
   (A) It arose from living systems brought to earth from another planet.
   (B) It required gaseous oxygen.
   (C) It has been duplicated in the laboratory.
   (D) It can be explained by using our knowledge of chemistry and physics.
   (E) It is still continuing.

Answers and Explanations to Model Questions: The Origin of Life

1. E According to the heterotroph-autotroph hypothesis, the first living organisms developed from complex molecules that were formed from simple organic compounds. These "organisms" depended on a continuous supply of fresh organic molecules for their existence.

   Incorrect Choices
   A, C Organisms could not arise before the necessary compounds were present.
   B This is a contradiction of the proposed theory, since the autotrophs are placed before the heterotrophs.
   D Molecules are needed before aggregation occurs.

2. B Oxygen is an oxidizing agent. As such, it immediately combined with substances, preventing a buildup in the atmosphere. After oxidizing reactions coated the exposed surface of the earth, oxygen could increase in concentration in the atmosphere. Photodissociation of water vapor also contributed to the production of atmospheric oxygen.

   Incorrect Choices
   A, C, D, E are all reducing substances.

3. B Fossils of blue-green algae, fungi, and bacteria have been found in rocks that are about 3 billion years old. The largest collection occurs in the Gunflint chert near the Minnesota-Ontario border. Since these forms are highly advanced, their predecessors must have been less complex in structure.

   Incorrect Choices
   A There is a scarcity of fossils from the Precambrian era. Those that exist are of blue-green algae.
   C The Cambrian era began at this time. This was the era in which there was an explosion of life forms.
   D The origin of human beings is estimated to have occurred over 3 million years ago.
   E Human recorded history began about 5,000 years ago.

4. D Experiments by Miller, Calvin, and Fox have shown how simple elements and molecules are involved in the formation of complex molecules that may possibly have been the forerunners of living systems. These experiments utilized apparatuses built on chemical and physical principles.

Incorrect Choices

A Outside of a visitation by space persons from another world, living systems would probably not have survived an interplanetary trip. There is no evidence to support this theory.

B The atmosphere was a reducing atmosphere.

C Experiments have succeeded only in producing complex organic molecules.

E The conditions that led to the origin of life are no longer present on the earth.

Genetics Explains the Evolutionary Process

In 1809, Jean Baptiste Lamarck proposed a theory to explain the origin of the species. His theory of evolution is known as the theory of *use and disuse.* Lamarck stated that organisms acquired traits through the continual use of a body part or organ in response to environmental conditions. Furthermore, he claimed that the *acquired characteristics* would be passed on to their offspring. Although the idea that acquired characteristics are inherited is incorrect, Lamarck's theory shifted the emphasis onto science to explain the origin of the various species of living things.

In 1858, Charles Darwin published *The Origin of the Species,* which shook the scientific world. In this book, Darwin proposed his theory of evolution by *natural selection,* which encompasses several ideas.

All species overproduce. Overproduction leads to a *struggle for existence* to obtain the resources necessary for survival. Many differences or *variations* occur among the individuals of a given species. Variations are the materials upon which environmental factors "choose" which organisms are best suited for the environment. Darwin termed this the *survival of the fittest.* The variations conferring an advantage to organisms are *inheritable.* Darwin's theory of descent with modification from ancestral species supplanted Lamarckism.

The concept of evolution is supported by evidence from various scientific fields. These include paleontology, comparative anatomy, comparative embryology, taxonomy, biochemistry, cytology, and molecular biology.

Paleontology, the study of fossils, has confirmed the existence of organisms in various geologic eras. It has also shown the relationship of those organisms to present-day living species.

Comparative anatomy, the study of structural similarities among organisms, provides the basics for taxonomy. The similarities in the stages and patterns of embryonic development is studied by *comparative embryology.*

Taxonomy groups organisms according to their homologous structures. Structural similarities resulting from common ancestry are known as *homologous structures. Vestigial organs* are remnants of structures that were once important to an organism. The human appendix is a vestigial organ, having no function in the body.

Biochemistry studies the chemical reactions in living things. *Cytology* examines the structure of cellular organelles and cell function. *Molecular biology* has elucidated the nature of the gene, genetic code, and protein structure. Organisms exhibiting common structural, functional, and chemical similarities share a common ancestry.

Despite the many evidences for evolution, it is really the study of genetics that adequately explains how the process of evolution occurs. Sexual reproduction provides the means for variation, while changes in the gene pool of a population provide the mechanism for evolution. A *gene pool* comprises all the genes available to a breeding population. A *population* is a group of freely and randomly interbreeding organisms living in a specified area.

It is difficult to determine directly the percentages or the frequencies of alleles of a specific gene in the gene pool of a population. However, from actual observation, it is possible to ascertain what percentage of a population shows a certain genotype as expressed phenotypically and, from this, to calculate the frequencies of allelic genes. The theoretical basis for these determinations is the Hardy-Weinberg Law.

Hardy-Weinberg Derivation

In a population of sheep where the trait for white wool is dominant over that for black wool, 25% of the sheep have black wool. The black individuals are homozygous recessive. The letter q represents the recessive allele in the gene pool. According to the laws of chance, the probability that two independent events will occur is equal to the product of their separate probabilities. Since two gametes each containing a q allele must combine to produce black offspring, the probability that this will occur is $q \times q$ or q^2. The 25% black sheep represent q^2 in the population. Therefore the square root of q^2 (0.5) represents the frequency of the q allele in the male and female gene pool. If the q allele makes up 50% of the alleles for coat color in the gene pool, then the allele p for white color makes up the remaining 50% ($p + q = 1$).

Although 75% of the population of sheep is phenotypically white, some are hybrid, while others are homozygous. The homozygous dominant individual is designated as p^2. The percentage of these individuals in the population depends on the probability of the combination of the gametes carrying the p allele ($0.5 \times 0.5 = 0.25$ or 25%). The hybrids are designated pq. Expansion of the binomial $(p + q)^2 = p^2 + 2 pq + q^2$ indicates that $2 pq$, or $2(0.5 \times 0.5) = 0.50$ or 50%. The mathematical relationship may be summarized by this equation: $p^2 + 2pq + q^2 = 1$.

Hardy and Weinberg observed that the gene frequencies in the gene pool remain relatively constant from generation to generation. Therefore the F_1 generation should show the same percentage of homozygous and heterozygous individuals as the parent generation. This means that the gene frequencies in the gene pool of the F_1 will be the same as in the parent generation. Actually, the gene frequencies in small populations fluctuate a little from generation to generation because of chance factors. This is known as *genetic drift*.

Conditions for Genetic Equilibrium

Any condition that contributes to the loss or gain of alleles in a gene pool may significantly alter the gene frequencies. In order for genetic equilibrium as proposed by the Hardy-Weinberg Law to be maintained, certain conditions that prevent the shifting of gene frequencies in a gene pool must be met.

First, the breeding population must be large. A large population can absorb the loss of individuals without any significant change in the gene frequencies of the gene pool. Also, there must be no accumulation of mutations and no selective mating among individuals. Genetic equilibrium maintains the status quo and does not lead to evolution or change.

Once the gene frequencies shift so that genes are lost from the gene pool, the population changes and evolution occurs. The force acting on the gene pool through the phenotypes within a population is *natural selection* or the environment. It is the environment that "chooses" the genes best suited for survival. Environmental pressures may affect the extremes or the middle of a population. In either case, the population changes.

Natural selection operates in three ways. It may be stabilizing, directional, or disruptive. Each can be represented by a graph (Figure 9.1).

Stabilizing selection affects the extremes of a population (Graph 1). The individuals that deviate too far from the average conditions are removed. The results are a decrease in diversity, maintenance of a stable gene pool, and no evolution.

Directional selection affects the extremes of a population (Graph 2). The eventual result is a shift in the gene pool toward the opposite extremes.

Selection that acts against individuals that have the average condition and for individuals at the extreme ends is called disruptive selection (Graph 3). The population is split into two.

Both directional and disruptive selection affect the gene pool. In both cases, the population changes and evolution occurs.

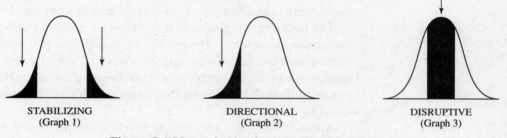

STABILIZING (Graph 1)   DIRECTIONAL (Graph 2)   DISRUPTIVE (Graph 3)

Figure 9.1 Natural selection operates in three ways

Evolution and Speciation

The direct result of evolution is *speciation,* resulting in genetically isolated, interbreeding organisms. When segments of a population are so isolated that gene flow ceases, the gene pool of each group is separated and subjected to particular selective pressures. Each group develops its own seasonal breeding pattern, courtship and mating behavior, subtle reproductive changes (genital and gametic), and habitat preference. Even if the isolating barrier is removed, gene flow between the groups may not be possible. New species may have been formed.

Speciation is best exemplified by a group of birds, known as Darwin's finches, on the Galapagos Islands. The ancestral finch probably came from Equador, on the mainland of South America, traveling in a flock that was blown off course. The small population experienced different selective pressures on each of the Galapagos Islands. Consequently, the finches of each island evolved their own gene pool. Isolation and genetic drift fostered adaptive radiation and thus divergent evolution.

Groups of finches from given populations spread out and occupied the many available habitats on each island. Selective pressures of adaptation to changed environments ultimately resulted in genetic isolation and speciation that prevented hybridization. *Adaptive radiation* is the evolutionary pathway that accounts for the existence of various species of finch on the same island.

Very often populations that are completely unrelated resemble each other—a condition known as *convergent evolution*. Unrelated species subjected to the same environmental factors may adapt in a similar fashion. For example, both the whale and the fish have streamlined bodies with fins although they do not share a common ancestry.

Gradualism Versus Punctuated Equilibrium

Traditionally, evolution was viewed as a slow, stepwise development of a species over a long period of time (millions of years). Many transitional fossil forms arose, forming a continuum from the ancestral species to the new species. This theory is known as *gradualism*. See Figure 9.2(a).

Speciation has resulted from a gradual change in the frequency of alleles in the gene pool due to selective environmental pressure. The fossil record of the elephant, camel, and horse support the theory of gradualism. However, even Darwin, whose theory of natural selection was based on gradualism, realized that major groups of organisms arose suddenly without evidence of transitional fossil forms.

In 1971, Niles Eldredge and Stephen Jay Gould proposed a theory of *punctuated equilibrium*. The theory proposes that species arose suddenly in a short period of time (thousands of years) after long periods of stability. Thus, evolution occurs in "spurts." The abrupt appearance of flowering plants without a fossil record of their origin is an example of punctuated equilibrium. See Figure 9.2(b).

The frequency of alleles in the gene pool changes rapidly in response to major environmental events. Polyploidy (a genetic event), natural catastrophes, and human-induced cataclysms are examples of major environmental events. The appearance of several species of banana-eating moths in Hawaii is an example of punctuated equilibrium resulting from a major environmental event induced by humans.

About 1,000 years ago, banana plants were introduced into Hawaii by Polynesian settlers. Today, several species of banana-eating moths exist that are closely related to other Hawaiian moths that feed on other plants.

Evolution is a dynamic force. The theories of gradualism and punctuated equilibrium are concerned with the tempo at which evolution occurs.

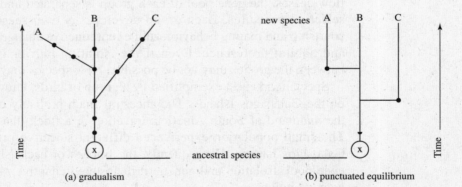

(a) gradualism   ancestral species   (b) punctuated equilibrium

Figure 9.2 (a) Gradualism and (b) punctuated equilibrium

Each point on the graphs represents an evolved species—each a little different from the one before. The letters at the top of the graphs represent the currently existing species. Compare the pathways of gradualism with the pathways of punctuated equilibrium.

Allopatric Speciation

Allopatric speciation occurs when a population that inhabits a particular area is divided into two or more geographically separated populations. A prime example of allopatric speciation is Darwin's finches, which, as stated above, migrated from mainland Equador to the Galapagos Islands. There 14 identifiable species evolved from the same finch ancestor. Each of the finch species is separate and distinct and cannot interbreed with any other species. Allopatric populations live in different geographical areas and come to differ genetically from one another. Allopatry is caused not only by migration of species but also by separation of population by a major environmental change, such as the division of a species range by a glacier.

Sympatric Speciation

Sympatric speciation is the appearance of a new species within a single population. Polyploidy, the multiplication of chromosomes, is not an unusual mechanism of sympatric speciation in plants. When chromosomes fail to segregate at meiosis, diploid gametes are formed. A polyploid stem may produce polyploid flowers that, in turn, produce diploid gametes. The diploid gametes will continue the cycle of producing polyploid progeny. Polypoidy is more common in plants than in animals because plants can self-fertilize. Occasionally, polyploidy occurs in certain species of earthworms, primitive fish, and frogs.

Reproductive Barriers

Gene flow between two closely related species occupying the same area is prevented by *reproductive barriers*. Reproductive barriers isolate gene pools.

Differences in habitat preference, mating, time (temporal isolation), genitalia (mechanical isolation), and failure of gametic fusion (gametic isolation) are *prezygotic* barriers. *Prezygotic* barriers deter fertilization.

Hybrid sterility and aborted hybrids due to genetic incompatibility are *postzygotic* barriers. *Postzygotic* barriers prevent the development of viable or fertile offspring.

Modern Explanation of Natural Selection

Clones and Natural Selection

A *clone* is a single-species population that has descended from one ancestor. All the individuals are genetically identical because they have been reproduced asexually. Seemingly, there is no variation in a clone. However, through random mutation variation can occur. A prime example of mutation in a clone is seen in the appearance of penicillin-resistant bacteria that thrive in the environment of this antibiotic. Usually, penicillin destroys the cell walls of bacteria, preventing them from carrying on the requisite metabolism. Through some slight variation in genetic makeup, however, a population of bacterial clones developed that are penicillin-resistant. At first, the population of the resistant types was relatively small but as penicillin kills off all of the susceptible bacteria, the population of penicillin-resistant organisms is increasing rapidly. Such has also been the case in the appearance and spread of penicillin-resistant gonorrhea bacteria.

In the summer of 1997, Dr. Ian Wilmut of the Roslin Institute in Scotland propelled to fame a sheep named Dolly. Dolly was cloned by Wilmut and his team of scientists from a mammary cell taken from an adult, pregnant sheep. The mammary cell is not an egg cell but a somatic cell in the tissue of a mammary gland. Cloning was accomplished by nuclear transfer. The chromosomes from the donor cell were removed and put into a denucleated egg cell of another sheep. Thus, the egg cell nucleus was replaced with the chromosomes of the donor cell. The egg cell with the transplanted chromosomes was implanted into the uterus of a surrogate mother sheep. The implanted cell went through cleavage and embryo formation, eventually developing into a lamb. This lamb grew to an adult sheep with the identical genetic makeup as the sheep from which the mammary cell chromosomes were taken.

The technique used is not an easy one. Many failures occurred before Dolly became a reality. It is believed that in order for cloning to be successful using the method of nuclear transfer, the production of RNA must be inhibited. It is also theorized that the regulatory molecules in the recipient egg act on the transferred nucleus to reprogram it.

Following the successful cloning of Dolly, five identical piglets were cloned by scientists in the year 2000 in a laboratory of PPL Therapeutics in the United States. In cloning these piglets, a different technique was used than the one that created Dolly. The ultimate goal of scientists who are carrying out cloning experiments is to develop a means of producing donor organs for human transplants.

Natural Selection and Environment

Before the Industrial Revolution in England, the moth *Biston betularia* appeared more frequently in its lighter, peppered form. The black form of the moth appeared rarely in the environment. With the Industrial Revolution, however, came layers of soot and carbon that blackened buildings and tree trunks. The changing environment with its blackened surfaces made the lighter morph of the peppered moth more visible to its bird predators, while the darkened surfaces provided protective coloration to the black moth. Consequently, the numbers of the peppered moth decreased dramatically with a corresponding increase in the numbers of the black moths. This is an example of environmental effect on natural selection.

Polygenic Effect

More often than not, natural selection is based upon the cumulative effects of numerous genes, each responsible for slight changes. When genes at more than one locus contribute to the same trait, the result is called a *polygenic effect*. In humans, polygenic traits are skin color, foot size, nose length, birth weight, height, and intelligence.

Genetic Drift

Genetic drift refers to random changes in the gene pool of a population. If a large and stable population were suddenly destroyed and only a few survivors remained, the population would be rebuilt by these survivors through reproductive means. However, the gene pool in this smaller and newer population might differ from that of the former population. As a result, mutations that were rare might become more concentrated and harmful genetic defects might show up with greater frequency.

There is an example in present day Afrikaaners. Afrikaaners are descendants from 30 families, a rather small population. A metabolic disorder called porphyria variegata is rare among most populations but occurs with great frequency among the South African descendants of Dutch settlers. This disease is characterized by excessive amounts of iron porphyrins in the blood. The urine turns red, and the afflicted person is very sensitive to light and ultimately suffers liver damage. The abnormal genes that cause this condition have been passed along to succeeding generations of the original settlers.

Model Questions Related to Genetics and Evolution and to Natural Selection

1. *The Origin of the Species by Means of Natural Selection* set forth the theory of evolution by means of natural selection as developed by
   (A) Charles Lyell
   (B) Thomas Malthus
   (C) Alfred Wallace
   (D) August Weissmann
   (E) Charles Darwin

2. Which of the following statements regarding the Hardy-Weinberg principle is INCORRECT?
   (A) Random mating sustains genetic equilibrium.
   (B) Selection upsets genetic equilibrium.
   (C) Nonrandom mating shifts the genetic equilibrium in a population.
   (D) New mutations upset genetic equilibrium.
   (E) Genetic equilibrium is maintained best in small populations.

3. The largest unit of population in which gene flow is possible and that is genetically isolated from other populations is the
   (A) kingdom
   (B) phylum
   (C) family
   (D) genus
   (E) species

4. Under certain conditions, the relative proportions of dominant and recessive genes in the gene pool of a given population tend to remain stable from one generation to the next. Which condition would upset such stability?
   (A) A large population of individuals capable of producing offspring
   (B) No mutations in the population
   (C) Random mating among members of the population
   (D) No significant emigration or immigration
   (E) Selective breeding among the members of a population

5. In the gene pool of a given population of rabbits, 80% of the gametes carry the dominant allele for gray coat. What percentage of the population is heterozygous gray?
   (A) 64%
   (B) 32%
   (C) 20%
   (D) 16%
   (E) 4%

6. Brown eyes are dominant over blue. In a population, 9% of the individuals have blue eyes. What percentage of individuals are hybrid brown?
   (A) 9%
   (B) 91%
   (C) 30%
   (D) 70%
   (E) 42%

7. In a study performed in Sweden, it was found that 16% of the population has type N blood. The N blood type results from a homozygous condition. What is the gene frequency for the N allele in the gene pool?
   (A) 16%
   (B) 40%
   (C) 84%
   (D) 60%
   (E) 24%

For Questions 8–11 choose the letter of the item that is MOST closely related to the numbered statement.

   (A) Adaptive radiation
   (B) Isolation
   (C) Natural selection
   (D) Stable gene pool
   (E) Convergent evolution

8. Members of the same species of moths are prevented from interbreeding because they live on opposite sides of a mountain range.

9. Darwin finches are a good example of this biological principle.

10. Members of a large population mate at random.

11. In the evolutionary history of the horse, eohippus (the dawn horse) was replaced by the modern one-toed horse.

For Questions 12–15 choose the letter of the item that BEST applies to the numbered statement.

(A) Embryology
(B) Fossils
(C) Homology
(D) Vestigial organs
(E) Biochemistry

12. The wing bones of a bird resemble the bones in the foreleg of a cow.

13. Indications of gill slits occur in the early development of mammals.

14. Few people can move their ears although they all possess muscles for that purpose.

15. Tests show that there are similarities in the DNA of chimpanzees and gorillas.

16. Modern theories of evolution propose that species may exist unchanged for a long period of time. Suddenly, an environmental change may cause a shift in the gene pool. This theory of evolution is known as
(A) natural selection
(B) sympatric speciation
(C) punctuated equilibrium
(D) parapatric evolution
(E) survival of the fittest

17. The species of wren found on the island of St. Kilda, which is off the coast of Scotland, closely resembles a wren species found on the Scottish mainland. These wren species support the theory of
(A) divergent evolution
(B) sympatric speciation
(C) natural selection
(D) allopatric speciation
(E) convergent evolution

18. Sympatric populations are those
(A) closely resembling species that live in the same place
(B) closely resembling species that live in different climes
(C) species that have traveled different lines of evolution
(D) species that have evolved from two ancestral lines
(E) closely resembling species that have migrated away from each other

19. Sympatric speciation usually occurs as a result of
    (A) polyploidy
    (B) geographic barriers
    (C) parapatric speciation
    (D) introgression
    (E) hybridization

Answers and Explanations: Genetics and Evolution and Natural Selection

1. E Charles Darwin's observations on his five-year trip around the world beginning in 1831 laid the foundation for his theory of evolution that was published in 1859 in the famous work *The Origin of the Species*.

   Incorrect Choices
   A Charles Lyell wrote *The Principles of Geology*, which suggested that the earth had undergone changes, a belief contrary to that of the times. This book influenced Darwin's thinking.
   B Thomas Malthus wrote a book entitled *Essay on Population* in 1798, which stated that food supplies can never keep pace with the rapid increase in human population.
   C Alfred Wallace, an English naturalist, came to the same conclusions about changes in living things as did Darwin; however, he was not the author of *The Origin of the Species*.
   D August Weissmann, a German biologist, proposed a theory of "continuity of the germplasm" in which he suggested that hereditary traits are carried from one generation to the next.

2. E In all populations, genes are lost and others are gained due to chance factors. In a large population, the loss or addition of genes to the gene pool does not significantly shift the gene frequencies in the pool. There will be only slight fluctuations of gene frequencies around a given value. However, in a small population, the loss or gain of genes shifts the gene frequencies so far away from the original equilibrium value that it cannot be reestablished.

   Incorrect Choices
   A In a population where mating among individuals is at random, the gene makeup of the population is in a state of balance.
   B The environment influences the genotypes through the phenotypes. Thus, certain genes may decrease while others may increase in the gene pool because of selection against or for particular phenotypes. This shifts the gene frequencies and upsets the equilibrium.
   C Nonrandom mating is selection. Item B above explains how selection will change a gene pool.
   D New mutations change the gene makeup of a population. Population shifts occur until a new point of equilibrium is reached.

3. E Members of a species share genes that allow for reproductive compatibility. This means that members of a species can interbreed with one another and produce fertile offspring. Biological species are reproductively isolated from other species.

Incorrect Choices
A, B, C, D are larger classification units in which there are diverse gene pools, not conducive to interbreeding as in the species.

4. E   See Answer 2.

5. B   A gene pool consists of all the genes available to a breeding population. 80% of the genes for rabbit coat color are gray alleles (p). Since the total alleles for coat color must equal 100%, the nongray color allele is 20% (q) of the total genes for color. According to the equation $p^2 + 2pq + q^2 = 1$, the hybrid or heterozygotes are designated $2pq$: $2pq = 2(0.8)(0.2) = 0.32$ or 32%.

6. E   Since the homozygous recessive condition results from the combination of two gametes carrying the recessive blue allele, the probability that the two gametes will combine is p^2 ($p \times p$). Because 9% of the population expresses this genotype, $\sqrt{0.09}$ or 0.3 represents the frequency of the recessive allele in the gene pool. The dominant allele comprises 70% of the gene pool. Then $2pq$ (hybrid) $= 2(0.3)(0.7) = 0.42 = 42\%$.

7. B   If 16% of the people are homozygous for the N allele, $\sqrt{0.16}$ (0.4 or 40%) equals the frequency of the N allele in the gene pool.

8. B

9. A

10. D

11. C

12. C

13. A

14. D

15. E

16. C   Punctuated equilibrium refers to a pattern of evolutionary change over millions of years. A long quiescent period (in terms of evolution) gives way to rapid changes in species caused by a shift in the gene pool.

Incorrect Choices
A Natural selection is the theory of evolution proposed by Charles Darwin, who observed that only the "most fit" organisms can survive.
B Sympatric speciation refers to the evolving of separate species from one ancestral line although the species live in the same geographical area.
D Parapatric evolution indicates that two adjacent populations have developed reproductive isolation between them, maintaining separate species.
E Survival of the fittest was one of Darwin's propositions in his theory of natural selection.

17. **D** Allopatric speciation occurs when two populations of a species become separated so that there can be no interbreeding between them.

    Incorrect Choices
    A Divergent evolution indicates that species from a common ancestor have branched off into different evolutionary directions making them separate species—unable to interbreed (among species).
    B For a discussion of sympatric speciation see 16B.
    C For a discussion of natural selection see 16A.
    E Convergent evolution is evolution of similar features by unrelated organisms, for example, a fish and a whale, which both live in the water but are otherwise unrelated.

18. **A** Sympatric populations are made up of species that closely resemble other species living in the same place.

    Incorrect Choices
    B, C, D, E, are incorrect. See discussion on allopatric populations, divergent evolution and convergent evolution.

19. **A** Sympatric speciation usually occurs as a result of polyploidy or mutation.

    Incorrect Choices
    B Geographic barriers result in allopatric speciation.
    C Parapatric speciation refers to two adjacent populations that have lost reproductive potential between them.
    D Introgression refers to the incorporation of the genes of one species into the gene complex of another as a result of hybridization.
    E Hybridization occurs more easily in plants through cross pollination. Genes of one species are incorporated into the gene complex of another species.

References for Area II: Heredity and Evolution

All of the following references are from *Scientific American*:

Genetics

Adleman, L. M. "Computing with DNA." August 1998

Anderson, W. F., and E. G. Diacumakos. "Genetic Engineering in Mammalian Cells." July 1988

Anderson, W. "Gene Therapy." September 1995

Benditt, J. "Genetic Skeleton." July 1988

Blaese, M. "Gene Therapy for Cancer." June 1997

Capecchi, M. "Targeted Gene Replacement." March 1994

Cavenee, W., and R. White. "The Genetic Basis of Cancer." March 1995

Collins, F., and K. G. Jegalian. "Deciphering the Code of Life." December 1999

Cunningham, P. "The Genetics of Thoroughbred Horses." May 1991

Dawkins, R. "God's Utility Function." November 1995

DeRobertis, E. M., G. Oliver, and C. V. Wright. "Homeobox Genes and the Vertebrate Body Plan." July 1990

Dickerson, R. E. "The DNA Helix and How It's Read." December 1983

Felgner, P. L. "Nonviral Strategies for Gene Therapy." June 1997

Friedman, T. "Overcoming the Obstacles to Gene Therapy." June 1997

Gehring, W. J. "The Molecular Basis of Development." October 1985

Gilbert, W., and L. Villa-Komaroff. "Useful Proteins from Recombinant Bacteria." April 1980

Greenspan, R. "Understanding the Genetic Construction of Behavior." April 1995

Holliday, R. "A Different Kind of Inheritance." June 1989

Iovine, J. "Genetically Altering Eschericia." June 1994

Kornberg, R. O., and A. Clug. "The Nucleosome." February 1981

Langer, R. S., and J. P. Vacanti. "Tissue Engineering: The Challenges Ahead." April, 1999

McKnight, S. L. "Molecular Zippers in Gene Regulation." April 1991

Miller, R. V. "Bacterial Gene Swapping in Nature." January 1997

Mirsky, S., and J. Rennie. "What Cloning Means to Gene Therapy." January 1997

Nathans, J. "The Genes for Color Vision." February 1989

O'Brochta, D., and P. W. Atkinson. "Building the Better Bug." December 1998

Parenteau, N., and G. Naughton. "The First Tissue-Engineered Products." April 1999.

Ptashne, M. "How Gene Activators Work." January 1989

Radman, M., and R. Wagner. "The High Fidelity of DNA Duplication." August 1988

Rennie, J. "Grading the Gene Tests." June 1994

Resenberg, S. A. "Adoptive Immunotherapy for Cancer." May 1990

Ross, J. "The Turnover of Messenger RNA." April 1989

Sapienza, C. "Parental Imprinting of Genes." October 1990

Stix, G. "Is Genetic Testing Premature?" September 1996

Tijan, R. "Molecular Machines that Control Genes." February 1995

Tsein, J. Z. "Building a Brainier Mouse." February 2000

Upton, A. C. "The Biological Effects of Low Level Ionizing Radiation." February 1982

Verma, I. M. "Gene Therapy." November 1990

Weinberg, R. A. "Finding the Anti-Oncogene." September 1988

Weiner, D. B., and R. C. Kennedy. "Genetic Vaccines." July 1999

Weintraub, H. M. "Antisense RNA and DNA." January 1990

White, R., and J. M. Lalouel. "Chromosome Mapping with DNA Markers." February 1988

Wilmut, I. "Cloning for Medicine." December 1998

Evolution

Abler, W. A. "The Teeth of the Tyrannosaur." September 1999

Agnew, N., and M. Demos. "Preserving the Lactoli Footprints." September 1998

Alexander, R. "How Dinosaurs Ran." April 1991

Allegre, C., and S. Schneider. "The Evolution of the Earth." October 1994

Alvarez, W., F. Asaro, and V. Corutillot. "What Caused the Mass Extinction?" October 1990

Badash, L. "The Age of the Earth Debate." August 1989

Bishop, J. A., and L. M. Cook. "Moth, Melanism and Clean Air." January 1975

Burgin, T., et al. "The Fossils of Monte San Giorgio." June 1989

Dalziel, I. "Earth Before Pangea." January 1995

Davies, N. B., and M. Brooke. "Coevolution of the Cuckoo and Its Hosts." January 1991

Doolittle, W. "Uprooting the Tree of Life." February 2000

Erickson, G. M. "Breathing Life into Tyrannosaurus Rex." September 1999

Erwin, D. "The Mother of Mass Extinctions." July 1996

Gould, S. J. "The Evolution of Life on the Earth." October 1994

Grant, P. R. "Natural Selection and Darwin's Finches." October 1991

Griffiths, M. "The Platypus." May 1988

Grimaldi, D. "Captured in Amber." April 1996

Hoffman, P. F., and D. P. Schrag. "Snowball Earth" January 2000

Knoll, A. "End of the Proterozoic Eon." October 1991

Leakey, M., and A. Walker. "Early Hominid Fossil Evolution from Africa." June 1997

Lovejoy, C. O. "Evolution of Human Walking." November 1988

Mayr, E. "Evolution." September 1978

McEvedy, C. "The Bubonic Plague." February 1988

Milner, R. "Charles Darwin: The Last Portrait." November 1995

Morris, S. C., and H. B. Whittington. "The Animals of the Burgess Shale." July 1979

Moxon, E., and C. Wills. "DNA Microsatellites: Agents of Evolution." January 1999

Novacek, M. "Fossils of the Flaming Cliffs." December 1994

Orgel, L. "The Origin of Life on the Earth." October 1994

Padian, K., and L. M. Chiappe. "The Origin of Birds and Their Flight." February 1998

Sagan, C. "The Search for Extraterrestrial Life." October 1994

Stebbins, G. L., and F. J. Ayala. "The Evolution of Darwinism." July 1985

Stringer, C. B. "The Emergence of Modern Humans." December 1990

Tattersal, I. "Once We Were Not Alone." January 2000

Turner, C. G., II. "Teeth and Prehistory in Asia." February 1989

Weinberg, S. "Life in the Universe." October 1994

Wellinhofer, P. "Archaeopteryx." May 1990

Wong, K. "Who Were the Neanderthals?" April 2000

PART FIVE

Area III of Advanced Placement Biology:

Organisms and Populations

CHAPTER TEN

Principles of Taxonomy, Including a Survey of Monera, Protista, and Fungi

Capsule Concept

To date, almost 2 million organisms have been identified and named. As early as the 1700s, Carl von Linne attempted to bring order to the vast diversity of living things by grouping them into categories based on their structural similarities. Modern taxonomy, however, transcends the simple aim of naming and classifying organisms. Taxonomy has become a useful tool in establishing evolutionary relationships among organisms.

Of the five kingdoms in the modern system of classification, the kingdom that contains the most primitive organisms, those most clearly related to ancestral types, is the Monera. Prokaryotic cells—bacteria and blue-green algae (cyanobacteria)—exhibit the simplest cellular organization.

The kingdom Protista consists of eukaryotic cells: protozoans and flagellated photosynthetic organisms that have been claimed by both the zoologists and the botanists. *Euglena* and *Volvox* are examples of two such controversial organisms.

The nonphotosynthetic plants—yeasts, rusts, mushrooms, and slime molds—have all been placed in their own kingdom, the Fungi.

The virus is a biochemical anomaly. Since it does not exhibit any of the characteristics attributed to living organisms, it is not classified in any of the kingdoms. The virus is merely a particle consisting of a protein coat that surrounds a core of nucleic acid, either DNA or RNA, but never both. It is speculative whether a virus arose from a primitive organism and eventually lost some of the life processes, whether it is an escaped gene surviving unattached, or whether it represents a primitive life form.

Vocabulary

Binomial nomenclature a system whereby each species is assigned a two-word Latin name, to distinguish it from every other type of organism

Class a classification category that is below the phylum and above the order

Division the major classification category, used in place of the phylum, for the plant kingdom

Genus (plural: **genera**) a classification category containing a group of similar species

Kingdom the largest and most inclusive classification category

Order a classification category that is below the class and above the family

Phylogeny the evolutionary history of a species

Phylum a major classification category that is just beneath the kingdom and is above the class

Species the basic unit of classification

Systematics the branch of biology that studies the history of life

Taxon (plural: **taxa**) a given classification category, such as a particular genus or species, in a formal system of nomenclature

Taxonomy

Taxonomy is the branch of biology that deals with the classification of living things. Using a hierarchial system of categories, the taxonomist groups organisms related to each other in biologically significant ways. The largest unit of classification is the *kingdom*, followed by six other categories that become smaller in descending order: *phylum*, *class*, *order*, *family*, *genus*, and *species*.

Under the system of *binomial nomenclature*, devised by the "father of taxonomy," Carolus Linnaeus (the Latinized version of Carl von Linne), each species is assigned a two-word Latin name, or binomial. The first word in this scientific name designates the genus; the second word, the species. Both words are italicized in print or underlined in handwriting. *Homo sapiens* is the scientific name of the human being.

Table 10.1 shows the classification of system of the human.

Systematics is the branch of biology that studies evolutionary relationships among species. The evolutionary history of a species or a group of species is known as *phylogeny*. Phylogenetic trees are used to show the descent of similar modern species from a common ancestor. The systematist studies and organizes the phylogenetic history of organisms, enabling the taxonomist to assign species to appropriate taxa.

Five-Kingdom System of Classification

As stated above, the kingdom is the largest unit of classification. For many decades, every living organism was thought to be either plant or animal and was grouped accordingly into the plant or the animal kingdom. Modern research techniques have elucidated a number of physiological, structural, and biochemical facts that indicate the need for additional kingdoms that permit more accurate classification. Currently, the five-kingdom system of classification is used. Table 10.2 provides a summary of this classification system.

TABLE 10.1   CLASSIFICATION OF THE HUMAN

Category	Taxon	Characteristics
Kingdom	Animalia	Multicellular organisms requiring preformed organic material for food and having a motile stage at some time in their life history
Phylum	Chordata	Animals having a notochord (embryonic skeletal rod), dorsal hollow nerve cord, gills in the pharynx at some time during the life cycle
Subphylum	Vertebrata	Backbone (vertebral column) enclosing the spinal cord, skull bones enclosing the brain
Class	Mammalia	Body having hair or fur at some time in the life, female nourishes young on milk, warm-blooded, one bone in lower jaw
Order	Primates	Tree-living mammals and their descendants, usually with flattened fingers and nails, keen vision, poor sense of smell
Family	Hominidae	Bipedal locomotion, flat face, eyes forward, binocular and color vision, hands and feet specialized for different functions
Genus	*Homo*	Long childhood, large brain, speech ability
Species	*Homo sapiens*	Body hair reduced, high forehead, prominent chin

TABLE 10.2   THE FIVE-KINGDOM SYSTEM OF CLASSIFICATION

Kingdom	Characteristics
Monera	All monera are single-celled organisms. Unlike other cells, the monerans lack an organized nucleus, mitochondria, chloroplasts, and other membrane-bound organelles. They have a circular chromosome. Examples are bacteria and blue-green algae.
Protista	Protists are one-celled organisms that have a membrane-bound nucleus. Within the nucleus are chromosomes that exhibit certain changes during the reproduction of the cell. Other cellular organelles are surrounded by membranes. Some protists take in food; other manufacture it by means of *photosynthesis*. Some protists can move from one place to another (motile); others are nonmotile. Examples of protists are *Amoeba*, *Euglena*, *Paramecium*, *Punnularia* (diatom).
Fungi	Fungi are nonmotile, plantlife species that cannot make their own food. The fungi (singular: fungus) absorb their food from a living or nonliving organic source. Fungi differ from plants in the composition of the cell wall, in methods of reproduction, and in the structure of the body. Examples of fungi are mushrooms, water mold, bread mold.
Plantae	The plant kingdom includes the mosses, ferns, grasses, shrubs, flowering plants, and trees. Most plants make their own food by photosynthesis and contain chloroplasts. All plant cells have a membrane-enclosed nucleus and cell walls that contain cellulose.
Animalia	All members of the animal kingdom are multicellular. The cells have a discrete nucleus that contains chromosomes. Most animals can move and depend on organic materials for food. After excluding the very simple species, most animals reproduce by means of egg and sperm cells.

Monera

Organisms grouped in the kingdom Monera are prokaryotes. A prokaryote is a cell that lacks an organized nucleus and membrane-bound organelles. The cell wall gives shape to the moneran cell, and the cell membrane controls the movement of substances into and out of the cell. Examples of monerans are blue-green algae and bacteria. Figure 10.1 shows a prokaryotic cell, and Figures 10.2 and 10.3 illustrate bacteria.

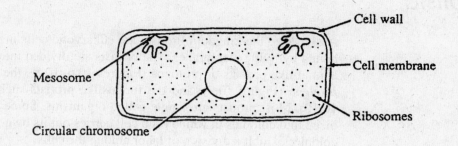

Figure 10.1 A prokaryotic cell

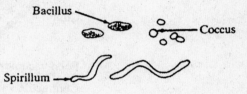

Figure 10.2   A flagellated bacterium   Figure 10.3   Three shapes of bacteria

Viruses

Viruses are nonliving particles. Viruses are not monerans and should not be placed within this grouping. Viruses are discussed here for convenience only.

Not possessing the enzymes necessary to carry out the biochemical activities of living cells means that viruses are inert. However, they have the ability to invade living cells. Once inside a cell, a virus reproduces itself many times over. The virus captures the host cell's nucleic acid and uses it to make virus particles. Figures 10.4 and 10.5 illustrate the basic structure of a virus.

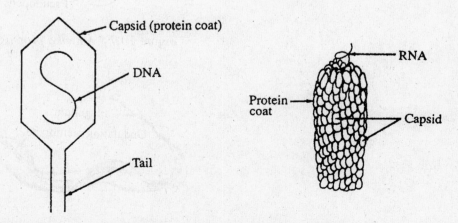

Figure 10.4   Structure of a virus   Figure 10.5   Tobacco mosaic virus

Protista

Assigned to the kingdom Protista are eukaryotic cells in which the organelles are bound by living membranes. The protists are divided into three major groups: the animal-like protists or protozoa (Figures 10.6–10.8), the algalike protists such as *Euglena* (Figure 10.9), and the funguslike protists such as slime mold (Figure 10.10). Most protists are single-celled organisms. Some protist species are organized into colonies in which each cell carries out its own activities. Certain protist colonies exhibit a division of labor among the cells.

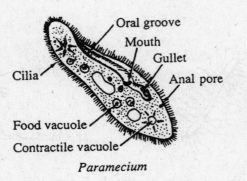

Oral groove
Mouth
Gullet
Cilia
Anal pore
Food vacuole
Contractile vacuole

Paramecium

Colpoda cucullus

Vorticella

Figure 10.6   Ciliated Protists

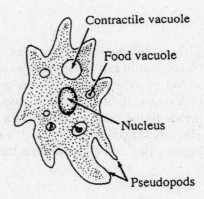

Contractile vacuole

Food vacuole

Nucleus

Pseudopods

Figure 10.7   *Ameoba proteus*

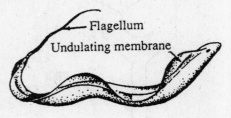

Flagellum
Undulating membrane

Figure 10.8   *Trypanosoma gambiense*

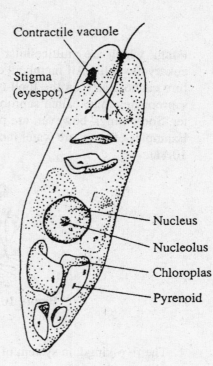

Figure 10.9   *Euglena*, an algalike protist

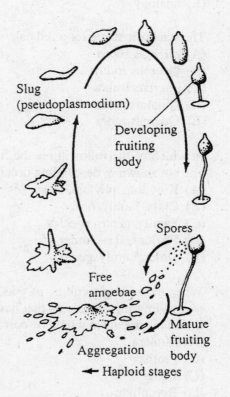

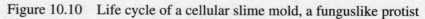

Figure 10.10   Life cycle of a cellular slime mold, a funguslike protist

Fungi

Fungi, which are multicellular and multinucleated organisms, are composed of eukaryotic cells. All members of the kingdom Fungi lack chlorophyll and therefore cannot carry on the food-making process of photosynthesis. Most fungi are saprophytes (also called saprobes), absorbing their food from dead organic matter. Some fungi, however, are parasites and absorb their food from a living host. Examples of fungi are water mold, bread mold, mushrooms, and yeast (see Figure 10.11).

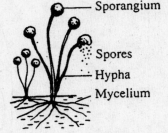

Figure 10.11   Bread mold and spores

Model Questions Related to Taxonomy

1. The five-kingdom system of classification is based on
   (A) morphology
   (B) taxonomy
   (C) phylogeny
   (D) fossil evidence
   (E) analogy

2. The scientific name for a red oak is properly written as
   (A) *quercus rubra*
   (B) Quercus *Rubra*
   (C) *quercus* Rubra
   (D) *Quercus rubra*
   (E) Quercus rubra

3. In which of the following is the listing of categories in the taxonomic heirarchy not shown in descending order?
   (A) Kingdom, phylum, class
   (B) Class, family, order
   (C) Family, genus, species
   (D) Phylum, class, order
   (E) Order, family, genus

4. When examining a culture of pond water through a microscope, you see some free-swimming organisms that have chloroplasts, a red eye-spot, and a flagellum. This organism would be correctly classified in the kingdom
   (A) Monera
   (B) Protista
   (C) Fungi
   (D) Bryophyta
   (E) Animalia

5. Of the following, which items are incorrectly paired (do not belong to the same kingdom)?
   (A) Palm tree/grass
   (B) Starfish/elephant
   (C) Blue-green algae/green algae
   (D) Toadstool/bread mold
   (E) Diatom/amoeba

6. The coyote, wolf, dingo, and jackal are doglike animals that share all of the same taxonomic categories EXCEPT the
   (A) family
   (B) genus
   (C) class
   (D) species
   (E) order

7. In the plant kingdom, the taxon known as division replaces the taxon in the animal kingdom known as
   (A) class
   (B) order
   (C) phylum
   (D) family
   (E) species

8. In which of the taxa listed below does its members display the greatest similarity?
   (A) Family
   (B) Class
   (C) Kingdom
   (D) Order
   (E) Phylum

9. A scientist who specializes in determining evolutionary relationships among organisms is called
   (A) an evolutionist
   (B) a biologist
   (C) a taxonomist
   (D) a systematist
   (E) a geneticist

10. Of the following, which characteristic has been most important in classifying the Monera?
    (A) Number of cells
    (B) Lack of membrane-bound organelles
    (C) Motility
    (D) Presence of chlorophyll
    (E) Method of reproduction

11. A characteristic of viruses is that they
    (A) contain a nucleus and cytoplasm
    (B) are motile outside a cell
    (C) can reproduce only in living cells
    (D) have a closed transport system
    (E) contain both DNA and RNA

Answers and Explantions to Model Questions: Taxonomy

1. **C** Phylogeny is the evolutionary history of an organism, which shows relationships useful in classification.

    Incorrect Choices
    A Morphology is the study of form.
    B Taxonomy is the science of classification.
    D Fossil evidence consists of the preserved remains of plants and animals, which establish their existence in a certain time period.
    E Analogy refers to structures of organisms that are used for the same purpose but have different evolutionary origins. An example is the wing of a bat and the wing of a fly.

2. **D** In a binomial, the first letter in the genus name is always capitalized. The species name is written entirely in small letters. The two-word binomial is either written in italics or underlined: *Canis familiaris*, the domestic dog.

    Incorrect Choices
    A, B, C, E are incorrect without explanation.

3. **B** This group is not in correct descending order because the family is a smaller taxon than the order. The descending taxonomic hierarchy is kingdom, phylum, class, order, family, genus, and species.

    A, C, D, E are in correct descending order.

4. **B** The protists are one-celled organisms with membrane organelles. Some are sessile, while others are free swimming. The organism described in the question has a flagellum, an eyespot, and chloroplasts. These characteristics tell you that this organism is free swimming and can carry on photosynthesis.

    Incorrect Choices
    A The Monera do not have membrane-bound organelles such as chloroplasts.
    C The Fungi are nongreen plantlike organisms.
    D The Bryophyta are mosses, liverworts, and hornworts, multicellular plants that live on land.
    E Members of the kingdom Animalia are multicellular and have neither chloroplasts nor flagella.

5. C Blue-green algae and green algae belong to different kingdoms. Blue-green algae are monerans, organisms that lack membrane-bound organelles. Green algae are variously classified as algal protists or as lower green plants. The chlorophyll of green algae is much like the chlorophyll of higher plants. Also, green algae have an organized nucleus.

Incorrect Choices
A Palm trees and grass belong to the kingdom Plantae.
B Starfish and elephants belong to the kingdom Animalia.
D Toadstools and bread mold are Fungi.
E Diagoms and ameobae are Protista.

6. D All of the animals named belong to the genus *Canis* but have different species names. The coyote is *Canis latrans*; the red wolf, *C. rufus*; the dingo, *C. familiaris dingo*; and the jackal, *C. adjustus*.

Incorrect Choices
A All of the animals belong to the family Tetropoda.
B All belong to the genus *Canis*.
C All belong to the class Mammalia.
E All belong to the order Carnivora.

7. C In the kingdom Plantae, the taxon "division" is used in place of "phylum," the taxon in kingdom Animalia.

Incorrect Choices
A, B, D, E Class, order, family, and species are used as plant taxa.

8. A The members of the family are more closely related than members in the other choices.

Incorrect Choices
B, C, D, E Remember the descending heirarchal order: kingdom, phylum, class, order, family, genus, species.

9. D The role of the systematist is to determine evolutionary relationships among organisms for classification.

Incorrect Choices
A An evolutionist studies the physical and biochemical evidences of evolution, fits organisms into a time period of emergence, and determines evolutionary links.
B A biologist studies the anatomy, physiology, and biochemistry of life.
C A taxonomist is responsible for assigning organisms into taxonomic groupings.
E A geneticist studies modes of inheritance.

10. B The lack of membrane-bound organelles (nucleus, mitochondria, chloroplasts, and other cellular organelles) sets the monerans apart from the protists.

Incorrect Choices

A The number of cells is not a definitive characteristic. A number of protists are single celled.

C The degree of motility (the ability to move from one place to another) is not a definitive moneran characteristic. Many protists are free swimming.

D The blue-green algae have chlorophyll, although it is different in composition from that of the true green plants. Of significance is the fact that in the monerans chlorophyll is not contained in membrane-bound chloroplasts.

E Monerans reproduce by binary fission, as do some protists.

11. C A virus reproduces itself by taking over the host's mechanisms for protein synthesis or nucleic acid synthesis.

Incorrect Choices

A Viruses are noncellular; they consist of nucleic acid inside a protein coat.

B, D Outside a host, the virus particle resembles a crystal, showing none of the characteristics of living organisms.

E Viruses contain either DNA or RNA, but not both.

Model Questions Related to Monera, Protista, and Fungi

1. Shown in the diagram is a pair of spirochetes as they appear through a microscope. In fact, spirochetes are so transparent that a technique known as darkfield is used to increase resolution. Spirochetes, which cause syphilis, are highly motile and nonucleated. Spirochetes are correctly classified in the same kingdom as
   (A) animal-like protists
   (B) blue-green algae
   (C) funguslike protists
   (D) *Euglena*
   (E) water mold

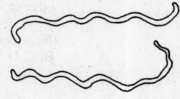

2. Toadstools are saprobes in that they
   (A) absorb nutritive material from living cells
   (B) absorb nutritive material from dead cells
   (C) ingest dead organic matter
   (D) invade living cells
   (E) act as phagocytes on smaller organisms

3. If it were possible to eliminate the organism *Trypanosoma gambiense* entirely, then it would be possible to eliminate the disease
   (A) AIDS
   (B) malaria
   (C) cholera
   (D) African sleeping sickness
   (E) botulism

4. In protists, all of the following are used for locomotion EXCEPT
   (A) pseudopodia
   (B) parapodia
   (C) gliding movements
   (D) cilia
   (E) flagella

5. Of the following, the most likely habitat for a parasitic species of amoeba would be
   (A) the hair in a dog's coat
   (B) decaying matter present on a forest floor
   (C) pond scum
   (D) the intestines of frogs
   (E) the north-side bark of trees

6. Slime molds resemble, in some of their characteristics, organisms of both
   (A) the kingdom Monera and the kingdom Plantae
   (B) the phylum Protozoa and the kingdom Fungi
   (C) the phylum Euglenophyta and the kingdom Protista
   (D) the kingdom Protista and the kingdom Monera
   (E) the bacteria and the blue-green algae

Answers and Explanations to Model Questions: Monera, Protista, and Fungi

1. B Spirochetes are prokaryotic cells. All unicellular organisms lacking an organized nucleus and other membrane-bound organelles are classified in the kingdom Monera. Blue-green algae also belong to Monera.

   Incorrect Choices
   A, C, D, E belong to the kingdom Protista. All of these organisms—the animal-like protists or protozoa, the funguslike protists, *Euglena,* and water mold—have organized nuclei and other membrane-bound organelles.

2. B Saprobes, or saprophytes, are organisms that absorb nutrients from dead organic matter. Dead cells are organic matter.

   Incorrect Choices
   A Organisms that absorb nutritive material from living cells are parasites. Parasites do harm to the host.
   C Toadstools cannot ingest materials.
   D Viruses, but not toadstools, invade living cells.
   E Phagocytes are white blood cells that ingest bacteria.

3. D *Trypansoma gambiense* is a parasite protist that causes African sleeping sickness. Humans are infected with the trypansome by the bite of an infected tsetse fly.

Incorrect Choices

A AIDS is caused by HIV infection, contracted through sexual intercourse with an infected person, through the use of shared "drug works" by intravenous drug users, or through transfusion with infected blood.

B Malaria is caused by the protist *Plasmodium vivax*. Humans contract the disease through the bite of an infected *Anopheles* mosquito.

C Cholera is caused by the bacterial pathogen *Vibrio cholerae*. Infection comes from drinking unclean water.

E Botulism is caused by the bacterial pathogen *Clostridum botulinum*. The common term for this infection is food poisoning.

4. B Parapodia are the numerous locomotive hairs present on the body of the sandworm, *Neries*, which is not a protist.

Incorrect Choices

A Pseudopodia are the "false feet" of the ameoba.

C The plasmodium of the slime mold moves by gliding motion.

D Cilia are the cytoplasmic hairlike structures that aid in the locomotion of *Paramecium*, *Vorticella*, and other ciliate protists.

E Flagella are longish, cytoplasmic hairlike structures that enable *Euglena* and other protist flagellates to locomote.

5. D Amoebae, whether parasitic or free-living, require a water environment, which is provided by a frog's intestines. Parasites live on or within the body of a host organism.

Incorrect Choices

A The hair in a dog's coat would not provide the water medium necessary for sustaining the life of an amoeba.

B Parasites must live inside or on a living host; decaying matter does not meet this requirement.

C Pond scum is not a living host.

E Tree bark is nonliving organic matter. Parasites must obtain their nutrients from the cells of a living host.

6. B Slime molds are fungal protists and have organized nuclei. One state in their life cycle consists of aggregate amoeboid forms that compose the plasmodium. Another stage consists of funguslike fruiting bodies.

Incorrect Choices

A Monerans lack organized nuclei. Plants have chlorophyll. Slime molds resemble neither.

C Slime molds do not resemble the flagellated *Euglena*, which also contains chloroplasts.

D Monera, as explained in A above, lack organized nuclei.

E Bacteria and blue-green algae belong to the kingdom Monera. See A above.

CHAPTER ELEVEN
Plants

Capsule Concept

The modern scheme of taxonomy groups all green plants in the kingdom Plantae. This kingdom exhibits a diversity of species. Some species of green plants are free-living cells, others live together in colonies, and still others form long filaments. Most species of Plantae are multicellular and are nonmotile, that is, anchored to one spot.

All members of the kingdom Plantae contain the green pigment chlorophyll, the life-giving substance of plants. Chlorophyll traps light energy, which is necessary in the food-making process of photosynthesis.

Vocabulary

Alternation of generations   the alternation of the haploid and diploid phases in sexually reproducing organisms

Bryophyte   a nonvascular green plant

Chlorophyte   a green alga

Gametophyte   the haploid phase of an organism that exhibits alternation of generations

Guttation   the loss of water from leaves caused by root pressure

Phloem   the nutritive conducting tissues in plants

Sporophyte   the diploid phase in the life cycle of an organism that exhibits alternation of generations

Tracheophyte   a vascular green plant

Tropism   the growth response of a plant to an external stimulus

Vascular bundles   veins in plants that consist of xylem tubules and phloem tubules

Xylem   the tubules that conduct water upward in plants

Survey of the Kingdom Plantae

All green plants belong to the kingdom Plantae. All species belonging to this kingdom have chlorophyll-containing cells in tissues such as modified leaves, true leaves, and in some species stems. Chlorophyll not only colors leaves green but also traps light energy in a process called photosynthesis.

Lower Plants

Approximately 50–90% of all photosynthesis is a result of the biochemical activity of algae. Most algae species live in water. Some algae are unicellular, other forms have cells that adhere together in colonies, while others have a flat, thallus-type structure. The thallus is multicellular.

The green algae belong to the Chlorophyta. A representative form is *Chlamydomonas* (Figure 11.1). Sea lettuce, *Ulva*, is another species of Chlorophyta (Figure 11.2). Patterns of reproduction in these species are shown in Figures 11.3 and 11.4.

Other groups of algae are the brown algae (Phaeophyta) and the red algae (Rhodophyta).

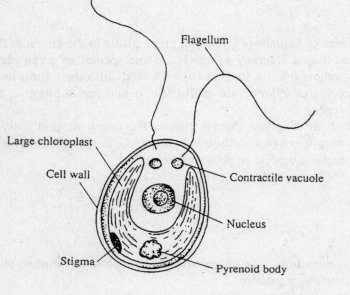

Figure 11.1   The green alga *Chlamydomonas*

Figure 11.2   The green alga *Ulva*, commonly called sea lettuce

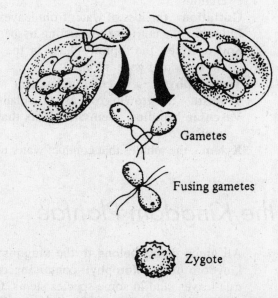

Figure 11.3   Gamete formation in *Chlamydomonas*

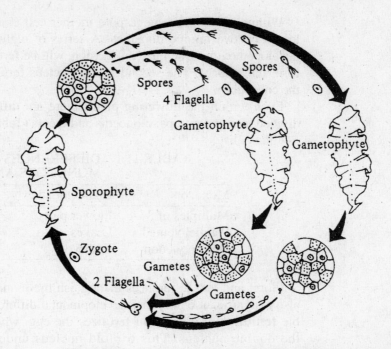

Figure 11.4   Alternation of generations in *Ulva*

Higher Plants

The higher plants have adapted to a land environment and are most conspicuous. They have been divided into two groups: bryophytes (mosses) and tracheophytes (ferns and seed-producing plants).

Member species of these groups exhibit *alternation of generations* (Figures 11.5–11.7). A major evolutionary trend from lower to higher plants has resulted in the reduction of the gametophyte generation to a few cells and the dominance of the sporophyte generation. Such a development is clearly observed in flowering plants, where the male gametophyte is the pollen grain and the female gametophyte is the ovule. From bryophytes to tracheophytes, the development of vascular or conducting tissue is also evident. Tracheophytes are further divided into gymnosperms and angiosperms.

The line of ascent in green plants is not easy to trace since fossils of these are rare. It is believed that a flagellated organism was the ancestor of the green algae. It is further theorized that mosses evolved from the green algae because the protonema of the former resembles a green alga.

The events that transpired to produce vascular plants are uncertain. However, the development of conducting tissues and an independent sporophyte made possible an increase in the kinds and numbers of land plants. Thus a new evolutionary step occurred.

The gymnosperms and the angiosperms top the evolutionary tree. The seeds of gymnosperms (conifers or evergreens) are borne in ovules that are in an exposed position on the cone, while the seeds of angiosperms are in ovules enclosed in an ovary. The flower is the reproductive organ of the angiosperm.

The stamen of a flower consists of a chamber filled with special cells that divide meiotically. A microspore mother cell gives rise to a tetrad of cells. Each cell in the tetrad divides mitotically. The two cells become the sperm cell nuclei, which are eventually surrounded by a hard protective coat, the pollen. When pollen reaches the stigma of the pistil, it germinates a tube that penetrates into the ovule.

Within the ovule, a megaspore mother cell undergoes a meiotic division followed by two mitotic divisions. A series of eight nuclei are formed within the developed ovule. One nucleus, the egg, will be fertilized by one sperm nucleus to form the embryo. The second sperm nucleus fertilizes two polar nuclei to form the endosperm.

Even within the flowering plants there are differences. The angiosperms are divided into two groups: monocots and dicots (Table 11.1).

TABLE 11.1   DIFFERENCES BETWEEN
MONOCOTS AND DICOTS

Monocots	Characteristic	Dicots
Multiples of 3	Flower parts	Multiples of 4 or 5
Parallel veined	Leaves	Net veined
One cotyledon	Embryo	Two cotyledons

Floral and leaf differences distinguish the monocots from the dicots. There is also a significant embryonic developmental difference. In the ovule there is double fertilization: one sperm fertilizes the egg, while the second sperm fertilizes the double nucleus. This triploid nucleus undergoes mitosis and forms the endosperm tissue. In monocots, the endosperm accumulates nutritive material and becomes a source of stored food. In dicots, the embryo leaves, or cotyledons, absorb the material of the endosperm, and they themselves become the major source of food for the emerging embryo.

While there are differences between monocots and dicots, there are also similarities in their basic structural parts: the roots, stems, and leaves.

ALTERNATION OF GENERATIONS

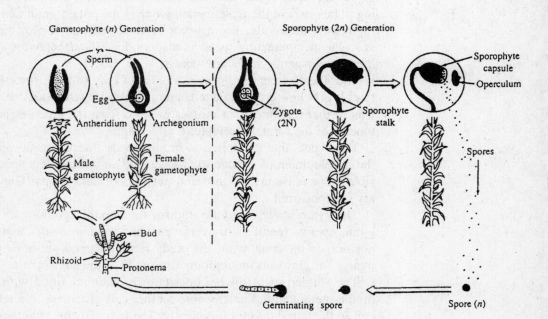

Figure 11.5   Life cycle of moss

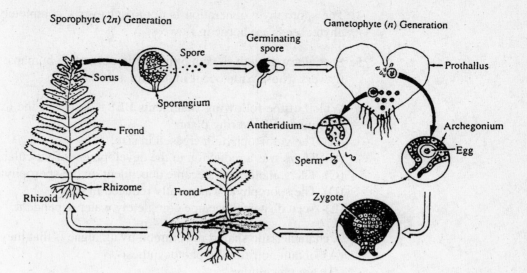

Figure 11.6   Life cycle of fern

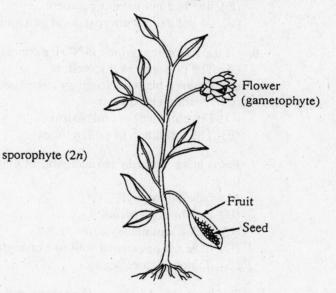

Figure 11.7   Life cycle of angiosperm

Model Questions Related to Plant Classification

For Questions 1–5 choose the word that BEST matches each numbered statement.

(A) Moss
(B) Fern
(C) Gymnosperm
(D) Angiosperm
(E) Dicot

1. It has a vascular system; both gametophyte and sporophyte are photosynthetically independent.

2. Seeds are borne on special leaves, sporophylls; has secondary growth.

3. It requires water for sexual reproduction; the gametophyte is the dominant generation.

4. The sporophyte generation is dominant over a completely parasitic gameto-phyte; seeds are borne in flowers.

5. Secondary growth is economically important to humans; seeds are borne in flowers whose parts occur in multiples of 4 or 5.

6. Which of the following statements BEST describes the trend followed in the evolution of flowering plants?
    (A) The gametophyte increased in size.
    (B) There was a reduction in the development of vascular tissue.
    (C) The gametophyte became dependent on the sporophyte.
    (D) The sporophyte was greatly reduced in size.
    (E) Seed dispersal became completely water dependent.

7. A characteristic shared in common by all algae is that they
    (A) obtain nutrition by photosynthesis
    (B) are unicellular
    (C) exhibit heterogamous sexual reproduction
    (D) live in a marine environment
    (E) do not exhibit alternation of generations

8. Of the following, which is NOT a characteristic of all flowering plants?
    (A) The presence of a scutellum
    (B) Embryo plant enclosed by a seed coat
    (C) Double fertilization
    (D) Development of endosperm
    (E) The formation of pollen tubes

9. Ferns have not been as successful as gymnosperms and angiosperms because they
    (A) lack chlorophyll
    (B) lack motile sperms
    (C) lack a vascular system
    (D) have not responded well to a changing environment
    (E) do not produce seeds

For Questions 10–14 choose the organism that BEST fits the description.
    (A) Spirogyra   (D) Fern
    (B) Fir tree   (E) Liverwort
    (C) Maple tree

10. Seeds enclosed in fruit

11. Seeds enclosed in a cone

12. Conjugation is a sexual mode of reproduction

13. Nonseed bearers with true roots, stems, and leaves

14. Nonseed bearers without true roots, stems, and leaves

Answers and Explanations to Model Questions: Plant Classification

1. B Ferns have vascular tissue. Although inconspicuous, the heart-shaped game-tophyte is photosynthetic and therefore independent.

2. C Seeds develop on special leaves that form a cone.

3. A The gametophyte is the obvious generation, forming the green carpet found on a forest floor. The sperms are motile but require water to swim to the egg.

4. D Angiosperms are plants that produce seeds in an enclosed flower.

5. E The term "secondary growth" implies the continual growth of vascular tissue after the formation of primary xylem and phloem. This condition exists in many types of plants. When the flower parts occur in multiples of 4 or 5, the plant is a dicot.

6. C The gametophyte is reduced in size to only a few cells in flowering plants, and is totally dependent on the sporophyte.

   Incorrect Choices
   A The gametophyte generation is reduced in size and relegated to a role of complete dependence on the sporophyte.
   B There was an increase in the development of vascular tissue, without which the height of plants would have been limited.
   D The sporophyte becomes the dominant generation and increases in structural complexity as one ascends the ladder of plant evolution.
   E In higher flowering plants, wind and animals are the chief agents of seed dispersal.

7. A All algae contain chlorophyll and therefore obtain nutrition by photosynthesis.

   Incorrect Choices
   B Not all algae are unicellular. The brown and red algae have multicellular sporophytes.
   C The algae display both heterogamous and homogamous sexual reproduction. Within the green algae, *Spirogyra* shows the homogamous type, where there is no distinction between the male and female gametes. In *Oedogonium* there are two distinctly different types of gametes (heterogamous).
   D Not all algae live in salt water (marine environment); a considerable number of alga species live in fresh water. Algae in a lichen partnership live in soil or on bare rock. A lichen is composed of two organisms: an alga and a fungus.
   E Some algae, such as *Ulva* (green algae), exhibit a marked alternation of generations.

8. A A scutellum is the name given to the single cotyledon or seed leaf in monocot seeds. The scutellum is not present in dicotyledons.

   Incorrect Choices
   Choices B, C, D, E are common to all flowering plants.

9. E Ferns reproduce by means of spores and must rely on these to spread the species from place to place. Angiosperms and gyrosperms are seed bearers. Seeds are less fragile than spores and are better protected from desiccation.

   Incorrect Choices
   A Ferns have chlorophyll.
   B Ferns produce motile sperms.
   C Ferns have an efficient vascular system.
   D Ferns are more ancient than the flowering plants and have responded to changing environments over time.

10. C

11. B

12. A

13. D

14. E

Plant Structures and Functions

Root Structure

The innermost cell layer of the cortex is a region known as the *endodermis*. The cells of the endodermis are thickened by suberin, a waterproofing material. This waterproof band is called the Casparian strip. The narrow band of endodermis surrounds the water-conducting *xylem* and *phloem* tubules. Just inside the endodermis is the pericycle; this gives rise to lateral roots, which force their way through the cortex and the epidermis.

Roots are vital plant structures. The movement of water, minerals, and gases into the plant occurs by both passive transport (diffusion and osmosis) and active transport (Na^+-K^+ pump and proton pump). Water entering the root hairs from the epidermal cells travels to the vascular tissue by movement along the cell walls (the apoplast pathway) or by passage through openings in the walls of cells (plasmodesmata) connected to each other by cytoplasm (the symplast pathway). From the roots, dissolved materials move to other plant structures. In addition to transport, roots have other functions. They anchor plants to the soil; they function in vegetative propagation; and they serve as storage areas.

A longitudinal section provides a different view of root cells and tissues. Four regions become visible: the root cap, the meristematic region, the region of elongation, and the zone of maturation (differentiation).

The root cap, as the name implies, protects the root as it pushes through the soil. Root cap cells secrete carbon dioxide, which dissolves in soil water to form carbonic acid. The downward movement of the root becomes possible as soil is dissolved. The thin-walled, protoplasm-filled cells of the meristem undergo frequent mitosis. This is the region in which the initial phase of growth takes place in roots. Cells from the meristem help to form the elongation region, where root cells increase in length, new protoplasm is formed, and vacuoles increase in size. In the region of maturation, differentiation occurs, forming the mature tissues, xylem, phloem, and the root hairs.

Stem Structure

A cross section of the dicotyledonous stem makes visible the following tissues: epidermis, cortex, and stele (including pericycle, phloem, cambium, xylem, rays, and pith). The arrangement of tissues is different in the cross sections of herbaceous monocot, dicot, and woody stems, as shown in Figure 11.8.

(a) Stem of herbaceous monocot with scattered vascular bundles

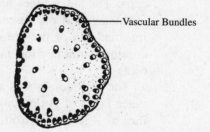

(b) Stem of herbaceous dicot with vascular bundles arranged in a ring

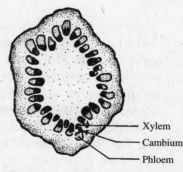

(c) Stem of a woody dicot with vascular tissue arranged in concentric circles

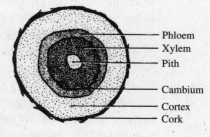

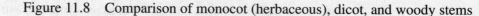

Figure 11.8   Comparison of monocot (herbaceous), dicot, and woody stems

In woody stems, the cambium layer, positioned between the xylem and phloem, differentiates into secondary xylem and secondary phloem. A continuous development of new vascular tissue during the growing seasons appears as annual rings in the stem. The overall effect of the active cambium tissue is an increase in the diameter of the stem, accompanied by compression of all the vascular tissue except the outermost layer. Examples of this stem structure are the conifers and the woody dicots.

Wood is xylem tissue. In conifers, the xylem is made up of cells known as tracheids. Water is conducted upward in the xylem through pits that interconnect the heavy, thick-walled, overlapping tracheids. Wood in which the tracheids are both supportive and conducting is known as soft wood.

In woody dicots, cells comprising the vessel element form the xylem. These cells are not as thick walled as tracheids and require the support of special fibers, the *sclerenchyma*. Both the soft wood of the conifers and the hard wood of the dicots are economically important.

Monocots are usually herbaceous and lack a cambium layer. There is no secondary growth. The vascular bundles are scattered throughout the *parenchyma*, which fills the stem.

Phloem tissue, which conducts materials downward, arises from *sieve tube* cells. The sieve tubes are the chief conducting cells of the phloem. Companion cells surround the sieve tubes and seem to mediate the movement of materials into and out of the phloem from adjacent tissue. The function of phloem is to conduct soluble food materials from the leaves downward into the stems and roots.

The bark consists of an epidermal layer above the cork, which arises from the outer layer of cortex cells. In woody plants, the cork, as well as the epidermis, dies as it is cut off from its food supply. The cork creates insulation and protection for the living tissue of the tree and is the most obvious part of a woody stem.

Water Transport

How water moves upward in the xylem is a question that persists. Two theories have been proposed, both having merit. The root pressure theory relies on the establishment of a water potential gradient by the active transport of ions into various tissues of the root. The cohesion or transpiration theory suggests that the evaporation of water (transpiration) from the leaves and the cohesion of water molecules in the xylem encourage upward movement.

Carbohydrates and other soluble food products are transported downward from the leaf through the phloem. It is believed that the companion cells which surround the phloem actively pump carbohydrates into the sieve cells of the phloem. Increased osmotic pressure within the sieve cells results in the diffusion of water out of the adjacent xylem and parenchyma cells and into the sieve cells. As the water pressure (turgor) increases in the rigid sieve cells, carbohydrate-laden water is forced downward into the lower sieve cells. A concentration gradient between the sieve cells of the phloem and other tissues determines the rate of movement of materials out of the phloem and into the adjacent tissue.

Leaf Structure

The photosynthetic organ of a plant is the leaf. Chloroplasts are located in the cells of the mesophyll (palisade and spongy layers). It is through the leaf that carbon dioxide is made available to the cells. Stomata (openings) in the leaf allow the gases, carbon dioxide and oxygen, to pass into and out of the cells.

Diffusion of gases within the mesophyll is made possible by its numerous air spaces and by the ever-present water vapor contained therein. Since the concentration of water is higher in the leaf than in the atmosphere, diffusion of water from the leaf is to be expected. Evaporation of water from the leaves (*transpiration*) occurs through the stomata. However, during nonphotosynthetic periods, excessive water loss would be most detrimental to plants. Water loss by transpiration is decreased when the stomata close at night.

Guard cells control the size of the stomata by an unusual mechanism. The walls of the guard cells are not equally rigid. The portion of the cellulose wall that borders the stoma is inflexible, but the other side of the cell wall is flexible. Water entering the cell causes the flexible portion to bulge; this, in turn, pulls the inflexible part. As the rigid portion is pulled away, the stoma opens, its size being increased. It is believed that water enters the guard cells in response to a change in osmotic pressure caused by an increase of sugar solution and the influx of K^+, which is taken up by the guard cells from the adjacent epithelial cells when blue-light receptors near the vacuole are activated by light. The receptors trigger a $K^+ - H^+$ pump. Increased sugar concentration comes about through reactions that change starch to sugar. The enzyme involved functions at a high pH. Since the photosynthetic process reduces the amount of carbon dioxide in the cell, the pH moves toward the basic range. Therefore the stomata are open during the periods of light when carbon dioxide is required for photosynthesis.

Model Questions Related to Plant Structures and Functions

1. Which of the following pressure systems increases the force of transpirational pull enabling the movement of water up through the xylem from roots to leaves?
   (A) Guttation and gravity
   (B) Cohesion and adhesion
   (C) Condensation and evaporation
   (D) Gravity and root pressure
   (E) Surface tension and root pressure

2. In many plant species the main factor assuring a seed's viability is
   (A) the loss of water before dormancy
   (B) enclosure in a hard cover
   (C) being surrounded by distasteful fruit
   (D) exposure to light and oxygen
   (E) the adaptation for seed masting

For Questions 3–7 choose the letter of the term that BEST matches each numbered statement:

   (A) Vessels   (D) Sieve tubes
   (B) Epidermis   (E) Parenchyma
   (C) Meristem

3. These cells are responsible for the transport of carbohydrates.

4. These cells serve as protective covering for roots, stems, and leaves.

5. These cells are mitotically active and also produce hormones.

6. These cells serve as storage cells.

7. These thick-walled cells form the wood of trees.

Your answers to Questions 8–12 are based on the diagram that follows:

Cross Section of a Dicot Root

8. Cells contain amyloplasts that store starch.

9. Cells in this layer do not differentiate and remain meristemic.

10. These single cells absorb water and minerals.

11. Conducts water and minerals upward from the root through the stem

12. Controls movement of substances between the root cortex and interior root

13. A student observed droplets of water hanging from the tip of the leaves of her house plant. The most likely explanation for the formation of the droplets is
    (A) transpiration pull
    (B) injury to the plant root system
    (C) root pressure
    (D) condensation of atmospheric water
    (E) water movement down the phloem

Answers and Explanations to Model Questions: Plant Structures and Functions

1. **B** Transpiration is the loss of water from the leaf that causes a pulling force moving water upward through the xylem. Transpirational pull on the xylem extends from the leaves to the roots. The cohesion of water (the sticking together of water molecules) makes it possible to pull up a column of sap without the water separating. Also helping to override the force of gravity is the adhesion of water to the walls of the xylem cells.

   Incorrect Choices
   A Guttation is the loss of water from blades of grass or the leaves of small plants caused by root pressure. Root pressure is not forceful enough to push water up the xylem of tall trees. Gravity is a force that pulls down.
   C Condensation is the changing of water vapor into liquid. Evaporation is the changing of water into the gaseous water vapor. Neither has any relationship to transpirational pull.
   D Root pressure is not strong enough to pull sap great distances upward in xylem tubules.
   E Surface tension helps to move water through the stomates but not upward from the roots.

2. A In many species, dryness is the main factor assuring a seed's viability. As a seed enters dormancy it loses approximately 95% of its water content. Viability refers to a seed's ability to break dormancy.

Incorrect Choices
B, C, D, E are unrelated to viability.

3. D Sieve tubes are composed of sieve cells. There are openings in the longitudinal axis of the cells to allow the passage of material from cell to cell.

4. B

5. C

6. E

7. A Vessels are tubes made from the interconnection of vessel cells. The vessels form the water-conducting tubes in woody dicots.

8. E Parenchyma

9. B Pericycle

10. A Root hairs

11. C Xylem

12. D Endodermis

13. C Root pressure causes guttation, that is, the exudation of water drops from the tips of leaves.

Incorrect Choices
A Transpiration pull is a theory that explains the movement of water up the xylem from the roots to the leaves.
B Injury to the roots usually causes leaf wilting since water movement to the leaves is impaired.
D Dew is the condensation of atmospheric water on the entire surface of the plant.
E Water moves up the xylem; phloem carries food downward from leaves to stem to root.

Plant Hormones

The growth and the development of plants are controlled by substances known as *hormones*. There are five classes of plant hormones, which affect division, elongation, and cell differentiation. The actions of the plant hormones are summarized in Table 11.2.

TABLE 11.2   FUNCTIONS OF PLANT HORMONES

Hormone	Function	Location in Plant
Auxin	Stimulates elongation of root and stem cells, differentiation of secondary leaves, xylem, development of fruit, phototropism, gravitropism	Meristematic tissue of roots, stems, buds, and seed embryo
Cytokinins	Stimulate cell division and differentiation, flowering inhibits leaf senescence	Actively growing tissues of roots, embryos, fruits
Gibberellins	Stimulate leaf growth, flowering, fruit development, seed germination	Meristematic tissue of roots, buds, leaves; embryo
Abscisic acid	Inhibits growth, closes stomates, maintains seed dormancy	Leaves, stems, green fruits
Ethylene	Stimulates fruit ripening, leaf abscission, and growth and development of roots, leaves, flowers	Ripening fruit, stem nodes, aging leaves

Although the mode of operation of some plant hormones has been determined, the way in which most hormones work remains unknown. Auxin initiates cell elongation by loosening the cell wall. This may be accomplished by the stimulation of a H^+ pump that alters the pH in the cell wall, thus activating cell-wall enzymes.

Auxin is transported from cell to cell by *polar transport*. This is an active transport by carriers located at specific ends of a cell.

Cytokinins stimulate RNA and protein syntheses. However, the presence of auxin is required for cytokinins to control cell division and differentiation.

Plant Movements

Tropisms

Plant movements termed *tropisms* are growth responses of a plant toward or away from a stimulus. There are three stimuli: light (phototropism), gravity (gravitropism), and touch (thigmotropism). Tropisms depend upon the distribution and concentration of plant hormones.

Statoliths are starch-storing plastids that accumulate at the bottoms of root cap cells and function in the gravitotropism of roots.

Turgor Movements

Turgor movements are plant movements that occur because of changes in turgor pressure within specialized cells in response to a stimulus. Unlike tropisms, turgor movements are rapid and reversible. The leaf movement of the sensitive mimosa and the sleep movements of many plants are examples of turgor movements.

Photoperiodism

Photoperiodism is the physiological response of an organism to the length of a day or night (Figure 11.9). In plants, flowering is an example of photoperiodism, a response to the length of darkness. It is believed that a hormone, florigen, is produced in the leaves that stimulates flowering when the plant is exposed to the proper period of darkness.

A *phytochrome*, a light-absorbing pigment, exists in two forms, P_r and P_{fr}, in the leaves. One form, P_{fr}, is required to produce florigen. Each species of plant requires a specific amount of florigen for flowering; therefore, a specific period of darkness is required.

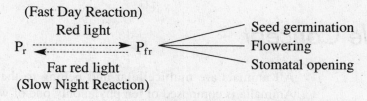

Figure 11.9   The mutual conversion of phytochrome between the inactive P_r form and the active P_{fr} form.

Model Questions Related to Plant Hormones and Movements

1. In an agricultural practice, plants are sprayed with a combination of auxin and gibberillins. The purpose of this spraying is to
   (A) stimulate fruit development
   (B) kill weeds
   (C) stimulate flowering
   (D) increase the height of the plants
   (E) stimulate the ripening of fruit

2. Statoliths are involved in
   (A) thigmotropism
   (B) phototropism
   (C) photosynthesis
   (D) gravitropism
   (E) flowering

Answers and Explanations to Model Questions: Plant Hormones and Responses

1. **A** The combination of auxin and gibberillins stimulates the development of fruit.

   Incorrect Choices
   B Neither auxins nor gibberillins are herbicides.
   C Cytokinins and ethylene stimulate flowering.
   D Gibberillins increase the length of the plant stem.
   E Ethylene stimulates the ripening of fruit.

2. **D** Statoliths are starch granules in cells that respond to gravity. In some unknown way, the accumulation of these granules affects hormone concentrations in the cell and thereby the growth response to gravity.

   Incorrect Choices
   A Thigmotropism is a growth response to contact.
   B Phototropism is a growth response to light.
   C Photosynthesis is the food-manufacturing process in plants.
   E Florigens control flowering in plants.

CHAPTER TWELVE
Animals

Capsule Concept

All animals are multicellular and belong to the kingdom Animalia. Kingdom Animalia is composed of ten phyla, only one of which includes animals that have backbones. This chapter will take you through the classification of animals, enabling you to grasp the concept of evolutionary relatedness. You will find that the organization of the animal body differs markedly from that of the plant body and therefore requires a different type of descriptive vocabulary.

Vocabulary

Analogous   denoting structures that have the same function but different evolutionary origins

Bilateral symmetry   the relative positions of body parts such that a line drawn through the center of the body lengthwise divides the body into mirror-image halves

Coelom   the body cavity; the space between two mesoderm layers

Homologous   denoting structures that have the same evolutionary origin

Larvae   immature animal forms

Pseudocoelomate   a body cavity not completely surrounded by mesoderm

Phylogeny   the line of evolutionary descent

Symmetry   the relative positions of body parts on opposite sides of a dividing line

Survey of the Kingdom Animalia

The kingdom Animalia contains multicellular animals that exhibit a division of labor. In general, characteristics of Animalia include the ability for self movement and sensory response to external stimuli. This kingdom is divided into ten phyla.

A phylum is a classification grouping to which organisms are assigned because of a distinctive characteristic. For example, all members of the subphylum Vertebrata have backbones. Each phylum is subdivided into classes. The Arthropoda, one of the most successful phyla, consists of five classes: insects, arachnids, crustaceans, millipedes, and centipedes. The classes of vertebrates include fish, amphibians, reptiles, birds, and mammals. A class is subdivided into orders, orders into families, families into genera, and genera into species. As stated in Chapter 10, this system of classifying organisms (taxonomy) has provided an orderly means for the study of the living world. The binomial system of nomenclature (genus + species), which was developed by Linnaeus, is used universally.

A species is a group of interbreeding, genetically similar organisms that resemble each other more closely than they do the organisms in any other group. The species is the basic unit of classification. The genus consists of groups of closely related species. *Felis concolor* (mountain lion), *Felis pardalis* (ocelot), *Felis leo* (lion), *Felis tigris* (tiger), and *Felis domesticus* (house cat) are examples of related species that belong to the same genus.

Taxonomy is based on homologous structures. Homology is used not only to classify animals but also to establish evolutionary relationships. Once an organism has been classified, its evolutionary kinship to other organisms becomes clear. The arms of a human being, the wings of a bat, and the flippers of a whale are homologous structures, indicating that these species share common ancestors. Like Darwin's finches, the ancestral species may have spread into different habitats. The basic forelimb structure was modified for each species through the processes of mutation and natural selection, resulting in divergent evolution. However, the bone pattern of the forelimb did not change.

Analogous structures are those that are similar in function. They may point to convergent evolution. Both flying insects and flying birds were capable of adapting to a similar habitat. Both have wings, but they are not related to each other.

The *phylogeny* (evolutionary tree) of the animal kingdom has been investigated rather extensively, and the findings are supported by fossil and morphological evidence. For example, biologists believe that the sponges arose from a colonial flagellated ancestor. Nor is it difficult to reason that flatworms and coelenterates have a common ancestor because the two organisms have structurally similar body cavities.

Many organisms hatch from eggs in an immature or larval form. The larval stage undergoes gradual changes terminating in adulthood. A similarity in larval development has been used to postulate that annelids, mollusks, and arthropods have a common ancestor. Similarly, echinoderms and chordates (tunicates and acorn worms) pass through larval stages that are remarkably similar to each other including the cleavage pattern of the eggs. These similarities in development seem to indicate that the chordates are related to the echinoderms.

In the late 1800s, E. H. Haeckel proposed the Recapitulation Theory (Biogenetic Law). He believed that every stage of development that an organism passes through corresponds to an ancestral adult. Hence each organism repeats its evolutionary history. Ontogeny (development) recapitulates (repeats) phylogeny (history). According to this theory, since mammals evolved from reptiles, which evolved from amphibians, which evolved from fish, each successive stage of mammal embryonic development corresponds to the evolutionary development of a fish, an amphibian, and a reptilian ancestor.

Today the theory has been modified. An individual's development does not correspond to the development of its adult ancestors. The similarity in the developmental stages of various species shows that the organisms are closely related to each other, not to ancestral types.

Definite evolutionary changes had to occur in vertebrates to make them a successful phylum. From fish to amphibian there had to be a development of limbs and lungs. In order for reptiles to become successful land-dwelling organisms, internal fertilization, the appearance of the amnion, and the development of an egg with a hard protective coat were essential. The most significant change in mammalian history was the internal development of the young, which ensured the survival of the group.

There is further evidence for the phylogenetic theory of development in the animal kingdom. Examination of the body plan within each phylum reveals a significant feature that distinguishes later evolutionary animal forms from earlier ones. A body plan constructed on a tube-within-a-tube system is an evolutionary advance. The inner body tube is separated from the outer body tube by a cavity known as the *coelom* (coelomate organisms). In order for this type of body plan to develop, three embryonic germ layers (ectoderm, mesoderm, and endoderm) must be present. The mesoderm layer separates into two parts and the space between the two mesoderm layers becomes the body cavity. The coelom is completely lined by the mesoderm.

The coelomates are further divided into two groups: *protostomes* (mollusks, annelids, arthropods) and *deuterostomes* (echinoderms, chordates). Each group is distinguished by a different pattern of cleavage, gastrulation, and coelomic development (Table 12.1).

TABLE 12.1   COMPARISON OF PROTOSTOMES
AND DEUTERSTOMES

Characteristic	Protostomes	Deuterostomes
Cleavage	Spiral cleavage Determinate cleavage	Radial cleavage Indeterminate cleavage
Blastopore fate	Mouth from blastopore	Anus from blastopore
Coelomic development	Split in mesodermal layer	Outpocketing of mesodermal layer

In pseudocoelomate organisms, the mesoderm does not split into two discrete sections. Parts of the mesoderm separate, but they remain attached to either the ectoderm or endoderm layer, resulting in the formation of a cavity that is not completely surrounded by the mesoderm. This is called a false coelom.

Although there is a mesoderm layer in the flatworm, the layer remains attached to both the ectoderm and the endoderm. No body cavity is formed. The flatworm is an acoelomate. In other acoelomates (sponges and coelenterates), there are only two germ layers, ectoderm and endoderm. There is no tube-within-a-tube body plan in acoelomate organisms.

Table 12.2 summarizes the classification of animals.

TABLE 12.2   CLASSIFICATION OF ANIMALS

Phylum	Characteristics	Examples	Class	Class Characteristics
Porifera (sponges)	Acoelomate, porous body, spicules, lime-silica skeleton	Sponge	—	—
Coelenterata	Acoelomate, radial symmetry, tentacles	Hydra Sea anemone, coral Jellyfish	Hydrozoa Anthozoa Scyphozoa	Polyp stage No medusa stage Medusa stage
Platyhelminthes (flatworms)	Acoelomate, bilateral symmetry	Planaria Fluke Tapeworm	Turbellaria Trematoda Cestoda	Free-living Parasitic Parasitic
Aschelminthes	Pseudocoelomate	Rotifer	Rotifera	Microscopic
Nematoda (round worms)	Pseudocoelomate, bilateral symmetry	Hookworm Trichina Pinworm	—	All parasitic
Annelida (segmented worms)	Coelomate, segmented	Earthworm Sandworm Leech	Oligochaeta Polychaeta Hirudinea	Few setae (bristles), no head Many setae, head Parasitic
Mollusca	Coelomate; soft, fleshy body; muscular foot	Oyster, clams Snails Octopus, squid	Pelecypoda Gastropoda Cephalopoda	Bivalve (2-part shell) Univalve (1-part shell) Head with tentacles, camera-type image-forming eyes
Arthropoda	Coelomate, jointed chitinous exoskeleton and legs, segmented body	Crayfish, lobster Grasshopper Spider, mite Millipede Centipede	Crustacea Insecta Arachnida Diplopoda Chilopoda	Two pairs of antennae, gills 3 body regions (head, thorax, abdomen), 6 legs, usually 2 pairs of wings, 1 pair of antennae 2 body regions, 8 legs, simple eyes, book lungs, no antennae 2 pairs of legs per segment 1 pair of legs per segment

TABLE 12.2 (Continued)

Phylum	Characteristics	Examples	Class	Class Characteristics
Echinodermata	Coelomate, radial symmetry, spiny skin, water-vascular system, marine dwelling	Starfish	Asteroidea	Central disk with 5 or more arms
		Sea urchin	Echinoidea	No arms
		Sea cucumber	Holothuroidea	Leathery body, tentacles around mouth
		Sea lily	Crinoidea	Attached to substratum by a long stalk
Chordata	Bilateral symmetry, notochord, pharyngeal gill slits, dorsal hollow nerve tube			
Subphyla (1) Hemi-chordata	Ciliated larval stage	Acorn worm	—	—
(2) Urochor-data	Notochord present only in larval form	Tunicate	—	—
(3) Cephalo-chordata	Notochord present throughout life	Amphioxus (lancelet)	—	—
(4) Verte-brata	Supporting backbone	Lamprey	Agnatha	Jawless Fish
		Shark, skate	Chondrich-thyes	Cartilaginous skeleton, cold-blooded
		Goldfish, bass	Osteichthyes	Scales, bony skeleton, cold-blooded
		Frog	Amphibia	Lungs, land and water, reproduction in water, moist skin, cold-blooded
		Snake, turtle	Reptilia	Dry skin, scaly skin, cold-blooded
		Robin, blue jay	Aves	Feathered bipeds, warm-blooded
		Human being, horse	Mammalia	Mammary glands, 4-chambered heart, warm-blooded, hair on body

Table 12.3 compares the life functions of members of different phyla.

TABLE 12.3   COMPARISON OF LIFE FUNCTIONS OF ORGANISMS IN VARIOUS PHYLA

Organism	Life Activities				
	Digestion	Respiration	Circulation	Excretion	Response
Hydra	Hollow digestive cavity	Diffusion into and from environment	Diffusion from cell to cell	Diffusion (ammonia)	Nerve ring
Earthworm	Digestive tract, special organs	Through moist skin	Closed system	Nephridia	Nervous system, ganglion structure
Grasshopper	Digestive tract, special organs	Tracheal tubes into air sacs	Open system	Malpighian tubules (uric acid)	Nervous system, ganglion brain
Starfish	Digestive tract, special organs	Skin gills	Closed system through skin gills	Amoeboid cells in coelom, diffusion through body wall	Nervous system, nerve ring, radial nerves, eyespot at tip of each arm
Human being	Digestive tract, special organs	Respiratory system—lungs	Closed system, 4-chambered heart	Excretory system—kidneys	Nervous system

Model Questions Related to Animal Classification

Questions 1–4 are based on the following list of criteria for indicating the possible common ancestry of organisms. Choose the letter of the item that BEST describes each numbered statement.

(A) Homologous structures
(B) Physiological similarities
(C) Biochemical similarities
(D) Embryonic similarities
(E) Genetic similarities

1. At one point in their lives, both human beings and birds have gill slits.

2. The wing of a bird is used for flying, while the flipper of a whale is used for swimming.

3. Uric acid is excreted through the kidneys of both birds and reptiles.

4. If the blood of a rabbit is sensitized to human blood, it will react to a chimpanzee's blood in very much the same way it reacts to human blood.

5. According to the binomial system of classification, to which organism is *Quercus rubra* most closely related?
(A) *Quercus alba*
(B) *Asclepias rubra*
(C) *Acer rubrum*
(D) *Juniperus communis*
(E) *Juniperus virginiana*

6. Of the following organisms, which have the closest taxonomic relationship?
   (A) Human and toad
   (B) Amoeba and paramecium
   (C) Whale and kangaroo
   (D) Crayfish and spider
   (E) Jellyfish and comb jelly

7. All of the following organisms have a true coelom EXCEPT the
   (A) clam
   (B) human being
   (C) earthworm
   (D) planarian
   (E) starfish

8. The development of an egg with a hard outer shell was an important advance in the history of vertebrates. The first class of vertebrates to reproduce in this way was the
   (A) Amphibia
   (B) Reptilia
   (C) Aves
   (D) Osteichthyes
   (E) Mammalia

9. All of the following pairs represent homologous structures EXCEPT
   (A) flipper of a whale; arm of a human being
   (B) wing of a bat; wing of a bird
   (C) flipper of a whale; lateral fin of a fish
   (D) arm of a human being; forelimb of a horse
   (E) forelimb of a horse; wing of a bird

10. All of the following are characteristics of chordates EXCEPT
    (A) a ventral nerve cord
    (B) segmented body musculature
    (C) a ventral heart
    (D) a notochord
    (E) gill slits in the anterior part of the body

11. Which of the following is TRUE of invertebrates that are best suited to a dry land environment?
    (A) They have a ciliated epithelium.
    (B) They obtain oxygen by way of gills.
    (C) They are supported by a firm exoskeleton.
    (D) They excrete free ammonia.
    (E) They move by contraction and expansion of the body.

Answers and Explanations to Model Questions: Animal Classification

1. D The early embryonic stages of vertebrates resemble each other. For example, a series of pouches from the pharynx push out and make contact with the outer body covering, forming slits. These slits become the gills in fish, but in other vertebrates, the slits close up and disappear.

2. A Although the uses of flippers and wings are different, the anatomic and developmental patterns are the same. Structures that are anatomically similar evolved from the same basic ancestral structure.

3. B Since the removal of uric acid from vertebrates requires a particular type of excretory organ, the excretory organs of birds and reptiles must be similar in structure and physiology. This similarity reflects a common ancestry.

4. C When a rabbit is injected with human blood, it forms specific antibodies that react against the blood protein only. If the serum from the rabbit reacts to the blood protein of a chimpanzee (antibody-antigen reaction), the chimpanzee's blood protein must be similar to human blood protein since the antibodies are specific for human blood.

5. A The binomial system of classification requires the genus and species name for an organism. Since *Quercus rubra* and *Q. alba* are in the same genus, they are more closely related to each other than to any other organism listed.

6. C The whale and the kangaroo are animals in the same class, Mammalia. They have the basic features of all mammals.

    Incorrect Choices
    A Although a human being and a toad are in the same subphylum, they belong to different classes (human being—Mammalia, frog—Amphibia).
    B Although the amoeba and paramecium are in the same phylum, Protista, they are in different classes (amoeba—Sarcodina, paramecium—Ciliata).
    D Even though the crayfish and spider are in the same phylum, Arthropoda, they belong to different classes (crayfish—Crustacea, spider—Arachnida).
    E Jellyfish belongs to Phylum Cnidaria. The comb jellies belong to Phylum Ctenophora.

7. D In order for a coelom to develop, there must be three germ layers. The mesoderm must divide to form the coelom. Although all of the choices include three germ layers, complete mesoderm separation is not observed in the planarian.

    Incorrect Choices
    A, B, C, E Complete mesoderm separation into two layers occurs. One mesoderm layer lines the outside of the endoderm, forming the gut. The other layer lines the inner part of the ectoderm.

8. B To be completely independent of a water environment, the embryos of organisms must have a covering that provides protection from drying out. The development of the reptile egg made complete land existence possible.

Incorrect Choices
A Amphibians return to the water to mate. The eggs develop in the water environment.
C Birds arose later than reptiles in evolutionary history.
D Fish are completely aquatic.
E Since mammals had reptile ancestors, the reptiles must have appeared first.

9. C Homologous structures are those that are anatomically similar. Although the flipper of the whale and the lateral fins of the fish are used for the same purpose (analogous structures), they are not homologous.

Incorrect Choices
A, B, D, E All are anatomically similar to each other.

10. A Chordates exhibit a dorsal nerve cord.

Incorrect Choices
B During the process of differentiation, the mesoderm divides into segments, or somites. Some of the somites, which are separated from each other by tissue, give rise to muscles.
C The hearts of vertebrates develop from the ventral mesentery in the area of the foregut. This pattern of development is characteristic of all chordates.
D A notochord is a stiff supporting rod of the nerve cord. It is characteristic of chordates only.
E Gill slits are observed in the embryonic development of all chordates. In some chordates, such as human beings, they quickly disappear.

11. C The phylum Arthropoda contains the largest number of invertebrate dry land dwellers. They all possess a chitinous exoskeleton.

Incorrect Choices
A Ciliated epithelium is associated with a respiratory system geared to filter the air. Arthropods breathe by means of gills or tracheal tubes.
B Arthropods that live in water have gills. Gills are thin areas embedded with capillaries. To facilitate gas exchange, there must be a constant supply of water to keep them moist. In land organisms, the water is absent.
D Ammonia is a highly toxic waste product excreted by organisms that can easily and quickly dilute it with water. Since water is limited in land environments, nitrogenous wastes are converted to a less toxic substance—uric acid or urea—that does not require large amounts of water for dilution.
E Mollusks move by the action of a large contracting foot muscle. Since the land mollusks breathe through their skin, they must live in a moist environment.

CHAPTER THIRTEEN
Modern Concepts of Human Physiology

Capsule Concept

Physiology is the science dealing with the functions of various parts and organs of living organisms. It is the branch of biology that provides explanations for the many processes that occur in the body's internal environment. The physiology of each body system depends on the maintenance of a steady state. Although the molecular exchange between the external and internal environments may vary, the conditions of the internal environment remain the same. For example, even if you eat several oranges in a single day, the pH of your blood remains between 7.35 and 7.45. If mechanisms for keeping the internal environment constant did not exist, human beings (and other animals) could not survive. The ability to monitor the internal environment, maintaining it in a steady state, is known as *homeostasis*. A major homeostatic mechanism is feedback control. Negative feedback tends to keep a system stable; positive feedback upsets stability. There are many body systems that utilize feedback control for homeostatic regulation, as you will become aware while studying physiological processes.

Vocabulary

Actin   a muscle protein
Adenohypophysis   the anterior pituitary
Adenyl cyclase   an enzyme in cell membranes
Corpus luteum   a temporary endocrine gland
Endocrine glands   glands of internal secretion
Erythrocyte   a red blood cell
Gonadotrophins   sex hormones
Homeostasis   steady-state control
Hormone   a chemical regulator
Hypophysis   the pituitary gland
Leukocyte   (leucocyte) a white blood cell
Metamorphosis   a change of body form
Myoneural junction   a neuromuscular junction
Myosin   a muscle protein
Nephron   the basic unit of kidney structure
Neurotransmitter   a secretion from a nerve cell
Neurohypophysis   the posterior pituitary
Neuron   a nerve cell

Ornithine cycle   biochemical processes that produce urea
Peristalsis   a wavelike contraction of muscle
Synapse   a junction between nerve cells
Tetany   a sustained muscle contraction

Hormones

In many organisms, there are dramatic changes from one stage of development to another. *Metamorphosis* (change of body form) is characteristic of insects and amphibians. Whether the metamorphosis is complete or incomplete (Figure 13.1), the changes are mediated by hormones.

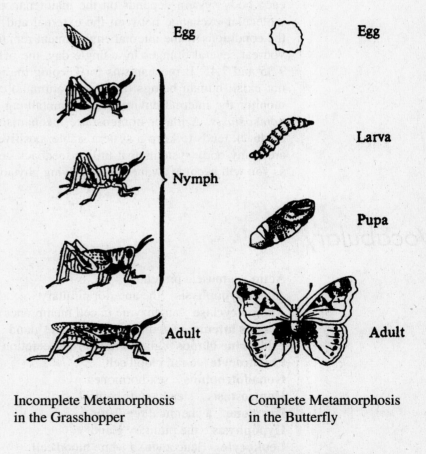

Incomplete Metamorphosis
in the Grasshopper

Complete Metamorphosis
in the Butterfly

Figure 13.1   Incomplete and complete metamorphosis

Hormones are chemical compounds that are produced in ductless glands and travel through the bloodstream to tissues or organs, on which they exert their effect. In higher animals, the hormones are produced by glands of internal secretion, which comprise the endocrine system (Figure 13.2).

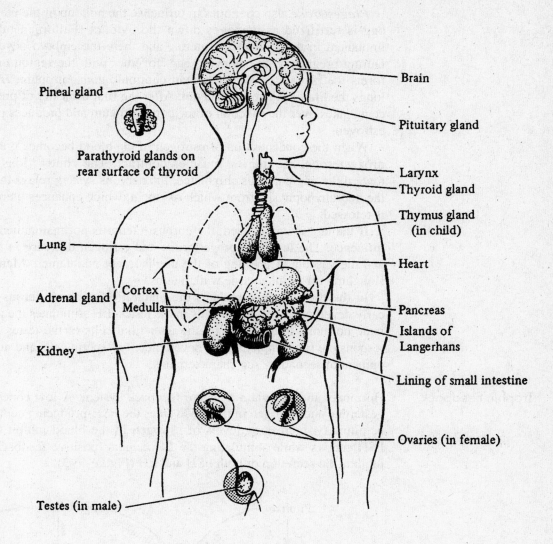

Figure 13.2   The human endocrine system

The pituitary, or hypophysis, a gland about the size of a green pea, is divided into two parts: the anterior pituitary (adenohypophysis) and the posterior pituitary (neurohypophysis). The anterior pituitary produces and secretes six protein hormones, including the gonadotrophins, which affect the human reproductive system.

Gonadotrophins

In females, *FSH* (follicle-stimulating hormone), produced in the anterior pituitary, stimulates the development of the ovarian follicle. As the follicle ripens, it produces and secretes the hormone estrogen. *Estrogen* induces the buildup of the endometrium, the mucous membrane lining the uterine wall, and it triggers the release of *LH* (luteinizing hormone) from the anterior pituitary.

When the concentration of LH exceeds the concentration of FSH, the ovarian follicle ruptures. Ovulation occurs. Under the influence of LH, the ruptured follicle forms scar tissue, the *corpus luteum*, which acts as an endocrine gland producing estrogen and progesterone.

Progesterone also continues to influence the buildup of the uterine lining. If an egg is fertilized as it moves down the oviduct (Fallopian tube), it becomes implanted in the wall of the uterus, and there the embryo develops until birth. During pregnancy, the cleaving egg, together with the region of the uterus that forms the placenta, secretes human chorionic gonadotrophin, *HCG*, which prolongs the life of the corpus luteum. After the first trimester of pregnancy, the placenta takes over the function of the corpus luteum and produces progesterone and estrogen.

When the concentration of estrogen in the blood becomes greater than that of progesterone, the uterine muscles contract. The contractions force the fetus toward the vagina. Thus stimulated, the nervous system relays the information to the neurohypophysis, from which oxytocin, which enhances uterine contractions, is released.

If the egg is unfertilized, the corpus luteum is not maintained beyond the LH influence. The levels of progesterone and estrogen decrease in the bloodstream, and the vascularized lining of the uterus is not maintained. Menstruation, or the sloughing off of the uterine wall, begins.

In the male, FSH stimulates the maturation of seminiferous tubules and the early development of sperm cells in the testes. LH stimulates the production of the male hormone testosterone by the interstitial cells of the testes. Testosterone is responsible for completing the development of sperm cells and also for the development of secondary sex characteristics.

Trophic Feedback

Hormones interact via a negative feedback system. A low concentration of progesterone and estrogen in the blood frees the FSH-producing centers in the anterior pituitary. Increasing levels of estrogen in the blood inhibit the hypophyseal FSH centers while stimulating the LH centers (positive feedback). Progesterone inhibits the secretion of both FSH and LH (Figure 13.3).

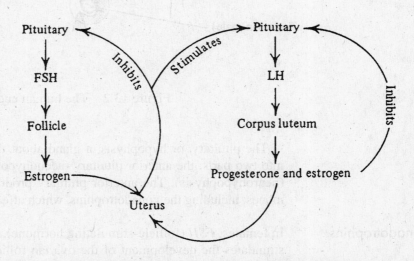

Figure 13.3   Negative feedback system of pituitary hormones

The anterior pituitary gland does not respond directly to the concentrations of hormones in the bloodstream. The gland is indirectly connected to a part of the brain, the *hypothalamus*, by means of a short stalk. Blood passes through a highly vascularized region of the hypothalamus, the median eminence. The median eminence responds to the low concentrations of blood hormone by releasing short polypeptide chains, called *releasing factors* (RF). Each of the six hormones of the adenohypophysis has a particular site for manufacture of RF in the median eminence. The RF travels to the anterior pituitary by way of the hypophyseal portal system, where it stimulates the production and release of pituitary hormones. Thus FSH is released from the pituitary only when a releasing factor travels from the median eminence to the pituitary. High concentrations of progesterone and estrogen will inhibit the production of FSH-releasing factor.

Other Hormones of the Anterior Pituitary

Other hormones, *TSH* (thyroid-stimulating hormone), *STH* (somatotrophic, or growth-stimulating, hormone), and *ACTH* (adrenocorticotrophic hormone), are produced and released from the hypophysis in the same manner. Under the influence of TSH, the thyroid gland produces *thyroxin*, a hormone that controls the metabolic rate of the body. Iodine is a component of thyroxin. Bone growth is regulated by STH. The electrolytic balance of the body and water reabsorption are controlled by the adrenocortical hormones, which are produced by the adrenal cortex under the influence of ACTH.

Hormones of the Posterior Pituitary

The posterior hypophysis releases two protein hormones. (1) *Vasopressin* (antidiuretic hormone, ADH) regulates water reabsorption or balance in the distal convoluted tubules of the kidneys; it also can increase blood pressure by causing the arterioles to constrict. (2) *Oxytocin* controls uterine contractions during childbirth.

The mechanism of hormone elaboration and release in the posterior pituitary differs from that in the anterior pituitary. Hormones of the neurohypophysis are secretions of nerve cells that are part of the hypothalamus. These hormone secretions travel down the axons, which, in turn, terminate in the posterior pituitary. Hormone release is initiated by the arrival of a nerve impulse from the hypothalamus. This indicates a close relationship between the endocrine system and the nervous system.

Other Endocrine Glands

Other endocrine glands synthesize proteinaceous hormones. The pancreas produces *insulin*, a hormone that stimulates glycogen formation and storage, stimulates carbohydrate oxidation, and inhibits the formation of new glucose. The pancreas also secretes *glucagon*, a hormone that stimulates the conversion of glycogen into glucose-6-phosphate. The parathyroids release *parathormone*, which controls the reabsorption of calcium and phosphate from bone; while *calcitonin*, a hormone secreted by the thyroid, controls the absorption of calcium and phosphate from the blood.

The various hormones are listed in Table 13.1 and their effects in Table 13.2.

Action of Hormones

How do hormones exert their effects on cells? Hormones may act on a cell's metabolic mechanism, on its genetic material, or on both. It has been determined that protein hormones act on the energy production machinery of a cell by making glucose-6-phosphate available to the cell. Unlike steroid hormones, protein hormones are molecularly too large to penetrate the cell membrane.

The cell has specific receptor sites for the various kinds of hormones. When a hormone (first messenger) combines with its specific receptor site, a complex is formed that activates the enzyme *adenyl cyclase*, normally present in cell membranes. Adenyl cyclase catalyzes the conversion of ATP to 3', 5'-cyclic AMP or cAMP (cyclic adenosine monophosphate). Then cAMP acts as a second messenger by activating another enzyme, glucagon, which converts glycogen to glucose-6-phosphate and lipids to fatty acids. Cells under hormonal influence are either engaged in synthesis or dividing. In either case, a large supply of energy is required. (See Figure 13.4.)

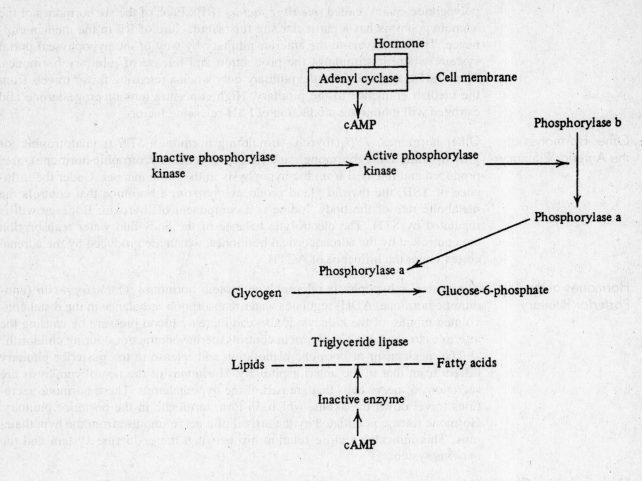

Figure 13.4   Hormone activation of second messenger cAMP

TABLE 13.1   TYPES OF HORMONES

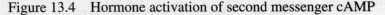

Protein	Steroid
FSH (follicle-stimulating hormone)	Cortisone group of hormones
LH (luteinizing hormone)	Estrogen
STH (somatotrophic hormone)	Progesterone
TSH (thyroid-stimulating hormone)	Testosterone
Prolactin	
Parathormone	
Glucagon	
Insulin	
Thyroxin	
Oxytocin	
ADH (antidiuretic hormone)	
RF (releasing factor)	
ACTH (adrenocorticotrophic hormone)	

TABLE 13.2   ENDOCRINE SECRETIONS AND THEIR FUNCTIONS

Gland	Hormone	Effect
Anterior pituitary	FSH	Stimulates growth of ovarian follicle and production of sperm in testes
	LH	Stimulates ovulation and secretion of sex hormones by ovaries and testes, maintains corpus luteum
	Prolactin	Induces milk secretions in mammary glands
	TSH	Stimulates thyroxin production by thyroid
	ACTH	Stimulates adrenal cortex to produce cortisone hormones
	STH	Stimulates growth
Posterior pituitary	ADH	Stimulates increased water reabsorption by kidneys and constriction of blood vessels
	Oxytocin	Stimulates uterine contractions and release of milk by mammary glands
Thyroid	Thyroxin	Controls oxidative metabolism
	Calcitonin	Regulates calcium and phosphate blood levels
Adrenal cortex	Cortisones	Regulate water reabsorption, electrolyte balance, protein and carbohydrate metabolism
Pancreas	Insulin	Stimulates glucose transport to cells
	Glucagon	Converts glycogen into glucose
Ovaries	Estrogen	Stimulates development of female secondary sex characteristics and buildup of uterus
	Progesterone	Stimulates female secondary sexual characteristics; maintains pregnancy
Testes	Testosterone	Stimulates male secondary sex characteristics
Parathyroid(s)	Parathormone	Controls reabsorption of calcium and phosphate from blood, stimulates release of calcium from bone
Thymus	Thymosin	Stimulates immunological response in lymphoid tissues
Pineal	Melatonin	Stimulates development of melanophores in vertebrates, maintains seasonal reproductive cycles

cAMP may also function in the activation of genes. The mechanism of gene stimulation is a positive control mechanism called catabolic repression.

The promoter site of an operon may be structurally incompatible with the RNA polymerase configuration. When a protein, CAP (combining activating promoter), combines with cAMP, the configuration of the protein is altered so that it "fits" into the promoter site, altering the configuration of the latter. RNA polymerase can then combine at the promotor site, and mRNA transcription begins. The availability of cAMP depends on the presence of a hormone that stimulates cAMP production. Protein hormones, therefore, exert an effect on the genetic apparatus of cells.

Since steroid hormones are small molecules and are lipid soluble, they penetrate the cell membrane. It is believed that a steroid hormone combines with a specific cytoplasmic protein, forming a complex. The hormone-protein complex penetrates the nucleus and interacts with the nonhistone portions of chromosomal proteins. The interaction frees the genes for transcription. This hypothesis is still under investigation.

The differences in the actions of protein and steroid hormones are listed in Table 13.3 and some common endocrine disorders in Table 13.4.

TABLE 13.3   DIFFERENCES IN HORMONE ACTION

Protein	Steroid
Acts on cell membrane nondiffusable	Acts in nucleus diffusable into cell
Activates energy production	—
Has possible action (cAMP) on genes	Forms protein complex that acts on genes

TABLE 13.4   SOME COMMON ENDOCRINE DISORDERS

Gland	Disease, Condition, or Symptom
Thyroid	Cretinism—undersecretion *in utero* Goiter—undersecretion in an adult Exophthalmic goiter—oversecretion in an adult
Pituitary	Dwarfism—undersecretion of STH in a child Giantism—oversecretion of STH in a child Acromegaly—oversecretion of STH in an adult Diabetes insipidus—undersecretion of ADH
Pancreas	Diabetes mellitus—undersecretion of insulin
Adrenal cortex	Addison's disease—undersecretion of ACH

**Model Questions
Related to the
Endocrine System**

Below is a diagram of organs in the endocrine system. For Questions 1–10, choose the number or numbers that best replaces "this gland" or "these glands" in each statement.

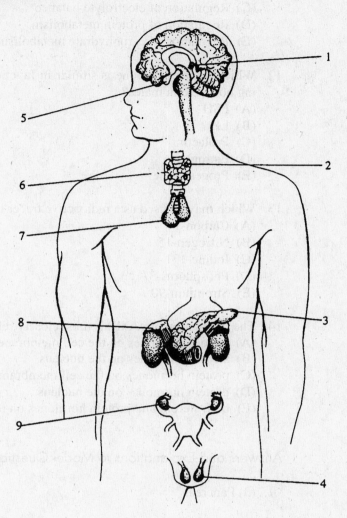

Organs of the Endocrine System

1. This gland is both an endocrine and an exocrine gland.

2. Steroid hormones are produced by these glands.

3. The accidental removal of this gland will upset the calcium-phosphate metabolism of the body.

4. An abnormal secretion from this gland results in Addison's disease.

5. Abnormal secretion from this gland results in acromegaly.

6. Cretinism is a result of hyposecretion of this gland.

7. Tetany results from the undersecretion of this gland.

8. The hypophyseal portal system is related to this gland.

9. Hypothalamic control is exerted over this gland.

10. These glands contribute to the metabolism of carbohydrates.

11. Which is NOT a function of adrenal cortical hormones?
    (A) Regulation of calcium absorption from the bloodstream
    (B) Regulation of water balance in the kidneys
    (C) Regulation of electrolyte balance
    (D) Regulation of protein metabolism
    (E) Regulation of carbohydrate metabolism

12. Which female hormone is similar in function to the interstitial cell-stimulating hormone in males?
    (A) FSH
    (B) LH
    (C) Prolactin
    (D) Estrogen
    (E) Progesterone

13. Which may be used as a radioactive tracer in studying thyroid function?
    (A) Carbon-14
    (B) Nitrogen-15
    (C) Iodine-131
    (D) Phosphorus-31
    (E) Strontium-90

14. The production of cAMP is most closely related to the activity of
    (A) steroid hormones on the cell membrane
    (B) steroid hormones on the nucleus
    (C) protein hormones on the cell membrane
    (D) protein hormones on the nucleus
    (E) both steroid and protein hormones on the nucleus

Answers and Explanations to Model Questions: Endocrine System

1. (3) Pancreas

2. (4), (8), (9) Testis, adrenal, ovary

3. (2) Parathyroid

4. (8) Adrenal

5. (5) Pituitary

6. (6) Thyroid

7. (2) Parathyroid

8. (5) Pituitary

9. (5) Pituitary

10. (3), (6), (8) Pancreas, thyroid, adrenal

11. A Regulation of calcium absorption is a function of the parathyroid gland.

   Incorrect Choices
   B, C, D, E The adrenal cortex secretes hormones that regulate water and electrolyte balance and the metabolism of proteins and carbohydrates.

12. B ICSH (interstitial cell-stimulating hormone) stimulates the secretion of testosterone by the interstitial cells of Leydig in the testes. Since LH injected into experimental animals produces the same effect, ICSH and LH are considered to be the same in the male and female, respectively.

   Incorrect Choices
   A FSH acts in the male to stimulate sperm production, while it stimulates follicular development in the female.
   C Prolactin stimulates the mammary glands.
   D, E Both hormones maintain the uterus.

13. C Since iodine is a major part of thyroxin, the fate of radioiodine can be easily followed and thyroid gland activity can be monitored.

   Incorrect Choices
   A, B, D, E Since the other elements are not major or special constituents of thyroxin, their use would not indicate how the thyroid gland is functioning.

14. C The union of a protein hormone and its receptor site in the cell membrane activates a series of reactions that make cAMP available to the cell. The hormone does not penetrate the membrane.

   Incorrect Choices
   A, B Steroid hormones penetrate the cell, combine with a special cytoplasmic protein, and accumulate in the nucleus. They are believed to free genes for RNA transcription.
   D Protein hormones indirectly influence genetic activity through catabolic repression, which has not been fully investigated.
   E Only protein hormones are known to be involved in cAMP production.

Overview of the Nervous System

The nervous system, responsible for integrating and coordinating activities of the body, consists of the central nervous system and the autonomic nervous system. The central nervous system coordinates skeletal muscle response to stimuli. The autonomic nervous system controls involuntary activity. The nerve cell is the basic unit of structure and function of the nervous system.

A nerve consists of a group of nerve cells (neurons or nerve fibers) supported by neuroglia and held together by connective tissue. Each individual nerve cell is approximately a few millimeters thick, but may be over a meter long. The nerve cell (Figure 13.5) is made up of the cyton (cell body), the dendritic processes, and the axon (which terminates in end branches). Neurons in the central nervous system (the brain and the spinal cord) are myelinated. The myelin sheath acts as an insulator, allowing an impulse to travel down the length of the axon at a very rapid rate.

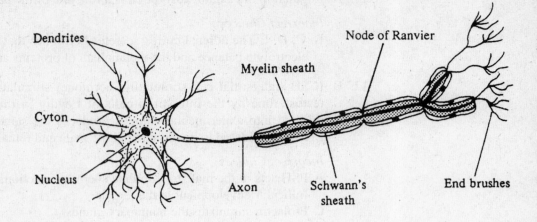

Figure 13.5   Structure of a motor neuron

The basic unit of response in the central nervous system (CNS) is the reflex arc. An impulse, initiated in a receptor, travels along the afferent or sensory fiber to the spinal cord, where it is transmitted across a synapse to the efferent or motor neuron. The final result is muscular contraction or movement. A simple stretch reflex (knee jerk) is an example of a reflex involving two neurons. A more complex reflex, such as the withdrawal or pain reflex, involves an associative neuron and two synapses in the spinal cord. The cytons of the efferent neurons participating in the reflex arc are located in the gray matter of the brain and spinal cord.

The Nerve Impulse

An impulse (Figure 13.6) is a wave of depolarization and repolarization along the cell membrane of a neuron. The mechanism causing the impulse can best be explained on the basis of the permeability of the cell membrane to ions. Because of selective permeability of the membrane, sodium ions are kept out of the cell. The cytoplasm of the cell has a potassium ion concentration 30 times greater than that of the fluid medium that bathes the cell. However, the charge on the inner surface of the cell membrane is negative in relation to its external positive charge, where there is a high concentration of sodium ions (Figure 13.6). An electric potential equal to 70–90 millivolts is established across the membrane.

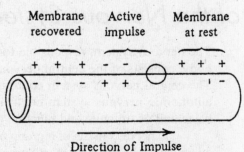

Figure 13.6   Impulse transmission

The membrane of the neuron consists of Na^+-K^+ ion exchange pumps and ion channels. The ion channels are regulated by specific proteins that act as ion gates. The sodium channel is regulated by two gates, the potassium channel by one gate.

When the permeability of the membrane is altered, sodium ions cross the membrane. An electric event (stimulus) changes the charges on both sodium gates and they open. The potassium gate opens very slowly. The influx of sodium through the gates changes the polarization of the membrane. The event is known as *depolarization*. When a maximum positive potential is reached inside of the cell, one sodium gate closes and movement of sodium ions cease. Potassium continues to flow out of the cell. The outflow of potassium ions reestablishes the membrane potential—*repolarization*—but not the ionic distribution. The Na^+-K^+ exchange pump reestablishes the ionic distribution (Figure 13.7).

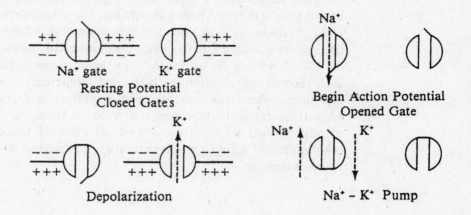

Figure 13.7   Sodium-potassium ion channels and pump

The graph of an action potential demonstrates the sequence of events involved in the transmission of an impulse (Figure 13.8). A threshold value (Ecrit) must be attained before depolarization occurs. Once the threshold value is reached, the sodium-potassium gates are affected. An impulse is generated throughout the nerve fiber.

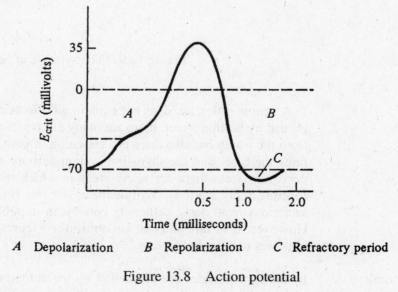

A Depolarization   *B* Repolarization   *C* Refractory period

Figure 13.8   Action potential

A nerve cell will transmit the impulse either totally or not at all. There are no graded responses. This maximal firing condition is referred to as an "all-or-none response."

In myelinated nerve cells, the impulse may travel 100 meters per second, or the length of a football field. Rapid conduction is possible because ion flow across the membrane occurs at the nodes of Ranvier. The impulse "leaps" from node to node (saltatory conduction). Although an impulse can theoretically flow in either direction along a nerve cell, the arrangement of the cells in relation to each other directs the flow in one direction. Nerve cells are arranged so that the terminal branches of one cell are in close contact with the dendrite end of another cell, at a junction called the *synapse* (Figure 13.9).

When an impulse reaches the synaptic knobs at the end of the axon, calcium ion gates are opened. The presence of the calcium ions causes the vesicles to be drawn against the presynaptic membrane. The vesicles release a *neurotransmitter.*

Neurotransmitters are chemical substances that diffuse across the synaptic gap and initiate a second impulse when chemoreceptors on the dendrites are stimulated. Not all nerve cells secrete the same neurotransmitters. Some cells are cholinergic (acetylcholine secreting); others are adrenergic (epinephrine secreting). The neurotransmitters may exert either an inhibitory or an excitatory effect on neurons. If neurotransmitters were not removed from the dendrite receptors, their effects would be continued and all control would be lost. Enzymes (cholinesterase acting on acetylcholine) are situated in the synapes to remove neurotransmitters.

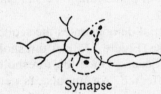
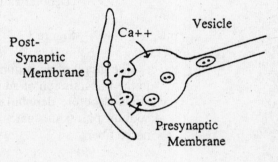

Figure 13.9   The synapse of nerve cells

A simple reflex arc does not explain all effects of a stimulus. If one's hand is burned by boiling water, there are many effects. Not only is the hand withdrawn from the water, but also there is a sensation of pain, the heartbeat and respiratory rates increase, and the digestive system activity slows down. Throughout the nervous system there are many areas in which masses of nerve cells synapse, forming relay stations. Within these stations (known as ganglia), excitatory secretions from nerve cells may convey an impulse to other areas of the body. However, it is not unusual for inhibitory secretions to prevent the flow of impulses to other body areas.

The Autonomic Nervous System

Involuntary muscles are controlled by the autonomic nervous system, which is separated functionally and structurally into two parts: the parasympathetic and the sympathetic systems. Table 13.5 summarizes the differences in these systems. Table 13.6 compares the central and autonomic systems, and Table 13.7 summarizes the action of the autonomic nervous system.

TABLE 13.5   THE AUTONOMIC NERVOUS SYSTEM

Parasympathetic	Sympathetic
Cranial and sacral nerves	Thoracic spinal nerves
Long premotor fiber	Short premotor fiber
Ganglia at or near effector	Ganglia near the spinal cord organ
Short postmotor fiber	Long postmotor fiber
Acetylcholine-motor neuron secretion	Epinenphrine-motor neuron secretion
Maintains status quo	Prepares body for action

TABLE 13.6   COMPARISON OF THE CENTRAL AND AUTONOMIC NERVOUS SYSTEMS

Central	Autonomic
Myelinated cells	Unmyelinated cells
Controls skeletal muscle	Controls smooth muscle
Voluntary actions	Involuntary actions

TABLE 13.7   ACTIONS OF THE AUTONOMIC NERVOUS SYSTEM

Parts of the Body Affected	Sympathetic System	Parasympathetic System
Heart	Speeds up beat	Slows down beat
*Arteries	Constriction, increases pressure	Dilates, decreases pressure
Digestive organs	Slows down peristalsis	Speeds up peristalsis
Urinary bladder	Relaxes	Constricts
Bronchial muscles	Dilates	Constricts
Sweat glands	Increases secretion	Decreases secretion

*Many sources cite this example. However, evidence from some experiments supports the idea that there is one vasomotor center. The frequency of signals passing from the vasomotor center over particular nerve pathways determines the diameter of the blood vessels.

In invertebrates and lower vertebrates, the brain functions as a ganglionic relay center, aiding in the coordination of body activities. In higher vertebrates, especially human beings, the brain has evolved into a highly specialized organ with functional parts.

The *medulla oblongata*, through the autonomic nervous system, controls involuntary and visceral activities. Body balance, muscular coordination, and equilibrium are controlled by the *cerebellum*.

The *hypothalamus*, located behind the thalamus, regulates body temperature, thirst, hunger, and metabolism and, in general, maintains the internal environment. The thalamic region is the center that sorts out the thousands of incoming and outgoing impulses and relays them to the various brain centers.

The most highly developed brain center is the *cerebral cortex*, which, in human beings, is the center for all voluntary muscular control and mental activity. It is in the *cerebrum* that the centers of analysis, coding and storage of information, recognition, memory, understanding, intelligence, and sense integration are located.

Sense Receptors

To gather information, special sensory receptors are necessary. Ninety percent of our information comes through the sense of sight. Through the arrangement of a lens and fluid in the eye, light is directed to a region on the retina known as the retinal disk. Here light energy is converted into electric impulses, which are then interpreted by the cerebrum.

Within the retina are receptor cells of two different types: rods and cones. The rods contain rhodopsin (visual purple), a light-sensitive pigment that consists of opsin (a protein) and retinene (a derivative of vitamin A). There are many isomers of retinene. Light striking the rods causes the *cis*-retinene to change to *trans*-retinene. This transformation results in an unstable rhodopsin, which then splits into opsin and retinene. In the process hydrogen ions are released, causing depolarization in the bipolar neurons that synapse with the rods. The depolarization spreads from the bipolar neurons through the complex web of synapses to the ganglion cells. The collection of axons of these cells forms the optic nerve, which carries the impulse to the cerebrum.

The cones, located in the fovea of the retina, also contain visual pigments. Whereas many rod cells synapse with one neuron, only one cone cell synapses with one neuron. This one-to-one arrangement produces visual acuity. The cone cells are responsible also for color vision.

Within the ear the cochlea, a fluid-filled chamber, is divided into two compartments. Each compartment consists of two membranous portions. One membranous section is lined with ciliated cells. When this membrane expands because of the fluid pressure within the cochlea, the ciliated cells rub against the opposite nonciliated membranous section. In some way the rubbing cilia cause depolarization in the nerve cells which are wrapped around the ciliated cells. This entire sensory portion of the cochlea is known as the organ of Corti.

The fluid pressure within the cochlea is affected by the pressure exerted by the three small bones of the inner ear against the membranous window of the cochlea. These bones, in turn, are moved by the tympanic membrane (eardrum), which is initially influenced by the compression of sound waves that enter the ear.

The senses of taste and smell depend on chemoreceptors which produce an action potential when stimulated. The sense of taste depends on stimulation of cells by dissolved chemicals; the sense of smell depends on stimulation of ciliated cells by dissolved vapors.

In order for an organism to respond to a stimulus, motor neurons must conduct the impulses to muscles or effectors. The transmission of impulses to muscle cells occurs in a region known as the *motor end plate* (or *neuromuscular junction*). This region consists of a membranous sheath that binds the nerve cell and muscle. However, there is no physical contact between the neuron and the muscle. The myoneural junction, commonly called the neuromuscular junction, is the gap between the axon of the motor nerve and its muscle (effector).

An impulse reaching the end of the axon causes the release of a neurohumor, acetylcholine. The neurohumor changes the permeability of the cells of the membranous sheath; all ions flow into the cells, and the potential changes. Since the membrane potential is different from the potential of the muscle cells, it may induce depolarization in the muscle cells. The final result is muscle contraction.

Model Questions Related to the Nervous System

Questions 1–5 are based on the following diagram of a reflex pathway. Choose the correct answer for each question.

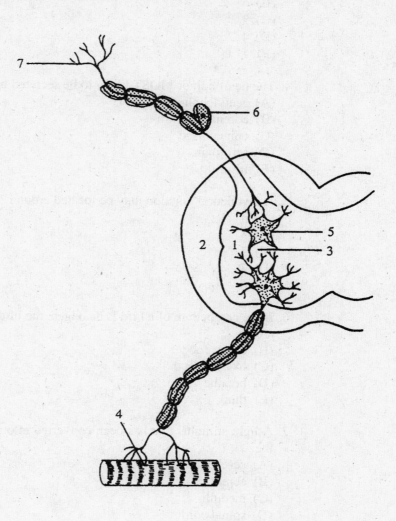

Reflex pathway

1. Number 7 refers to which of the following?
   (A) Dendrites from the motor neuron
   (B) Dendrites from the sensory neuron
   (C) The terminal portion of an axon from a sensory neuron
   (D) The terminal portion of an axon from a motor neuron
   (E) Dendrites from an associative neuron

2. Which number refers to a synapse?
   (A) 7
   (B) 6
   (C) 5
   (D) 3
   (E) 1

3. Which number represents the cell body of a sensory neuron?
   (A) 7
   (B) 6
   (C) 3
   (D) 4
   (E) 1

4. The neurohumor MOST likely to be secreted by number 4 is
   (A) acetylcholine
   (B) adrenaline
   (C) epinephrine
   (D) serotonin
   (E) insulin

5. A myoneural junction may be located around
   (A) 4
   (B) 3
   (C) 5
   (D) 7
   (E) 6

6. If the cerebellum of a bird is damaged, the bird is not able to
   (A) fly
   (B) digest food
   (C) oxidize food
   (D) breathe
   (E) think

7. A light stimulus that has been converted into a nerve impulse is interpreted by the
   (A) cerebellum
   (B) cerebrum
   (C) medulla
   (D) spinal cord
   (E) thalamus

8. A person under the increasing influence of alcohol loses first the ability to talk; second, the ability to walk straight; third, the ability to breathe normally. From this evidence, in what order may it be shown that alcohol affects the central nervous system?
   (A) Medulla, cerebellum, cerebrum
   (B) Cerebrum, cerebellum, medulla
   (C) Cerebellum, medulla, cerebrum
   (D) Cerebellum, cerebrum, medulla
   (E) Cerebrum, thalamus, medulla

Questions 9–11 are based on the following experiment.

> One galvanometer (an instrument that detects electric currents) was attached to the auditory nerve of a frog, and another galvanometer to the optic nerve. When the frog's eye was stimulated by light and its ear by sound, the galvanometer needles were equally deflected.

9. Compared to the current flowing in the auditory nerve, that which flows in the optic nerve appears to be
   (A) the same
   (B) larger
   (C) smaller
   (D) more intense
   (E) less intense

10. When the frog reacts to a sound, the nerve impulse started by the sound appears to be interpreted in the
    (A) auditory nerve
    (B) optic nerve
    (C) brain
    (D) synapses
    (E) rods and cones

11. The impulse that travels in the optic nerve most closely resembles
    (A) light rays
    (B) radio waves
    (C) electric currents
    (D) cosmic rays
    (E) water waves

12. The most accurate description of a nerve impulse is that it is a flow of
    (A) neurotransmitters across a synapse
    (B) electrons like an electric current
    (C) protons across the cell membrane
    (D) sodium ions back and forth across the cell membrane in response to a gradient
    (E) sodium ions in and out of the cell due to a change in the permeability of the cell membrane

13. Rhodopsin (visual pigment) is found in the
    (A) cochlea
    (B) cone cells
    (C) rod cells
    (D) taste buds
    (E) convolutions of the cerebrum

14. The organ of Corti is located in the
    (A) retina
    (B) cochlea
    (C) synapse
    (D) cone cells
    (E) cerebrum

15. Neurohumors are involved in all of the following activities EXCEPT
    (A) synaptic transmission
    (B) transmission at the myoneural junction
    (C) effector organ stimulation
    (D) thalamic activity
    (E) stimulation of the cochlea

Answers and Explanations to Model Questions: The Nervous System

1. **B** Since a stimulus is applied to the finger, the neuron indicated in the finger is a sensory neuron. Sensory neurons usually have their cell bodies located at a distance from the dendritic process, which is the direct recipient of the stimulus.

   The other numbered structures in the diagram are as follows:
   6 represents the cyton of the cell body of the sensory neuron.
   5 represents the associative neuron.
   3 represents one synapse in the spinal cord.
   4 represents the terminal portion of the axon of the motor fiber.
   2 represents the white matter of the spinal cord.
   1 represents the gray matter of the spinal cord.

2. **D** Refer to Answer 1.

3. **B** Refer to Answer 1.

4. **A** In the central nervous system, acetylcholine is the neurohumor secreted at the junction of the motor neuron and the effector organ.

   Incorrect Choices
   B, C Adrenaline is another name for epinephrine. This is a secretion of the autonomic nervous system as well as a secretion from the adrenal medulla.
   D Serotonin is the neurohumor primarily associated in the synapses in the brain.
   E Insulin is a hormone secreted by the pancreas.

5. **A** The myoneural junction is the space between the muscle and the motor neuron.

6. **A** The cerebellum controls activities involving muscle equilibrium, posture, and balance. Bird flight involves all of the above.

   Incorrect Choices
   B, D All involuntary activities are controlled by the medulla.
   C Oxidation of food involves the endocrine and the autonomic nervous systems.
   E Mental activities are controlled by the cerebrum of the brain (assuming a bird can think).

7. **B** Mental activities, such as the interpretation of light, are controlled by the cerebrum.

   Incorrect Choices
   A The cerebellum controls balance and coordination.
   C The medulla controls all involuntary actions.
   D The spinal cord is a center for reflex actions.
   E The thalamus is an integrating center in the brain. It directs incoming impulses to the appropriate brain areas.

8. **B** The cerebrum controls the ability to talk; the cerebellum, coordination and balance; and the medulla, breathing.

9. **A** Since both galvanometer needles were deflected equally, the electric currents they registered were the same.

10. **C** Interpretation of impulses is the function of the brain.

   Incorrect Choices
   A Auditory nerves merely conduct the impulse to the brain.
   B Optic nerves conduct the impulse initiated in the eye to the brain.
   D Synapses are gaps between neurons.
   E Rods and cones are special cells that can convert a light stimulus into an electrochemical event.

11. **C** As shown by the galvanometers, which measure electric current, the impulse that travels through the nerves is capable of moving electric particles.

   Incorrect Choices
   A Light rays consist of beams of photons released from incandescent bodies.
   B Radio waves are electromagnetic waves.
   D Cosmic rays are beams of high-energy particles traveling from deep space.
   E Water waves are the movements of water caused by a disturbance.

12. **E** An impulse is a wave of depolarization along the length of a cell membrane. Polarity is established by the impermeability of the cell membrane to sodium ions. For depolarization to occur, the permeability of the membrane must be altered.

   Incorrect Choices
   A Neurotransmitters initiate an impulse in neurons or effector organs once they are released by stimulated axons.
   B Ions, not electrons, move across the membrane.
   C A proton is the nucleus of an atom. In living systems, the hydrogen ion is the proton.
   D The cell membrane is impermeable to sodium ions. Those that leak across the membrane are removed by active pumps. The sodium is unable to flow across the membrane until the permeability of the membrane changes.

13. **C** Rhodopsin is the protein that is found in the rod cells of the retina of the eye.

   Incorrect Choices
   A The cochlea is a chamber in the ear.
   B Cone cells contain different kinds of proteins that are essential for color vision and the clarity of vision.
   D Taste buds do not require a protein to convert a chemical stimulus to an impulse.
   E The convolutions of the cerebrum extend along the surface area and are associated with increased intellectual capacities.

14. B The organ of Corti is an area in the cochlea that is necessary for converting the pressure produced by sound waves into electric impulses.

Incorrect Choices
A, D The retina contains the cone and rod cells necessary for vision.
C Refer to Answer 10.
E The cerebrum interprets impulses arriving from receptors.

15. E The fluid in the cochlea is affected by the pressure exerted against the eardrum; the eardrum responds to sound vibrations.

Incorrect Choices
A, D The thalamus is a highly complex network of synapses. The movement of impulses across a synapse depends on the secretion of neurohumors, which initiate a new impulse in the adjacent neurons.
B, C Effector organs and motor neurons are separated by a gap that forms the myoneural junction. The impulse terminates at the end of the motor neuron. Neurohumors released at the end of the axon must stimulate the effector if movement is to occur.

The Structure and Function of Muscle

Gross skeletal movement, delicate and precise eye movement, and slow peristalsis depend on the contraction of muscles. Muscles are bundles of muscle cells or fibers that are bound together by connective tissue. Three basic types of muscle tissue are present in vertebrates: smooth, cardiac, and skeletal. Each type is structurally different from the other.

Smooth (or visceral or involuntary) muscle is composed of spindle-shaped cells, each containing a single nucleus in the widest region. The individual muscle cells, or fibers, are not striated. They interlace to form sheets of muscle tissue, rather than converging into bundles. Smooth muscle forms the muscle layers in the walls of the digestive tract, bladder, various ducts, and other internal organs; it is also the muscle of the arteries and veins. Smooth muscle is stimulated by the autonomic nervous system.

Cardiac muscle (the tissue of which the heart is composed) consists of multinucleated, striated, branching cells attached to each other like a jigsaw puzzle.

Skeletal (or voluntary or striated) muscle (Figure 13.10) is composed of heavily striated multinucleate cells, or fibers.

The fibers are usually bound into bundles rather than sheets. A muscle fiber is an elongated cell that has many nuclei scattered in its semiliquid cytoplasm. Many long, parallel muscle fibrils marked with alternating light and dark bands are the basic units of the muscle fiber. The electron and phase-contrast microscopes have afforded a closer look at the striation of muscle fiber. The striations are caused by a special arrangement of the muscle proteins—myosin, actin, and tropomyosin. Skeletal muscle produces the movements of the limbs, trunk, face, jaw, and eyeballs; it is the most abundant tissue in vertebrates; and it is innervated by the somatic nervous system.

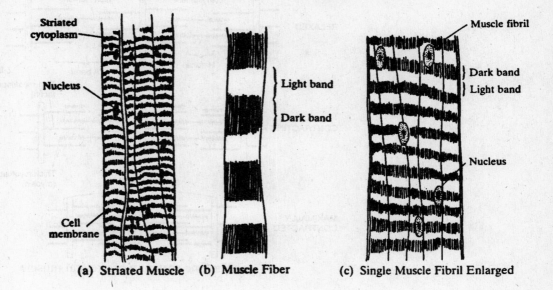

Figure 13.10   The Structure of muscle

Myosin is a heavy, rodlike protein molecule that has two active heads projecting from one end of the rod. These rodlike molecules are grouped together (several hundred) in a bundle to form a thick filament. Myosin filaments are located in the A and H regions of the fiber. The A region is darker in appearance than the H region because of the overlap of another protein filament, actin. Actin filaments, composed of light, helical-structured protein molecules, extend through the I regions. This entire pattern is repeated throughout the muscle fiber. Z lines form the boundary between the repetitive units of striated muscle fibrils, the *sarcomeres.*

Experimental data reveal that, when the I regions shorten, the Z lines are pulled closer together during muscle contraction. These observations plus visualization of the particular arrangement of protein filaments led to an explanation of muscle contraction known as the sliding-filament theory. It is reasoned that actin filaments slide over and under the myosin filaments. The sliding is aided by the myosin heads, which form cross bridges with the actin filaments, creating an actomyosin complex. When energy is released (after an actomyosin complex has been formed), the cross bridges swivel, pulling the actin filaments toward the center of the sarcomere. (Figure 13.11.)

All myosin heads exist in close association with ATP molecules. When the ATP-myosin complex couples with active sites on the actin filament, ATP is hydrolyzed and energy is released. The cross bridges swivel. Thus the actomyosin unit acts as an ATPase, as well as the structural component required to slide the actin filament over the myosin.

Other proteins function in muscle fibers. Tropomyosin, a thin fibril, coils around the actin filament. Troponin, a small protein molecule, is present at the end of the tropomyosin fibril that is contiguous to the active sites of the actin filament. Calcium is an element vital to the biochemistry of contraction. According to current theory, calcium combines with the troponin. The union of this mineral and protein encourages the movement of the tropomyosin away from the active sites of the actin filament, allowing the formation of the actomyosin complex. Calcium is stored in sacs (cisternae) of the sarcoplasmic reticulum, which adjoins a network of tubules known as the transverse system (T-system).

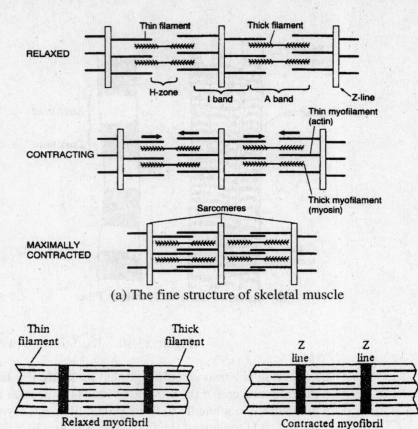

(a) The fine structure of skeletal muscle

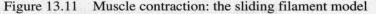

(b) Sliding filament mechanism of muscle contraction

Figure 13.11   Muscle contraction: the sliding filament model

The T-system communicates with the sarcoplasmic membrane (cell membrane). A wave of depolarization, which originates in the motor neurons, moves along the sarcoplasmic membrane and down to the cisternae by way of the channels of the T-system. The release of calcium initiates contractility of muscle, whereas the reabsorption of calcium into the sacs by active pumps results in muscle relaxation. The entire process of contraction and relaxation requires a few thousandths of a second. The pattern of striations in cardiac muscle indicates that the contractile mechanism is probably similar to that of skeletal muscle. However, the mechanism for the contraction of smooth muscle has not been demonstrated sufficiently to make knowledgeable comparisons.

Muscle contraction requires ATP, but the skeletal muscle cells of mammals are incapable of storing ATP. However, they do store another high energy phosphate compound, namely, creatine phosphate (phosphocreatine), which is hydrolyzed to transform ADP into ATP. It is believed that the enzymes for this process are located in the cisternae.

Under conditions of strenuous exercise, the quantity of energy produced through mitochondrial activity is limited by the amount of oxygen that the bloodstream can transport to the cells; it takes time for the mitochondrial system to mobilize. Anaerobic respiration (glycolysis) of glucose to lactic acid must provide the ATP when the mitochondrial process is too slow. However, the buildup of lactic acid impedes the operation of the glycolytic process. Therefore muscle contraction ceases until the oxygen debt is paid back by the reconversion of lactic acid to pyruvic acid.

The patterns of contraction by the three kinds of muscles differ (Figure 13.12). Smooth muscle contracts slowly and maintains its contractions over a long period of time; whereas each contraction of cardiac muscle is followed by a rest period, thus varying the rate of contraction.

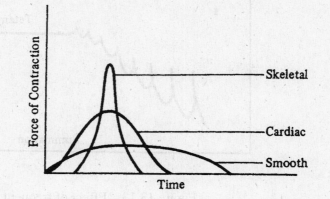

Figure 13.12   Types of muscle contraction

Both smooth muscle and cardiac muscle depend on mitochondrial activity to provide the energy-rich ATP molecules. Unlike these two, skeletal muscle may be required to contract rapidly and to remain contracted for long periods of time. Both aerobic and anaerobic respiration are essential in ATP production.

The special properties of skeletal muscle contraction are revealed by kymography. When stimuli are applied, there is a latent period before contraction. This delay is due to the time required for depolarization at the myoneural junction and along the muscle membranes. The latent period is followed by the contractive period, which, in turn, is followed by a relaxation period. The three phases of muscle contraction make up a single muscle twitch. If the strength (intensity) of the stimulus is increased, the strength of the muscle contraction is also increased, up to a point where the maximal contractive response level is reached.

If the frequency of the stimuli is increased, the strength of the contractions also increases. As the frequency of the stimuli increases, the muscle does not have time to relax fully from one contraction before the next stimulus arrives and causes it to contract again. This results in *summation*—contractions that are stronger than any single twitch (Figure 13.13).

If the stimuli are very frequent and the impulses continue to reach the muscle cells before they can relax, one contraction fuses with another until there is sustained contraction of the muscle without intervening periods of relaxation (*tetany*). Eventually the muscle fatigues because of the accumulation of lactic acid and the depletion of ATP. The muscle contraction weakens, and the muscle will require recovery time before any further contractions are possible.

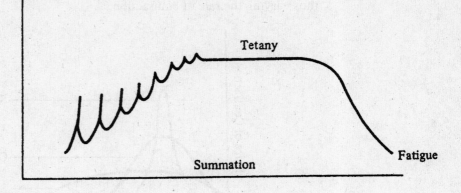

Figure 13.13   Effect of frequent stimuli on muscle

Unlike nerve tissue, muscles exhibit a graded response. The same muscle may be required to lift an 8-ounce glass of milk or a 20-pound bag of sand. Graded responses are possible because a single muscle is innervated by many nerve cells from a single nerve. Each nerve cell innervates a group of muscle cells within the muscle. This group is called a *motor unit*. A weak stimulus causes the excitation of a few motor units; a strong stimulus excites all or most of the motor units. A stimulus greater than that which engages all the motor units has no further effect on the muscle. Thus simple muscle twitches (Figure 13.14) depend on the strength of the stimulus. Many skeletal muscle activities depend on the tetanic contractions rather than on simple twitches. Muscles that enable the body to maintain its posture rely on tetanic contractions.

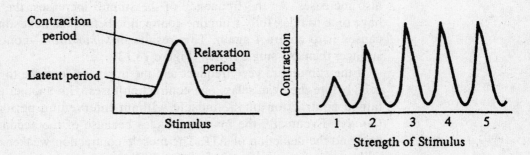

Figure 13.14   Phases of muscle twitch

**Model Questions
Related to Muscle
Physiology**

1. To time the phase of a single muscle twitch in a leg of a frog the instrument
   that is most useful is called a
   (A) manometer
   (B) kymograph
   (C) phase contrast microscope
   (D) chromatograph
   (E) sphygmomanometer

2. The maximum peak of a vertical curve produced by a muscle twitch on a
   kymograph tracing is associated with the
   (A) latent period
   (B) contraction period
   (C) relaxation period
   (D) recovery period
   (E) refractory period

3. The strength of a muscle twitch depends on the
   (A) strength of the impulse
   (B) number of muscle fibers stimulated
   (C) frequency of the impulse
   (D) amount of neurohumor secreted
   (E) number of sarcomeres stimulated

4. Muscle contraction occurs when myosin combines with
   (A) oxygen
   (B) actin
   (C) ADP
   (D) lactic acid
   (E) ATP

5. Experiments show that muscle fatigue is accompanied by an accumulation of
   (A) ATP
   (B) lactic acid
   (C) pyruvic acid
   (D) glycogen
   (E) creatine phosphate

6. During strenuous contraction of skeletal muscle, which is MOST likely to be
   found in high concentrations in the muscle cells?
   (A) Creatine phosphate
   (B) Lactic acid
   (C) Oxygen
   (D) ATP
   (E) Glycogen

7. The protein that acts as an enzyme as well as a contractile protein is
   (A) actin
   (B) myosin
   (C) tropomyosin
   (D) troponin
   (E) calcium

8. A man hanging from a ledge of a roof eventually loses his grip. The causes of his loss of grip include all of the following EXCEPT
   (A) there was a depletion of ATP
   (B) an accumulation of lactic acid impeded the glycolytic pathway
   (C) his muscles sustained a condition of tetanus for a brief period
   (D) his cells suffered from an oxygen debt
   (E) there was an insufficient store of glycogen in his liver

9. The element MOST essential in muscle contraction is
   (A) calcium
   (B) potassium
   (C) sodium
   (D) magnesium
   (E) iodine

Answers and Explanations to Model Questions: Muscle Physiology

1. B The contractile activity of a stimulated muscle may be recorded on a rotating drum. This device is known as a kymograph.

   Incorrect Choices
   A A manometer is a pressure gauge for gases and liquids.
   C A phase-contrast microscope provides contrasting backgrounds of microscopic specimens. It cannot measure muscle twitches.
   D A chromatograph records results obtained from the process of separating closely related substances on the basis of their different solubilities in a solvent.
   E A sphygmomanometer is used to measure blood pressure.

2. B The muscle is attached to a recording needle on the kymograph in such a way that contraction, or shortening, of the muscle pulls the needle upward. Thus the highest point on the graph represents the strength of the contraction of the muscle produced by a given stimulus.

   Incorrect Choices
   A The brief time lag between the application of the stimulus and the muscle contraction is called the latent period. Usually it is not visible on ordinary kymographic recordings.
   C The relaxation period is marked by the return of the muscle of its original length.

D Recovery is a brief period of rest when the oxygen debt is repaid. It is not a normal part of an actual muscle twitch.

E The refractory period occurs when another impulse cannot be initiated because the membrane potential is below the normal resting potential.

3. B The more muscle fibers stimulated, the more the muscle shortens. This allows for graded responses which are essential if the same muscle is to do light and heavy work.

Incorrect Choices

A, C, D While the impulse initiates muscle contraction by stimulating different numbers of motor units, it is the actual shortening of the fibers that produces the contraction.

E Sarcomeres are repetitive units within a fiber. If a fiber contracts, the sarcomeres are engaged.

4. B Once an actomyosin complex is formed, ATP is hydrolyzed and energy for the contraction is released.

Incorrect Choices

A Oxygen is utilized to produce ATP.

C ADP is the low-energy compound that must be restored to ATP if further contractions are to occur.

D Lactic acid is a product of anaerobic respiration, which occurs if contractions are sustained over a long period of time.

E Myosin filaments exist in combination with ATP. However, no work is done unless the ATP is split. This process is accomplished when myosin combines with actin.

5. B Muscle fatigue occurs when muscle contraction is sustained over a period of time. Since aerobic respiration is too slow in producing the required energy, anaerobic respiration occurs. Lactic acid accumulates from anaerobic respiration and impedes the glycolytic pathway. The muscle fatigues.

Incorrect Choices

A, C, D, E All of these substances are essential to muscle contraction and are utilized in the process.

6. B Refer to Answer 5.

Incorrect Choices

A, C, D, E Refer to Answer 5.

7. B The myosin heads that form the cross bridges give myosin its enzymatic and contractile properties.

Incorrect Choices

A Actin contains the active sites for the attachment to the cross bridges.

C, D Both proteins are on the actin molecule at or near the active sites.

E Calcium is the mineral that is essential in muscle contraction.

8. E The inability to sustain grip is due to muscle fatigue. Muscle fatigue occurs when there is a buildup of lactic acid and a depletion of ATP. Glycogen is not ordinarily a limiting factor.

   Incorrect Choices
   A, B See Answer 5.
   C Tetanus (sustained contraction of a muscle) no doubt existed since the muscle fibers were not able to relax during the entire episode.
   D Since anaerobic respiration enabled him to sustain his grip, lactic acid increased. Oxygen debt is lack of oxygen; oxygen is necessary to reconvert lactic acid into glycogen.

9. A Calcium begins the entire process of contraction by freeing the active sites on the action molecule.

   Incorrect Choices
   B Potassium is required for nerve function and acid-base balance.
   C Sodium is required for nerve function, acid-base balance, and water regulation.
   D Magnesium is required to activate enzymes for protein synthesis.
   E Iodine is a component of thyroxin, a hormone.

Circulation, Excretion, and Digestion

Circulation and Heartbeat

If the heart of a frog is removed from its body and placed into a saline solution, a rhythmic beating is maintained for several hours thereafter. Cardiac muscle exhibits the myogenic property of innate contractions.

Within the right atrium of the heart (Figure 13.15) near the entrance of the superior vena cava, there is a special tissue that depolarizes spontaneously. This tissue is the sinoatrial node (S-A node), or the pacemaker. It establishes the rhythm of cardiac contraction. A wave of depolarization spreads out across the atrium, resulting in synchronous contractions. Next, the impulse spreads to the atrioventricular node (A-V node), specialized tissue located in the septum between the right atrium and ventricle. From there the impulse travels through the bundle of His to the Purkinje fibers to the entire ventricular muscle. An electocardiogram (EKG) reading reveals the electric changes that occur during contractions (Figure 13.16).

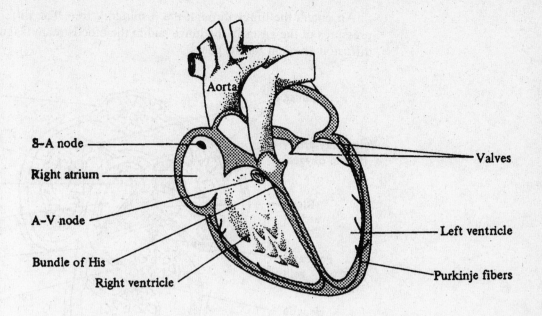

Figure 13.15   Diagram of the heart

Cardiac muscle is subjected to the control of the autonomic nervous system. Otto Loewi, through a series of experiments, proved that hormone-like substances, neurohumors, are secreted by nerves. These secretions slow down or speed up the rate of heartbeat. Acetylcholine slows down the heart rate, and epinephrine accelerates it. The impulses releasing the neurohumors from the nerves originate in the medulla of the brain in response to stimulation of stretch receptors in the atria and aorta.

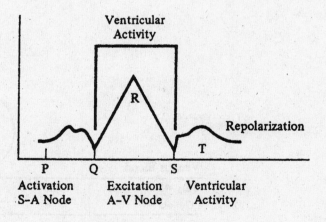

Figure 13.16   EKG graph

Heart Function

Unlike skeletal muscle, cardiac muscle does not contract before its relaxation phase is competed. This brief rest period is essential for the normal functioning of the organ.

The function of the heart is to move blood through the lungs to all parts of the body (Figure 13.17). The right ventricle pumps blood to the lungs, where oxygenation occurs. Alveoli—thin, moist, membranous sacs surrounded by capillary networks—are sites within the lung where gas exchange occurs.

Air enters the lungs through the respiratory tree. The difference in the partial pressures of the gases in the lungs and in the bloodstream (Figure 13.18) creates a diffusion gradient for the gases.

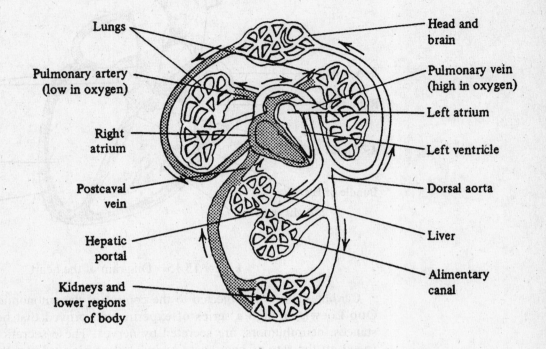

Figure 13.17   Path of the blood

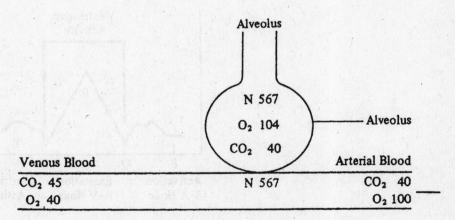

Figure 13.18   Partial pressures of gases in the blood

Oxygenation of Blood

The movement of the diaphragm, a thick skeletal muscle controlled by the autonomic nervous system and responding to hydrogen ion concentrations and partial pressures of gas, increases and decreases the gas pressure within the lungs. Atmospheric gas moves into the lungs (inhalation) as pressure in the lungs is decreased by the expansion of the thoracic cavity, which is induced by the contraction and downward movement of the diaphragm. Exhalation is a passive process and is caused by the relaxation of the diaphragm. Table 13.8 summarizes the functions of the respiratory organs.

TABLE 13.8   THE HUMAN RESPIRATORY SYSTEM

Structure	Function or Description
Nasal cavity	Warms, humidifies, filters air (ciliated mucous membranes)
Epiglottis	Prevents food from entering the air passages
Pharynx	Permits passage of air and food
Larynx	Produces speech (voice box), permits passage of air (ciliated epithilial membrane)
Trachea	Ciliated and mucus-covered cells for filtration of air
Bronchi	Convey air to and within the lungs (tubes in the trachea leading to the lungs)
Bronchioles	Small tubes leading to the alveoli
Alveoli	Exchange gas between the lungs and the blood

Composition of Blood

The part of the blood that carries oxygen is the red blood cell (erythrocyte). Each erythrocyte contains hundreds of molecules of hemoglobin, an iron protein compound similar to chlorophyll. Oxygen binds loosely to hemoglobin, forming oxyhemoglobin. In an environment of low partial pressure and low pH, the oxyhemoglobin releases the oxygen. The affinity of hemoglobin decreases as the pH is lowered. This is known as the *Bohr shift*. The pH shift occurs as carbonic acid is formed from the release of carbon dioxide by metabolizing cells.

In contrast, carbon dioxide is not efficiently carried by the hemoglobin molecule. Only about 35% of the carbon dioxide can combine with another portion of the hemoglobin molecule, forming carbaminohemoglobin. About 65% of the carbon dioxide is converted into carbonic acid by an enzyme, carbonic anhydrase, that is present in the red blood cells. Carbonic acid ionizes to hydrogen and bicarbonate ions. Through a buffering system, the hydrogen combines with the potassium salt of the hemoglobin molecule to form acid hemoglobin and potassium bicarbonate. The entire process is reversed in the lungs. The remaining 5% of the carbon dioxide is dissolved in the plasma.

White blood cells, or leukocytes, of which there are five types, function to protect the body against foreign proteins. *Neutrophils* and *monocytes* are amoeboid and actively phagocytize foreign particles. *Eosinophils* detoxify histamine-like secretions. *Lymphocytes* are involved in immune responses, while *basophils* produce anticoagulants.

Platelets are cell fragments that are essential in the complex blood-clotting mechanism (Figure 13.19).

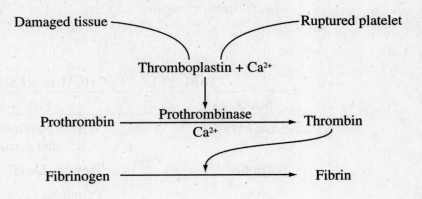

Figure 13.19 Blood-clotting process

Plasma, the liquid part of the blood, is mostly water. Nutritive materials, minerals, hormones, and antibodies are dissolved and carried by the plasma to all the cells.

One of the plasma antibodies responds to the presence of foreign red blood cells. As shown in Table 13.9, transfusions are possible only when the red blood cell antigens are compatible with the plasma proteins.

The thick walls of the left ventricle pump the oxygenated blood through the aorta to all parts of the body. The aorta is the largest of the elastic, muscle-walled arteries. Veins, which carry blood to the heart, are less muscular and lack the recoil properties of arteries. Between the arteries and veins are the capillaries. Because the capillaries are thin walled (one cell thick), substances can enter and leave the bloodstream.

Two forces operate to control the flow of material across the capillaries.

TABLE 13.9   BLOOD TYPES, ANTIGENS, AND ANTIBODIES

Type of Blood	Antigen	Antibody
A	A	b
B	B	a
AB	AB	—
O	—	a and b

Because of the pumping action of the heart, blood exerts hydrostatic pressure against the blood vessel walls. As the blood moves further from the heart, the pressure decreases. Osmotic pressure, caused by the colloidal action of blood in the vessels, is constant throughout the system. When the hydrostatic pressure is greater than the osmotic pressure, materials leak out of the blood into the tissues near the arteriole end. At the venule end, where the hydrostatic pressure is less than the osmotic pressure, materials move from the tissues into the bloodstream. In Figure 13.20 the circled numbers represent the hydrostatic pressure in a small section of capillary bed.

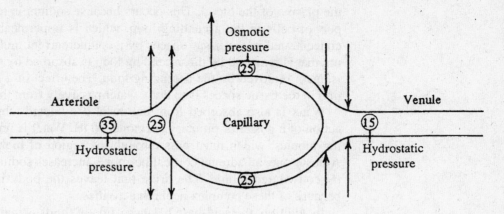

Figure 13.20   Pressure in a capillary bed

Not all of the fluid that leaks out of the vascular system reenters the blood-stream. It becomes *lymph*, which bathes the cells. Lymph returns to the circulatory system by way of vessels that comprise the lymphatic system. The fluid is moved through the lymph vessels by contraction of the surrounding muscles. Along the path of the lymph vessels, glands, or nodes (which contain lymphocytes), remove particles from the lymph. The lymphatic system joins the circulatory system at the subclavian veins.

Excretion

One of the nitrogenous waste products, urea, is carried by the bloodstream. Urea is formed when ammonia from deamination enters into the ornithine cycle (Figure 13.21) in the liver. Urea and other wastes are removed when blood passes through the kidneys.

Within each kidney there are filtering units called nephrons (Figure 13.22). Although the nephrons remove toxic wastes, they also maintain an electrolyte-water balance, control the pH of body fluids, and effect homeostasis of cells, tissues, and organs.

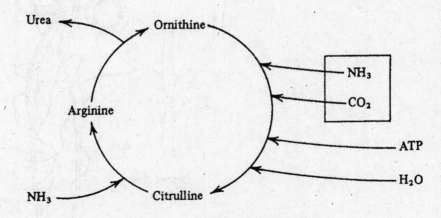

Figure 13.21   Ornithine cycle

A high hydrostatic pressure forces most substances (except red blood cells and large proteins) out of the glomerulus. Eighty percent of the water and most of the nutrients are reabsorbed by the proximal convoluted tubes (PCT). The filtrate that passes down the loop of Henle becomes increasingly hypotonic in relationship to the plasma of the blood. This occurs because sodium is removed by active transport of cells in the ascending loop, which is impermeable to water. Sodium is concentrated in the tissue spaces by a countercurrent multiplier. Part of the sodium that is removed by the ascending loop is absorbed by the descending loop and is again removed by the ascending loop. The effect of a high sodium concentration in the tissue spaces is to draw water passively from the descending loop.

Water is also absorbed from the distal convoluted tubules (DCT) through the action of a posterior pituitary hormone, ADH. Water is removed from the collecting tubules, which must pass through the region of high sodium concentration. Aldosterone, an adrenal cortical hormone, increases sodium absorption in the distal convoluted tubule. The urine that leaves the body is hypertonic to plasma because of these complex nephron activities.

In addition, the acid-base balance of body fluids is maintained by the secretion of hydrogen and potassium ions by the tubule cells in response to the hydrogen ion concentration of the blood.

Digestion

In order to maintain itself, an organism must obtain nutrients. The food that is ingested is changed to soluble products by the various digestive organs. Table 13.10 lists the organs that digest food and their functions.

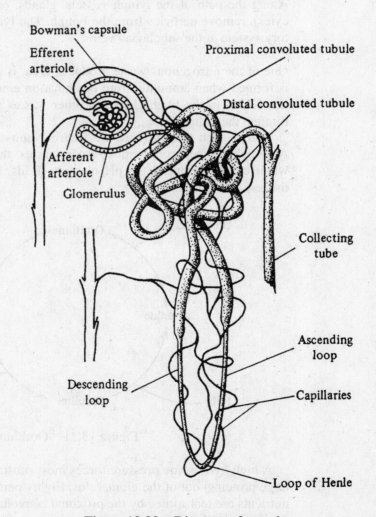

Figure 13.22   Diagram of a nephron

TABLE 13.10   DIGESTIVE ORGANS

Organ	Function	Enzyme
Mouth	*Mechanical digestion*: food broken down into small particles	—
	Chemical digestion: starch converted into sugar	Amylase of saliva
Stomach	*Mechanical digestion*: churning action—chyme	—
	Chemical digestion: proteins converted into peptones—casein of milk	HCI and pepsin of gastric juice; rennin
Small intestine	*Chemical digestion*: sugars converted into monosaccharides;	Lactase, sucrase, maltase
	peptones, into amino acids;	Peptidases
	fats, into fatty acids and glycerol	Lipase

There are also three accessory organs of digestion (Table 13.11).

TABLE 13.11   ACCESSORY ORGANS OF DIGESTION

Accessory Organ	Function	Secretions
Pancreas	Converts sugars into mono-saccharides	Amylase
	Converts peptones into amino acids	Trypsin, chymotrypsin
	Converts fats into fatty acids	Lipase
	Neutralizes acid pH of stomach	$NaHCO_3$
Liver	Emulsifies fats	Bile salts
Gallbladder	Stores bile	

Absorption of the soluble nutrients occurs in the small intestine through *villi*, which increase the surface area for absorption. Absorption of amino acids and glucose occurs through active transport, while the absorption of fatty acids occurs through passive transport. Fatty acids are absorbed into the lacteals, which are part of the lymphatic system. All nutrients carried by the bloodstream must pass through the liver by means of the hepatic portal system.

The removal and storage of excess nutrients are two of 200 functions of the liver. Deamination, production of urea, release of nutritive materials, destruction of old red blood cells, detoxification of substances, and production of blood proteins are some of the other functions.

Materials are moved along the alimentary canal by *peristalsis*, waves of muscle contractions. Each organ is stimulated to secrete its enzymes by a discharge of hormone. In fact, one of the first hormones investigated by W. M. Baylis and E. H. Starling was *secretin*, a secretion released by the small intestine, which stimulates the flow of pancreatic juice and bile.

Whereas external environmental conditions may force an organism to adapt or perish, the internal environment is maintained in a constant state by various homeostatic mechanisms.

Model Questions Related to Circulation, Excretion, and Digestion

1. A clean toothpick was dipped into a sample of human blood. The blood was then mixed with a drop of anti-A serum. No agglutination was observed. The blood type was
   (A) A or B
   (B) A or O
   (C) B or O
   (D) AB or B
   (E) B only

2. When using blood from the same sample, the procedure described above was repeated using anti-B serum. Agglutination was produced. The blood type was
   (A) O
   (B) A
   (C) B
   (D) A or O
   (E) B or O

3. All of the following statements about arteries are true EXCEPT
   (A) they are thick walled
   (B) they pulsate
   (C) they contain much elastic fiber tissue
   (D) they carry blood away from the heart
   (E) they contain valves

4. The liquid that bathes the cells of the body is called
   (A) fibrinogen
   (B) lymph
   (C) plasma
   (D) fibrin
   (E) blood

5. All of the following veins carry deoxygenated blood EXCEPT the
   (A) superior vena cava
   (B) inferior vena cava
   (C) pulmonary vein
   (D) renal vein
   (E) hepatic vein

6. All of the following statements about white blood cells are true EXCEPT
   (A) they are formed in lymph glands
   (B) they are formed in bone marrow
   (C) they move like paramecia
   (D) they are phagocytic
   (E) they are nucleate

7. The pacemaker of the heart is the
   (A) S-A node
   (B) A-V node
   (C) septum
   (D) Purkinje fiber
   (E) bundle of His

8. Below is a diagram representing a capillary bed.

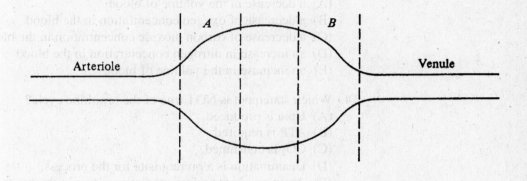

Capillary bed

Which statement is INCORRECT?
(A) More oxygen diffuses at A than at B.
(B) More CO_2 diffuses from the tissues at B than at A.
(C) More fluid leaves at A than at B.
(D) The hydrostatic pressure is greater at A than at B.
(E) Blood moves faster in area A than in area B.

9. Which represents the correct sequence in the clotting of blood?
   (A) Platelets, fibrinogen, prothrombin, fibrin, thrombin
   (B) Platelets, prothrombin, thrombin, fibrinogen, fibrin
   (C) Platelets, thrombin, fibrinogen, prothrombin, fibrin
   (D) Platelets, fibrin, fibrinogen, thrombin, prothrombin
   (E) Platelets, fibrinogen, prothrombin, fibrin, thrombin

10. Most of the CO_2 is carried in the bloodstream by
    (A) plasma
    (B) carbaminohemoglobin
    (C) HCO_3^-
    (D) myoglobin
    (E) CO_3^{2-}

11. Oxyhemoglobin is formed in the
    (A) lungs
    (B) right auricle
    (C) spleen
    (D) kidney
    (E) liver

12. After being in a small, poorly ventilated room for an hour with eleven other persons, a student noticed that his rate of breathing had increased. The most probable reason for this is that the
    (A) air in the room had become hot
    (B) carbon dioxide concentration in his blood had increased
    (C) oxygen concentration in his blood had increased
    (D) excess water in his body had to be eliminated
    (E) eleven people bored him

13. Thickening of the alveoli membranes may cause
    (A) a decrease in the volume of blood
    (B) a decrease of oxygen concentration in the blood
    (C) a decrease of carbon dioxide concentration in the blood
    (D) an increase in nitrogen concentration in the blood
    (E) an increase in the volume of blood

14. Which statement is NOT true of the ornithine cycle?
    (A) Urea is produced.
    (B) ATP is required.
    (C) CO_2 is consumed.
    (D) Deamination is a prerequisite for the process.
    (E) Arginine is not an intermediate product in the process.

15. All of the following are activities of the nephron EXCEPT
    (A) maintenance of pH balance
    (B) maintenance of electrolyte balance
    (C) control of sodium balance
    (D) control of red blood cell reabsorption
    (E) secretion of hydrogen and potassium ions

16. The active transport of sodium occurs in the region of the nephron known as the
    (A) glomerulus
    (B) proximal convoluted tubules
    (C) Bowman's capsule
    (D) ascending loop of Henle
    (E) collecting tubules

For Questions 17–21, choose the word that BEST fits each statement.

    (A) Stomach
    (B) Esophagus
    (C) Small intestine
    (D) Large intestine
    (E) Liver

17. Absorbs large quantities of water.

18. Secretes pepsin and hydrochloric acid.

19. Converts glucose to glycogen.

20. Is the major organ for the absorption of the end products of digestion.

21. Secretes a fat-emulsifying substance.

Answers and Explanations to Model Questions: Circulation, Excretion, and Digestion

1. C Anti-A is a serum antibody that reacts with antigen A. Since O blood has no red blood cell antigen and since there is no agglutination with antigen B, the blood type is either B or O.

2. C Since there was agglutination in the serum that contained anti-B, the blood cell antigen is B. The blood type is B.

3. E Because of the elastic recoil of arteries and the driving force of the heart, the blood will constantly move in one direction. There is no need for valves to prevent the backward flow of blood.

   Incorrect Choices
   A, C The thick walls are due to a well-developed muscular middle layer that contains extra elastic material.
   B Since the arteries receive blood directly from the heart, they alternately expand and contract in response to cardiac systole and diastole.
   D Arteries carry blood away from the heart.

4. B Lymph is the fluid that leaks out of the circulatory system by way of the capillaries and returns to the circulatory system through lymphatic vessels.

   Incorrect Choices
   A Fibrinogen is the plasma protein necessary for blood clotting.
   C Plasma is the liquid part of the blood.
   D Fibrin is the elastic fiber produced as the end product of the clotting reaction.
   E Blood is the transport medium of an organism.

5. C The pulmonary veins return blood to the heart from the lungs. In the lungs the blood is oxygenated.

   Incorrect Choices
   A, B, D, E Blood passes through the capillaries before it enters the veins. In the capillaries there is an exchange of gas. The blood passing from the capillaries to the veins has been deoxygenated. With the exception of the pulmonary veins, veins carry deoxygenated blood.

6. C White blood cells are phagocytic. The engulfing of particles occurs in a fashion similar to that of amoeboid, not paramecium, activity.

Incorrect Choices

A Lymphocytes, which give rise to antibody-producing plasma cells, are formed in the lymphatic tissue, lymph glands, spleen, liver, thymus, and bone marrow.

B Bone marrow is the active site for red and white blood cell production.

D The word "phagocytic" means "engulfing." White blood cells, like amoebae, form pseudopods that enable them to ingest foreign particles.

E The different shapes of the nuclei of white blood cells distinguish the various types of these cells.

7. A The S-A node is specialized for spontaneous depolarization and repolarization. Its rhythm regulates the heartbeat.

Incorrect Choices

B The A-V node directs ventricular contractions.

C The septum is a wall in the heart that separates the right side from the left side.

D Purkinje fibers conduct impulses throughout the ventricles of the heart.

E The bundle of His is a special tissue that joins the A-V node to the Purkinje fibers.

8. E The diameters at both ends of the capillary bed are the same. Blood should flow through both ends at the same rate.

Incorrect Choices

A, B The partial pressure of the gases at the different ends of the capillary bed influence the diffusion of gases from the blood to tissues and vice versa.

C The high hydrostatic pressure forces more fluid out at *A* than at *B*.

D The arteriole end of the capillary is affected by the beating heart. The blood entering the capillary from the arteriole end does so under greater pressure than the blood leaving the capillary. The further away from the heart action, the lower the hydrostatic pressure.

9. B Many chemical reactions are involved in the clotting reaction. The sequence shown in Figure 13.19 summarizes the main clotting reactions.

10. C Carbon dioxide is converted by an enzyme into carbonic acid, which quickly ionizes. A buffer system neutralizes the excess hydrogen ions by forming acid hemoglobin, a weak acid.

Incorrect Choices

A Plasma carries only 5% of the CO_2.

B Thirty-five percent of the CO_2 is carried by the hemoglobin molecule as carbaminohemoglobin.

D Myoglobin is a protein in muscles that stores oxygen.

E The carbonate ion does not enter into CO_2 transport.

11. A Oxyhemoglobin is formed when oxygen combines with hemoglobin. This occurs in the lungs.

Incorrect Choices

B, C, D, E There is no gas exchange with the external environment in these organs.

12. **B** Since the room was poorly ventilated, the concentration of CO_2 was probably very high. The partial pressures of CO_2 in the lungs and in the room were very similar; therefore the exchange of gas was not efficient—CO_2 increased in the bloodstream. Chemoreceptors detecting the high partial pressure in the bloodstream stimulated the respiratory centers.

Incorrect Choices
A The hot air would make it difficult to cool the body but would not diminish or increase the O_2 content of the air.
C If O_2 increased, the respiratory centers would not be stimulated, but rather would be depressed.
D Kidney function would have to be speeded up for the removal of excess water.
E A state of boredom would not increase the breathing rate because the consumption of O_2, due to the lack of mental or physical exertion, would decrease.

13. **B** Diffusion occurs across thin, moist membranes of the alveoli. If they thicken, the process is greatly slowed and CO_2 increases as O_2 decreases in the bloodstream.

Incorrect Choices
A If there is any effect on blood, it will be an increase in the number of red blood cells to compensate for the lack of O_2 in the bloodstream.
C CO_2 concentration increases because of the failure of the diffusion process.
D N_2 concentrations remain constant since N_2 does not diffuse into the bloodstream.
E If blood volume increases, it is in response to low O_2 concentration and high CO_2 concentration.

14. **E** Refer to Figure 13.21.

15. **D** Red blood cells do not leave the bloodstream under normal conditions; therefore there is no reabsorption.

Incorrect Choices
A, E The pH is maintained through secretion of hydrogen and potassium ions.
B, C Sodium is one of the most important electrolytes regulated through reabsorption of the filtrate. Without sodium, nerve impulses would not be possible.

16. **D** The loop of Henle is an area specialized in the absorption of sodium (active pumps).

Incorrect Choices
A The glomerulus is another name for the capillary tuft found in Bowman's capsule.
B The PCT are located in the region of the nephron closest to Bowman's capsule. All of the nutrient and 90% of the water reabsorption occurs in this region.
C Bowman's capsule is the area in which filtration occurs.
E Collecting tubules are located in the nephron and lead into the pelvis of the kidney.

17. D

18. A

19. E

20. C

21. E

Immunology

Special Vocabulary for Immunology

Acquired immune deficiency syndrome a collection of symptoms due to loss of helper T-cell function, believed to be caused by infection of a retrovirus

Antibodies immunoglobulins that are produced by lymphocytes and function by binding to antigens

Antigen any foreign molecule that stimulates the production of antibodies

B-cell lymphocyte an antibody-producing cell

Cell-mediated immunity immune responses using T cells and not requiring antibody

Complement plasma proteins that can destroy the antigen when activated by antibodies bound to antigen

Cytotoxic T cells specifically sensitized T lymphocytes that are able to kill target cells that have specific antigens attached to their surface

Fab region the variable portion of an antibody that binds antigen

Hybridomas fused cells of B lymphocytes and B-cell lymphocyte myeloma cells that produce monoclonal antibodies

Interferon a cellular substance that inhibits replication of viruses in virus-infected cells

Lymphocyte a nucleated cell produced by the stem cells in the bone marrow; lymphocytes play an important role in providing immunity

Macrophages phagocytic cells that seek out foreign cells and engulf them

Memory cell a type of B cell that stores instructions to build antibodies

Monoclonal antibodies specific antibodies (clones) produced from genetically engineered cells

Phagocytes macrophages and neurophils that ingest foreign cells and particles

Plasma cells antibody-producing B cells

T-cell lymphocyte a lymphocyte that matures in the thymus

Nonspecific Immunity

The purpose of the immune system is to protect the body against infection. Skin is the body's first line of defense against infection. As long as the skin is unbroken, no disease-causing organisms (pathogens) can enter the body. The enzyme lysozyme in tears and saliva and the stomach's hydrochloric acid destroy harmful bacteria. The mucous membranes that line the walls of the trachea and the lungs also prevent bacteria from entering cells. The gastrointestinal tract has many ways of keeping harmful bacteria confined, preventing their entry into body cells and tissues. All of these ways of preventing infection are called *nonspecific immunity* because they nonspecifically prevent any germs from entering the body.

Cells of the Immune System

The immune system is able to recognize and act upon foreign proteins or invading microorganisms that enter the body. Foreign substances that cause the immune system to react are called *antigens*. The immune system is capable of recognizing and acting upon at least 100 million different antigens. Antigens are usually proteins, glycoproteins, or carbohydrates. Fats are generally not antigenic.

The *immune system* is loosely organized and dispersed throughout the body. It consists of lymphoid tissue, fluid called lymph, and special cells (white blood cells) that act against foreign protein that invades the body. It is intimately involved with the blood circulatory system.

The cells of the immune system collect in the lymphoid tissues. These tissues are the adenoids and tonsils in the head region, the thymus gland in the thoracic cavity, bone marrow in the center of long bones, the spleen just below the heart, lymph nodes in regions under the arms and in the groin, and Peyer's patches in the small intestine.

When nonspecific immunity methods break down, infection results. Infection means that harmful organisms such as viruses, bacteria, protozoa, fungi, worm parasites, or foreign proteins have entered the body and begin to reside where they do not belong. The cells of the immune system are called into action in response to invading pathogens or proteins.

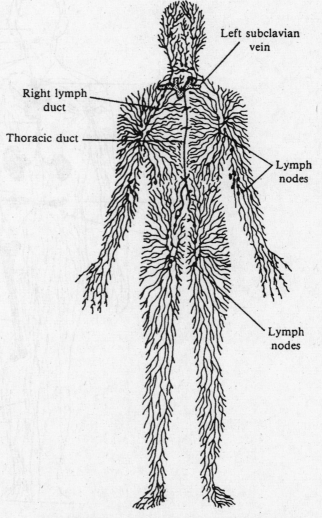

Figure 13.23   Network of lymphatic vessels

Phagocytes

Two types of cells that phagocitize (engulf) invading microorganisms provide the human body with a very effective first line of defense. They are macrophages and neutrophils.

Huge amoeboid *macrophages* and smaller *neutrophils* are white blood cells that are specialized for attacking invading microorganisms. Some macrophages remain fixed in the spleen and lymph nodes (Figure 13.23), where they ingest the microorganism invaders that pass their way. Other macrophages and the neutrophils patrol the body in search of invaders. Circulating in the blood are about 20 complement proteins that are called into action at different times during the immune response. Complement assists phygocytes in the destruction of foreign cells by coating them. Phagocytes are thus stimulated to ingest the complement-coated pathogens. Chemicals released from damaged blood platelets attract the macrophages and neutrophils, which then congregate at the site of infection, engulfing the foreign bacteria. When macrophages ingest large numbers of bacteria, these phagocytes are killed by bacterial toxins and a pileup of their dead bodies forms *pus*.

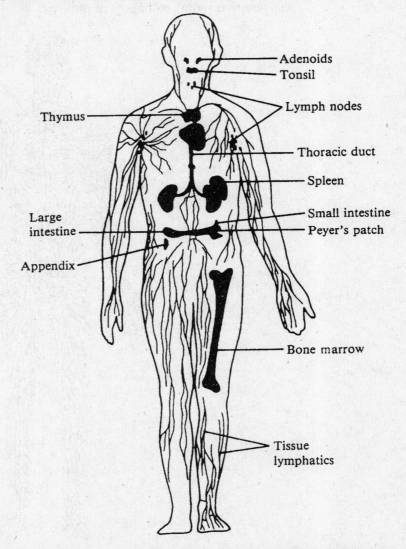

Figure 13.24   The immune system: location of lymphoid tissue

Lymphocytes

The second line of defense in the body is carried out by a kind of white blood cell known as a lymphocyte. There are two kinds of lymphocytes, the B-cell lymphocyte and the T-cell lymphocyte. Both of these cells carry out a complex series of events that is known as the *immune response*.

The human immune response consists of two major biochemical events: the *humoral immune response* and the *cell-mediated response*. The humoral immune response involves the production of protein molecules called antibodies, while the cell-mediated response evokes direct cellular action. Each of these responses is driven by a specific type of lymphocyte, either the B-cell or the T-cell lymphocyte.

Lymphocytes are produced by the stem cells which reside in the red marrow of long bones (Figure 13.24). Stem cells produce all of the blood cells, and by some mechanism differentiation takes place, effectively producing cells that carry out specific functions. The B-cell lymphocytes mature in the bone marrow (thus the name B cell), and the T-cell lymphocytes mature in the thymus (which accounts for the T designation). In the inactive state, the B cells and T cells look much alike, but they differ markedly when activated. When activated, B cells exhibit a very rough endoplasmic reticulum, indicating that an enormous number of ribosomes are attached to these membranes. Activated T cells, in contrast, have large concentrations of free ribosomes in the cytoplasm.

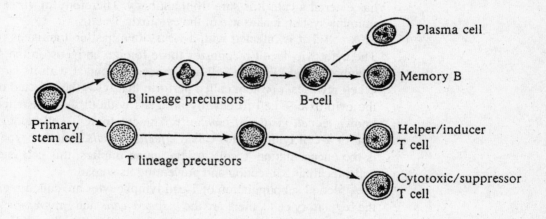

Figure 13.25   Development of B-cell and T-cell lymphocytes from a primary stem cell

Primary Immune Response

B-cell lymphocytes carry specific antigen-recognition proteins on the outer surface of their cell membranes. Each cell is a specialist carrying only one kind of recognition protein. When a virgin B-cell lymphocyte meets with a matching antigen, the B cell becomes activated and the antigen becomes attached to the recognition site on the membrane of the B cell. This B cell with its attached antigen becomes much larger and goes through mitosis and differentiation, producing two new kinds of cells—*plasma cells* and *memory cells* (Figure 13.25).

The *plasma cells* become active and produce an antibody that matches the captured antigen. The antibody production by these plasma cells is enormous; they produce about 2,000 molecules of antibody per minute. The antibodies attach to and immobilize the invading antigen. Now the phagocytes—macrophages and neutrophils—engulf the antigen-antibody complexes. This early reaction of the immune system is called the *primary immune response*.

Secondary Immune Response

The *memory cells* store all the instructions necessary to build the antibodies that the short-lived plasma cells make during the primary immune response. Memory cells have very long lives and can remain in existence for many years. Should a second attack by the same antigen occur, the memory cells go into action, producing huge numbers of antibodies through very rapid reproduction of plasma cells. The plasma cells are produced in even greater numbers than during the primary immune response. The secondary immune response is often so rapid that a person may not know that infection has taken place. The secondary immune response is far more effective than the primary immune response.

The Work of the T-cell Lymphocyte

Antibodies are effective against most pathogenic microviruses. Antibodies can inactivate viruses and prevent them from entering cells. However, once a virus has entered a cell, it is safe from antibody. Therefore for virus-infected cells, the immune system makes use of the cytotoxic T cell.

A cell that is infected with a virus often has viral antigens on its cell surface. The cytotoxic T cell recognizes those foreign antigens on the surface of the cell membrane and kills that cell as it comes in contact with its target. The cytotoxic T cell releases a protein called perforin that lyses the infected cell. The ability of the cytotoxic T cell to destroy antigens without the assistance of antibodies is known as *cell-mediated immunity*. The cytotoxic T cell belongs to a subpopulation of T-cell lymphocytes called *effector T cells*. Another type of effector T cell is the encapsulation T cell. As the name implies, this cell encapsulates foreign cells, localizing infection and preventing its spread.

A second subpopulation of T-cell lymphocytes are called *regulatory T cells*. Of the regulatory cells, there are the *helper T cells* and *suppressor T cells*. The helper T cells, also called the *inducer cells*, assist other T cells and B cells in responding to antigens. The helper T cells also activate some macrophages.

B-cell lymphocytes have antibodies interlaced in their cell membranes in which the antigen-bind region (of the antibody) faces outward and the fragment-crystallizable region is embedded in the membrane. In order to synthesize and secrete antibody, B cells must receive two signals. The first signal occurs when antigen binds to the Fab region of the membrane-bound antibody. The second signal occurs when a soluble protein secreted by the helper T cells alerts the B cells. Once activated by these two signals, the B cells will enlarge, divide, and secrete antibody.

Recognition of Self by the Immune System

The task of the immune system is to destroy foreign antigens but not antigens or cells that belong in the body where the immune system resides. Therefore, the immune system must be able to distinguish between self and nonself. To recognize self is the work of T-cell lymphocytes.

In the body there are two major groups of self antigens, and they differ in tissue allocation and function. The class I *histocompatibility antigens* (self antigens) are present on all body cells except the red blood cells. These antigens are responsible for the rejection of tissue grafts. Class II antigens are on the surfaces of T cells and B cells and some macrophages. Class II antigens are found only on cells in the immune system and are needed for the response of helper T cells. Helper T cells are not activated by class I self antigens, having been made tolerant to them in fetal life. Thus they cannot stimulate B cells to make antibodies against normal self tissues nor can they help cytotoxic T cells kill the self. T cells can be activated by foreign antigen in association with class II self antigens as they bind to the surface of a macrophage. Helper T cells prevent the accidental destruction of body cells by the immune system.

About Antibodies

Antibodies are protein molecules (immunoglobins) that are produced by the plasma cells of the B-cell lymphocyte lineage. Antibodies have a four-part structure composed of two identical long (heavy) chains of 450 amino acid molecules and two identical short chains (light) of 220 amino acids. Thus an antibody molecule consists of 1,340 amino acids (Figure 13.26). The heavy chains have a hingelike region that gives flexibility to the molecule. The chains are bound together by disulfide linkages.

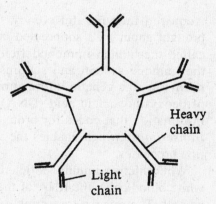

Figure 13.26   Structure of an antibody molecule

Antibody molecules have two functional regions. One end of the molecule binds to antigen, the other end to effector cells and proteins. The fragment antigen-binding region (Fab), which binds to a specific antigen, is the variable part of the molecule. Each antibody molecule has two identical Fab regions. There are as many antigen-binding regions as there are antigens.

The other functional region is called the Fc (fragment-crystallizable) region. This end of the molecule binds to specific receptors on effector cells or on other plasma proteins. There are only five kinds of regions, one for each of the five kinds of antibodies (see Table 13.12).

When antibodies bind to a specific pathogen, such as a bacterial cell, the cell is marked for destruction. A complement protein forms a bridge between the two antibodies. The combination of antibodies and complement causes rupturing of the bacterial cell membrane resulting in the lysis of the cell.

TABLE 13.12   ANTIBODIES AND THEIR FUNCTIONS

Kinds of Antibodies	Functions
IgA	This is a dimer, a double immunoglobin molecule that acts as a single molecule (monomer) in the blood. It is found in mother's milk, the intestinal tract, respiratory mucus, saliva, and tears. It functions as a first line of defense in the secondary immune response.
IgD	This is a monomer (a single immunoglobin) occurring as a receptor site on the surface of B cells.
IgE	This is a monomer that acts in allergic responses, releasing histamine.
IgG	This is a single immunoglobin (monomer) appearing in blood during late primary infections and throughout secondary infections.
IgM	This is composed of five immunoglobins and is thus called a pentamer. It functions in the blood during the primary immune response.

AIDS and the Inoperative Immune System

Acquired immune deficiency syndrome (AIDS) is an infectious disease brought about by a suppressed immune system. It is known that a *retrovirus*, called the human immunodeficiency virus (HIV), escapes the safeguards of the immune system and suppresses the normal activity of the T cells. In a retrovirus, the genetic information is coded in ribonucleic acid (RNA) instead of deoxyribonucleic acid (DNA). The usual role of RNA is to carry genetic information that codes for proteins. Retroviruses produce the enzyme *reverse transcriptase*, which enables the virus to make more of itself through replication processes.

When human immunodeficiency virus infects a T4 lymphocyte (a type of white blood cell), the RNA of the retrovirus takes over the DNA of the invaded cell. The reverse transcriptase makes the DNA of the infected white blood cell follow the coded instructions contained in the viral RNA. In this way, HIV viruses are produced very rapidly within the infected T4 white blood cell. When the infected white blood cell can hold no more of the newly replicated HIV viruses, the cell membrane ruptures, releasing great numbers of the HIV virus into the bloodstream. The process continues: invading of a T4 lymphocyte—replication of the virus—release of numerous viruses into the blood.

Once deactivated, the helper T cells cannot stimulate reproduction of B cells, which are the progenitors of the plasma (antibody-producing) cells and/or the memory cells. In effect, AIDS shuts down a person's entire immune system.

Until recently, the condition of full-blown AIDS meant certain death. Today, however, medications have been developed that prolong the lives of those suffering from AIDS. People with AIDS usually die from the opportunistic diseases with which they are infected because their immune systems do not work. One such opportunistic disease is *Pneumocystis carinii* pneumonia (PCP). Drugs that prevent and work against the killer PCP have been developed, thus reducing the death rate of AIDS patients.

Model Questions Related to Immunology

1. The most important mediators of nonspecific immunity are the
   (A) stem cells and the T-cell lymphocytes
   (B) macrophages and the T-cell lymphocytes
   (C) macrophages and the neutrophils
   (D) neutrophils and the B-cell lymphocytes
   (E) leucocytes and lymphocytes

2. A true statement about antibodies is that
   (A) they are protein molecules produced by lymphocytes
   (B) they bind to invading microorganisms
   (C) together with macrophages and blood proteins kill or incapacitate pathogens
   (D) all of the above
   (E) none of the above

3. A chemical produced by a living organism that kills or inhibits bacterial growth is known as an
   (A) antiseptic
   (B) antibiotic
   (C) antibody
   (D) antigen
   (E) anticodon

4. An antibody is best described as
   (A) a particle
   (B) an organism
   (C) a molecule
   (D) a fragment
   (E) a pathogen

5. Complement destroys bacterial cells by
   (A) preventing reproduction by binary fission
   (B) breaking apart the singular circular chromosome
   (C) depositing antigen on the bacterial cell surface
   (D) stimulating reproduction of the helper T cells
   (E) perforating the cell wall and the cell membrane

6. Antibodies are not effective against
   (A) intracellular viruses
   (B) extracellular viruses
   (C) retroviruses
   (D) all of the above
   (E) none of the above

7. Effector cells are a subpopulation of
   (A) T-cell lymphocytes
   (B) B-cell lymphocytes
   (C) helper T cells
   (D) suppressor T cells
   (E) regulatory T cells

8. The osmotic gradient between the internal and external environments of bacterial cells can be destroyed by
   (A) leukocytes
   (B) stem cells
   (C) antibodies
   (D) antigens
   (E) complement

9. The light and heavy chains of antibodies are bound together by
   (A) weak hydrogen bonds
   (B) strong amide bonds
   (C) carboxyl groups
   (D) disulfide bonds
   (E) dipeptide bonds

10. Interferons are proteins that act against
    (A) certain types of bacteria
    (B) natural killer (NK) cells
    (C) virally infected cells
    (D) hybridomas
    (E) B-cell lymphocytes

11. T-cell lymphocytes are able to distinguish between self and nonself by means of
    (A) antibody levels in the blood
    (B) histocompatibility antigens
    (C) enzyme-substrate complex
    (D) macrophage-antigen complex
    (E) plasma cell levels in the blood

12. Myeloid stem cells produce
    (A) cytotoxic T cells
    (B) memory cells
    (C) B-cell lymphocytes
    (D) inducer cells
    (E) macrophages

13. Monoclonal antibodies are most closely associated with
    (A) hybridomas
    (B) leukocytes
    (C) neutrophils
    (D) eosinophils
    (E) interferons

14. AIDS is a collection of diseases that infect the victim because his/her
    (A) antibody titer is low
    (B) immune system shuts down
    (C) genes manifest a predisposition
    (D) tolerance level for infection is low
    (E) macrophages do not work

15. If the primary immune response failed to generate memory cells, then
    (A) the initial work of plasma cells would be inhibited
    (B) the T-cell lymphocytes would not produce the inducer cells
    (C) the secondary immune response would generate too few antibodies
    (D) the T-cell lymphocytes could not distinguish self antigens
    (E) the DNA coding in the stem cells would mutate

Answers and Explanations to Model Questions: Immunology

1. **C** Macrophages and neutrophils are phagocytic cells that move to sites of infection and ingest immobilized bacterial cells. These cells are "generalists" and do not act specifically against a particular species of bacterium.

   Incorrect Choices
   A Stem cells are the progenitors of blood cells. T-cell lymphocytes act specifically in the immune system as helpers or suppressors of other lymphocyte groups.
   B, D, E See explanation above.

2. **D** All of these statements are true about antibodies.

   Incorrect Choices
   E See explanation above.

3. **B** An antibiotic is an organic compound produced by a living microorganism that kills or inhibits bacterial growth.

   Incorrect Choices
   A An antiseptic is a chemical manufactured to destroy or inhibit bacterial growth on the human body.
   C An antibody is a protein produced by B-cell lymphocytes that attacks foreign bodies.
   D An antigen is any substance which stimulates the immune system.
   E An anticodon is a triplet base at one end of a tRNA molecule.

4. **C** An antibody is a protein molecule that contains more than 1,300 amino acids.

   Incorrect Choices
   A, B, D, E Incorrect without explanation.

5. **E** Complement is a group of blood proteins that function in immune responses. Complement digests the cell wall and membrane of bacterial cells.

Incorrect Choices
A, B, C, D Incorrect without explanation.

6. **A** Antibodies are not effective against viruses once they have entered a cell.

   Incorrect Choices
   B, C Antibodies can be effective against viruses before they enter cells.
   D, E Incorrect without explanation.

7. **A** Effector cells are a subpopulation of T-cell lymphocytes. There are two types of effector cells: cytotoxic T cells and encapsulating cells.

   Incorrect Choices
   B, C, D, E Incorrect without explanation.

8. **E** The osmotic gradient between the internal and external environments of bacterial cells can be destroyed by complement, which destroys the cell membranes.

   Incorrect Choices
   A, B, C, D Incorrect without explanation.

9. **D** The light and heavy chains of antibodies are bound together by disulfide bonds.

   Incorrect Choices
   A, B, C, E Incorrect without explanation.

10. **C** Interferons are proteins that act against virally infected cells.

    Incorrect Choices
    A Interferons are produced by cells but are ineffective against bacteria. Interferons are produced by cells that are infected with viruses and inhibit the reproduction of intracellular viruses.
    B Interferons are not cells, but protein molecules.
    D Hybridomas are genetically engineered cells that produce a single line of clone antibodies.
    E Interferons are not cells.

11. **B** Histocompatibility antigens are "self" antigens and assist T-cell lymphocytes in recognizing the body's own cells and tissue. Otherwise, the immune system would destroy its own cells and tissues.

    Incorrect Choices
    A, C, D, E Incorrect without explanation.

12. **E** Myeloid stem cells produce macrophages.

    Incorrect Choices
    A Cytotoxic T cells are produced from T lineage precursors, which come from the multipotential stem cell.
    B Memory cells are produced by B lineage precursors, which come from the multipotential stem cell.

C See B above.
D See A above.

13. A Monoclonal antibodies are most closely associated with hybridomas, which are fused cells of B-cell lymphocytes and B-cell tumor cells that are genetically engineered to produce a single line of antibodies.

Incorrect Choices

B, C, D are white blood cells and do not produce antibodies.
E Interferons are cell substances that inhibit the replication of intracellular viruses.

14. B AIDS is characterized by the clamping down of the immune system because of malfunctioning of the helper T-cell lymphocytes. It is believed that a retrovirus HTLV III is responsible for rendering T cells inactive.

Incorrect Choices

A, C, D, E Incorrect without explanation.

15. C The memory cells are B cells that store instructions about the structure of antibodies. Enormous numbers of antibodies are produced during the secondary immune response because memory cells are capable of synthesizing antibodies at fantastic rates. Without memory cells, the synthesis and secretion of antibodies would not be possible.

Incorrect Choices

A, B, D, E Incorrect without explanation.

Human Reproduction and Development

Fertilization

As previously mentioned, the meiotic process of oogenesis produces an egg cell, or female gamete, while spermatogenesis produces a sperm cell, or male gamete. The union of one female and one male gamete (or the nuclei of the gametes) is called *fertilization*. In humans, fertilization occurs in the fallopian tubes when one sperm cell is able to penetrate the dense layer of cells that surround the egg cell. The impinging of a sperm cell on the egg cell membrane initiates a series of events that prevent the penetration of other sperm cells. A fertilization membrane forms around the egg cell, the sperm cell is drawn into the egg, and the nuclei of the two cells fuse. Fertilization has then occurred.

The *zygote*, or fertilized egg, undergoes a series of mitotic divisions known as *cleavage* (described in the next section). When the embryo, consisting of 32 cells, arrives in the uterus, implantation commences.

Extraembryonic membranes—yolk sac, allantois, amnion, and chorion—develop (Figure 13.27). The *yolk sac* gives rise to blood cells. The *allantois* serves as the tissue for blood vessel formation. The *amnion* forms a fluid-filled sac that provides the embryo with protection and moisture. The *chorion* contributes to the formation of the placenta. The *placenta* is the organ that provides for the exchange of materials between the embryo and the mother.

Gastrulation (described on pages 309-310) continues in the uterus.

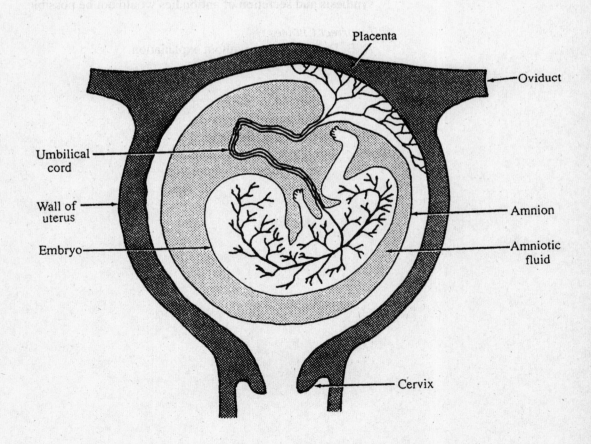

Figure 13.27 Human embryo in uterus

Mitosis and Cleavage

Mitosis plays an important role in the process of cleavage. After fertilization, the single-celled zygote undergoes many changes until an organism is ultimately formed. Cleavage is a series of cell divisions, without growth, that the fertilized egg undergoes (Figure 13.28).

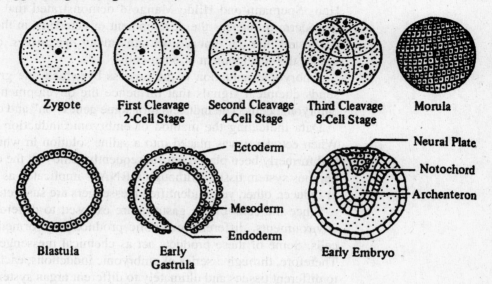

| Zygote | First Cleavage 2-Cell Stage | Second Cleavage 4-Cell Stage | Third Cleavage 8-Cell Stage | Morula |

Blastula   Early Gastrula — Ectoderm, Mesoderm, Endoderm   Early Embryo — Neural Plate, Notochord, Archenteron

Figure 13.28   Stages of cleavage in the frog embryo

A single cell becomes two, two become four, four become eight, and so on, geometrically. Eventually the cells are arranged in a hollow ball known as the *blastula*. Even though the nucleus of each daughter cell divides equally, the cytoplasm divides unequally. The cells are not all of the same size or of the same cytoplasmic composition. The explanation for this is that the egg cell does not contain an equal distribution of cytoplasmic or nutritive material. Within the egg cell there are two recognizable poles: animal and vegetal. The vegetal pole contains most of the nutritive material.

The first two cleavage divisions occur through the poles, but the next division occurs perpendicularly to the poles. The cells at the vegetal end, which are usually heavier and larger, divide at a slower rate than the cells at the animal end. Although the cells of mammals have virtually no stored yolk, they still follow the same pattern of cleavage as do cells of organisms that contain yolk.

Differentiation

Hans Driesch showed that the blastomeres from the first and second cleavages of sea urchin eggs, when separated, are capable of developing into normal embryos. When the separation occurs at the eight-cell stage, however, the blastomeres do not develop into normal embryos. The eight cells from the third cleavage do not contain equal distribution of cytoplasmic material. Experimentation on the green alga *Acetabularia* and on frog embryo nuclei transplants reveals that there is an interaction between the nucleus and cytoplasm of cells. Although the cleavage cells are genetically alike, they differ biochemically. The cells differ in the concentration of proteins and mRNA. This sets the stage for differentiation. Complex cellular movements result in the formation of a two-layered, cup-shaped gastrula.

The heavy cells at the vegetal end of the blastula begin to invaginate during a process called *gastrulation*. Although the pattern of *gastrulation* varies in different organisms, the result is the same. The invagination proceeds to distribute the cells into two layers. A third layer of cells, the mesoderm, forms in the region of the dorsal lip.

The development of the mesoderm creates many cell and position differences. Hans Spemann and Hilde Mangold demonstrated that transplanted dorsal lip mesoderm influences the development of ectoderm in the nervous system. They suggested that the embryo contains many *organizers*, cells that influence and induce the development of other cells.

Embryonic induction is the process by which one group of embryonic cells sends chemical signals that influence the development of a second group of embryonic cells. The inducers turn some genes "on" and other genes "off."

Data indicating the method of embryonic induction were supplied by Nue. When ectoderm was placed into a saline solution in which dorsal lip mesoderm had formerly been placed but subsequently removed, the ectoderm developed into nervous system tissue. Although mRNA is implicated as the chemical messenger or inducer, other, yet unidentified messengers are suspected.

Since the cells of the gastrula are exposed to different external and internal environments, different metabolic products are formed by various clusters of cells. Some of these products act as chemical messengers on neighboring cells. Therefore, through a series of embryonic inductions, each layer of cells gives rise to different tissues and ultimately to different organ systems (Table 13.13).

TABLE 13.13   DERIVATIVES OF EMBRYONIC GERM LAYERS

Ectoderm	Mesoderm	Endoderm
Skin	Muscles	Digestive system
Nervous system	Skeleton	Pancreas
tissue	Gonads	Liver
Hair	Excretory system	Thyroid
Nails	Circulatory	Lungs
Eye lens	system	Bladder

Model Questions Related to Human Reproduction and Development

1. In the process of cleavage, the mesoderm appears during
   (A) the blastula stage
   (B) differentiation
   (C) the gastrula stage
   (D) the zygote stage
   (E) organ formation

2. Which structure does NOT develop from the ectoderm layer of embryonic cells?
   (A) Skin
   (B) Lens of the eye
   (C) Brain
   (D) Hair
   (E) Stomach

3. In a series of experiments performed by Spemann and Mangold, an embryo containing two nervous systems was produced by implanting the mesoderm from the dorsal lip into the gut region of the embryo. The term that explains this occurrence is
   (A) preformation
   (B) induction
   (C) differentiation
   (D) regeneration
   (E) parthenogenesis

4. What is the correct sequence of events in mammalian embryonic development after cleavage begins?
   (A) Differentiation, gastrulation, blastula formation, mesoderm formation
   (B) Gastrulation, blastula formation, mesoderm formation, differentiation
   (C) Blastula formation, mesoderm formation, gastrulation, differentiation
   (D) Mesoderm formation, differentiation, blastula formation, gastrulation
   (E) Blastula formation, gastrulation, mesoderm formation, differentiation

Base your answers to Questions 5 and 6 on the diagram below of the female reproductive system.

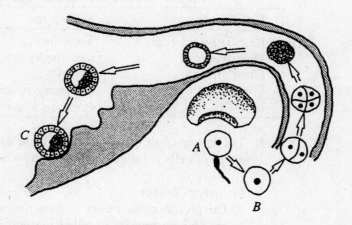

5. What process is occurring between A and B?
   (A) Fertilization
   (B) Spermatogenesis
   (C) Oogenesis
   (D) Cleavage
   (E) Ovulation

6. Where does the process shown at C normally occur?
   (A) Oviduct
   (B) Uterus
   (C) Vagina
   (D) Ovary
   (E) Follicle

Answers and Explanations to Model Questions: Human Reproduction and Development

1. **C** The stage of development in which invagination and the appearance of the mesoderm occur is termed gastrulation.

   Incorrect Choices
   A The blastula stage is the hollow stage of embryonic development.
   B Differentiation does not occur until after the mesoderm develops.
   D The primary germ layers do not appear during the zygote stage.
   E Organ formation is the result of specialization of tissue, which cannot occur until the third germ layer is present.

2. **E** The stomach is part of the digestive system, which develops from the endoderm.

   Incorrect Choices
   A, B, C, D See Table 13.14 below.

TABLE 13.14 DIFFERENTIATION OF PRIMARY TISSUE

Endoderm	Mesoderm	Ectoderm
Respiratory system	Skeleton	Nervous system
Digestive tract	Circulatory system	Skin
Gland: liver,	Muscles	
pancreas	Excretory system	
	Reproductive organs	

3. **B** The influence of one layer of cells on the development of another layer or group of cells is called embryonic induction.

   Incorrect Choices
   A The preformation theory implies that all the essential parts of an embryo are already present in either the egg or the sperm and that fertilization and time are required for growth.
   C Differentiation does not occur until the mesoderm develops.
   D The regrowth or replacement of a lost body part is termed regeneration.
   E Parthenogenesis is the development of an egg without fertilization by a sperm.

4. **E** A blastula is the hollow-ball stage of development. A gastrula is the cuplike stage of embryonic development in which two germ layers are formed. Mesoderm formation is the development of a third or middle germ layer. Differentiation is the specialization of tissues and organs.

   Incorrect Choices
   A, B, C, D In all of these choices, at least one stage is out of the proper sequence of mammalian development.

5. **A** The process occurring between A and B is fertilization, the union of egg and sperm cell.

Incorrect Choices
B Spermatogenesis is the production of sperm cells.
C Oogenesis is the formation of an egg cell.
D Cleavage is the process by which a single fertilized egg cell becomes a multicellular embryo.
E Ovulation is the release of an egg cell from the ovary.

6. B The process shown at C usually occurs in the uterus. This process is implantation.

Incorrect Choices
A Fertilization occurs in the oviduct.
C The fetus is expelled through the vagina.
D An egg cell is produced in the ovary.
E The follicle is the area within the ovary in which the egg develops.

Behavior

The nervous system is an integral aspect of the behavior of an organism. The action or series of actions of an organism in response to a stimulus in the environment is known as *behavior*. Behavior may be *inborn* or *learned*. Simple reflexes and instincts are inborn, while conditioned reflexes and habits are learned. The response of an organism to environmental changes allows it to communicate, recognize, cooperate and mate with others of its own species, defend territory, avoid danger, rear its young, and perform all the activities essential for its own survival and ultimately the survival of its species.

Instinctive or Innate Behavior

The response of an organism to a *stimulus* may trigger a set of complex activities known collectively as a fixed action pattern (FAP). An FAP is a prewired genetic program, a series of actions in response to a stimulus. The stimulus, or the factor that evokes or *releases* instinctive patterns, is called a *releaser*. This type of behavior is often called *instinctive* or *innate behavior*. Instinctive behavior is divided into two stages, appetitive and consummatory.

The *appetitive stage* involves a series of actions, which may be extremely variable, that enables an organism to secure and obtain food. The *consummatory stage*, the taking of food into the digestive system, depends upon FAP.

Learning

Learning is a modification of behavior. There are three types of learning: habituation, classical conditioning, and operant conditioning. *Habituation* is the loss of sensitivity to a stimulus. *Classical conditioning* is the training of an organism to respond to a substitute stimulus. The response of Pavlov's dogs to flashing light is an example of classical conditioning. *Operant conditioning* is trial-and-error learning.

The relationship of learning and instinctive or innate behavior is seen in *imprinting*, a special kind of learning. The response of an organism to a stimulus is instinctive. However, the response must be directed to an object within a critical period. Thus geese will follow the first object, such as another goose or a human being, that they observe to move within the first two days of their hatching. Movement is the stimulus. The following of a specific object (for example, a parent), however, is not innate.

Foraging behavior, competitive interactions (territoriality, dominance hierarchies), courtship behavior, altruistic behavior, and recognition and communication behavior are essential for the survival of a species. Behavior is an important adaptation for survival on the population and community level.

B. F. Skinner (learning behavior of rats), Wolfgang Kohler (insight into learning activities), Konrad Lorenz (imprinting), Karl von Frisch (bee behavior), and Ivan Pavlov (conditioning) are some of the individuals who have contributed to the study of animal behavior.

Model Questions Related to Behavior

1. The viceroy and monarch butterflies look alike. However, the monarch contains a toxic substance that causes blue jays to regurgitate. After a few experiences of eating a monarch, a blue jay will no longer capture either type of butterfly. This behavior is an example of
   (A) imprinting
   (B) classical conditioning
   (C) habituation
   (D) operant conditioning
   (E) instinct

2. Salmon will return to the same stream in which they were hatched. It has been determined that they are able to follow a chemical scent peculiar to the stream of their birth. This behavior is an example of
   (A) imprinting
   (B) classical conditioning
   (C) habituation
   (D) operant conditioning
   (E) learning

3. Belding ground squirrels are vulnerable to predators when they are foraging above ground. If a predator approaches, one of the squirrels will emit a high-pitched alarm call, alerting the others but putting itself in jeopardy of being discovered. This is an example of
   (A) imprinting
   (B) territoriality
   (C) conditioning
   (D) kin selection
   (E) monogamy

Answers and Explanations to Model Questions: Behavior

1. D Operant conditioning is trial-and-error learning. After a few unpleasant experiences the blue jay learned not to eat either viceroy or monarch butterflies.

   Incorrect Choices
   A Imprinting is a response that involves both learning and instinctive behavior.
   B Classical conditioning is "training" an organism to respond to a substitute stimulus.
   C Habituation is the loss of sensitivity to a stimulus.
   E Instinct is an innate response to a stimulus.

2. **A** Imprinting is a response that involves both learning and instinctive behavior.

   Incorrect Choices
   B Classical conditioning is "training" an organism to respond to a substitute stimulus
   C Habituation is the loss of sensitivity to a stimulus.
   D Operant conditioning is trial-and-error learning.
   E Learning is a modification of behavior.

3. **D** Kin selection is a concept in altruistic behavior. Altruistic behavior is a response that may endanger a particular individual but save another of the species. This behavior is often manifest among organisms that are genetically related to each other.

   Incorrect Choices
   A Imprinting is a response that involves both learning and instinctive behavior.
   B Territoriality is the defense of a specific area or territory.
   C Conditioning is training an organism to respond to a specific stimulus.
   E Monogamy is a type of behavior in which an organism remains with one mate for a period of time.

CHAPTER FOURTEEN
Modern Concepts of Ecology

Capsule Concept

Ecology is a biological science that explores the relationship between living things and their physical environments. The ecological unit of structure and function is the *ecosystem*, which includes plants and animals together with the nonliving environment.

Of greater significance than the heterotrophs, autotrophs, and decomposers is the flow of energy through an ecosystem. Energy takes many forms and can be changed from one form to another. A principle basic to ecology is that matter is cycled, but energy is lost, in ecosystems.

Vocabulary

Biomass   the total amount of living organisms at each level in a food web

Biome   a geographical area (or major ecologic community) distinguished by climax plants and animals and having an overall type of climate

Biosphere   the part of the earth in which living forms exist

Climax community   a stable, mature community

Consumer   an organism that cannot produce its own food

Ecological niche   a feeding pattern

Food chain   an arrangement of the organisms of an ecologic community according to the order of predation in which each uses the next, usually lower, member as a food source

Food web   the totality of interacting food chains in an ecologic community

Producer   a species that can use inorganic materials to synthesize organic nutrients

Succession   the orderly process in which one biotic community is replaced by another; a unidirectional change in the composition of an ecosystem as the available competing organisms, especially the plants, respond to the environment and modify it

Nutritional Relationships

Four types of nutritional relationships between species are defined below.

Commensalism   a relationship between two kinds of species in which one species obtains food or other benefits from the other species without damaging or benefiting that other species.

Mutualism   a relationship between two kinds of species in which each provides benefit for the other

Parasitism   a relationship in which one species lives inside or on another species, usually doing harm to the host

Symbiosis   a nutritional relationship in which two dissimilar species live together

Community and Succession

All living organisms interact with both the living and nonliving environments. The study of life patterns as related to the environment is called *ecology*. The physical area in which the interacting organisms live is called the *environment*. The *ecosystem* consists of abiotic (nonliving) and biotic (living) elements.

A *community* is a group of interacting species. In each community, a basic trophic (food) relationship exists among organisms. Autotrophs provide the energy for the maintenance of the entire community. They are the producers that support many consumers: primary consumers (herbivores) and secondary or tertiary consumers (carnivores). Each species occupies a particular niche (position in the trophic levels). A pyramid is used to show the energy flow in a community or ecosystem.

At each upward trophic level, the amount of available energy decreases. The producers can utilize only about 6% of the total radiant energy available because of dissipation of light into space. Nevertheless, producers are very efficient, converting 95% of the radiant energy they utilize into chemical energy. Even though green plants require energy to live, about 50% of the energy they make is left over for the herbivores. In the process of cellular respiration, much of the herbivores' energy is lost as heat, thereby reducing the amount of energy available for the next trophic level. The energy loss continues at each ascending level of the pyramid. Ecologists use the pyramid to describe levels of biomass distribution and to depict the numbers of individuals at each trophic level.

In addition to the producers and consumers, each community possesses decomposers. These are saprophytes: bacteria and fungi. Saprophytic organisms decompose dead organic material, thus returning the valuable nutrients, minerals, and elements to the environment.

A community is often named after its dominant form of vegetation. A gradual succession of events occurs in a region until its biological potential is reached and one form of vegetation predominates. *Succession*, the orderly change of plants and animals in a community, is visible on two levels: primary and secondary. Primary succession begins in a place devoid of soil. Once soil is formed through weathering and lichen activity, mosses inhabit the area. The mosses are followed by grasses and ferns, which precede shrubs and trees. Each serial stage enriches the environment, enabling another group of plants to become established. The final form of vegetation in a region, the *climax community*, is a stable community in which one or two plant species predominate. For example, a beech-maple forest is a climax community. Abiotic factors, such as temperature, precipitation, and mineral content of the soil, influence the type of climax community. A large climax community in a geographic region of one kind of climate is called a *biome*.

Tables 14.1 and 14.2 show the important land and marine biomes.

TABLE 14.1 LAND BIOMES

Biome	Characteristics	Vegetation	Location
Tundra	Permanently frozen subsoil, frozen desert, swamps, ponds	Mosses and lichens	Alaska, Siberia
Taiga	Defrostable subsoil	Conifers (spruce, firs)	Canada
Deciduous forest	Moderate rainfall, cold winters, warm summers	Trees that shed leaves (oak, hickory)	Eastern United States
Tropical rain forest	Warmth with constant rain	Broadleaf trees that do not shed leaves, epiphytes	Central America, South America, Asia
Grassland	Low rainfall	Grasses	Rocky Mountain area
Desert	Small amount of rainfall	Cactus, xerophytes	Arizona

TABLE 14.2 MARINE BIOMES

Biome	Characteristics	Vegetation
Intertidal	Area alternately covered and exposed by tides	Algae and barnacles
Littoral	Continental shelf, muddy bottom	Eelgrass
Pelagic	Open sea	Phytoplankton, zooplankton
Benthic	Floor of the ocean	?

Although fairly stable, a climax community can be destroyed. As succession begins again, several serial stages may be skipped, in what is known as secondary succession. Fields, ponds, and lakes also undergo succession. Succession is a part of the natural order that may be accelerated or impeded by human beings. *Eutrophication* is the accumulation of organic matter in lakes due to overfertilization by minerals from land.

Population Size

If a population in a community is left unchecked, its size can increase astronomically. Population size is controlled by various factors: competition, size of the population itself, predation, environmental factors.

The competition among organisms of the same species is known as *intraspecific competition*. Intraspecific competition weeds out the least fit in a population. The competition between two different species is known as *interspecific competition*. The species that is best adapted to the environment will survive while the other species may perish or move into an empty niche.

An S-shaped (sigmoid) curve (Figure 14.1) is used to describe the pattern of growth of a population when organisms move into an empty niche. Unlimited resources allow for a rapid reproductive phase. Eventually, the *carrying capacity*, the number of individuals that can be supported by the environment, is reached. The population size may stabilize or fluctuate around the carrying capacity (Figure 14.2). When the birth rate equals the death rate, an equilibrium is attained.

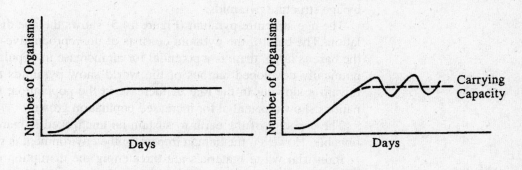

Figure 14.1 S-shaped growth curve   Figure 14.2 Fluctuation around carrying capacity

Stress due to overcrowding produces psychological and behavioral changes in a population. Difficulty in mating, high infant mortality, and inability to nurture the young reduce the population. Parasitism increases as resistance is weakened by stress. When the death rate exceeds the birth rate, the density (size) of the population is reduced.

Predation not only removes the very old, the very young, and the sick from the population but also reduces the population of the prey. If the predators do not keep the prey population in balance, the carrying capacity is exceeded and the prey may starve (Figure 14.3). Predator and prey are closely interdependent.

Populations of algae, annual plants, and insects with short life spans are controlled by seasonal and nutritional environmental changes. A J-shaped growth

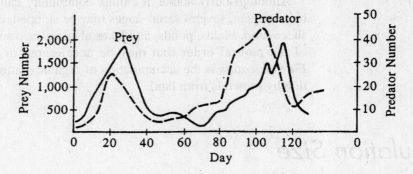

Figure 14.3 Predator-prey relationship

curve (Figure 14.4) shows a plunge in population as the reproductive potential declines because of environmental changes. The following year there is an exponential increase in the population.

Other relationships (*symbiosis*) exist between species. *Mutualism* is an association of two species in which both derive benefits: for example, the protozoans that live in the intestines of termites. *Commensalism* is a relationship between two species from which one species benefits and the other is unaffected: for example, barnacles attached to a whale.

The human population has followed a J-shaped curve for 4,000 years. In this time period, there have been three surges in population growth. Although the rate of increase has slowed, the human population is still in the third phase of the growth surge. A good indicator of future trends in population growth is furnished by age structure pyramids.

The age structure pyramid (Figure 14.5) shows the age distribution of a population. The base of the pyramid consists of nonreproductive-aged females. When the base is large, there is a potential for an increase in population. While the economically developed nations of the world show pyramids that indicate a trend toward a slowing in the rate of increase in the population, the lesser developed nations show a potential for increased population growth.

The capacity of the earth to sustain an unchecked human population is questionable. However, the human impact on the environment is well documented.

Industrial waste materials are threatening the disruption of the earth's greenhouse effect and have produced acid rain. Biomes such as the tropical rainforest are disappearing because of extensive lumbering. The agricultural base of some

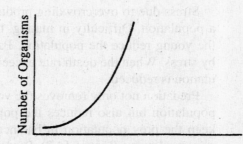

Figure 14.4 J-shaped growth curve

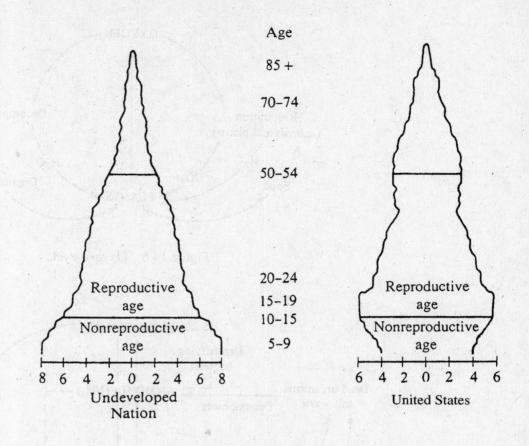

Figure 14.5   Age structure pyramids

nations has been eroded by the removal of soil due to the denuding of the countryside. Wetlands and farmlands are being used for home development. The transport and storage of toxic and nuclear wastes are making some areas uninhabitable. The future success of the human species may ultimately depend upon controlling environmental pollution.

Biogeochemical Cycles

No discussion of ecology is complete without consideration of the biogeochemical cycles. For communities to survive, there must be constant recycling of vital materials provided by the oxygen (Figure 14.6), carbon dioxide, and nitrogen (Figure 14.7) cycles.

In addition, the hydrologic (water) and sulfur cycles contribute to the conservation of vital materials in the environment. The water cycle involves the movement of water from the oceans to the atmosphere and back to the seas. Autotrophic bacteria (the sulfur bearers) cycle sulfur between green plants and the soil.

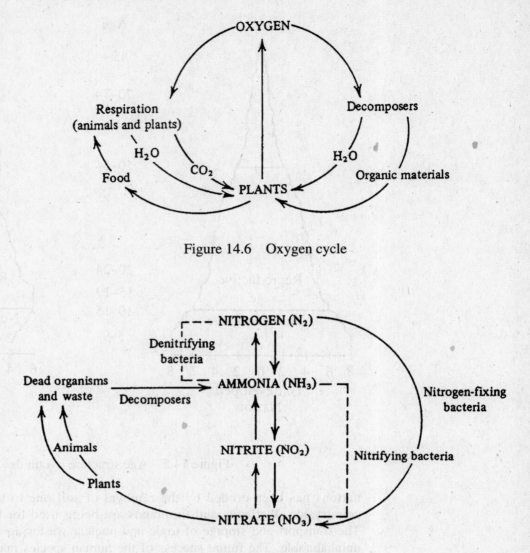

Figure 14.6   Oxygen cycle

Figure 14.7   Nitrogen cycle

The Ecology of Disease Transmission

Indoor Pollutants

Although environmental regulations have improved the quality of outdoor air, little attention has been paid to the quality of indoor air. Dr. Wayne Ott and his team of research scientists have been assessing the quality of indoor air. After carrying out more than 6,000 studies in individual homes, Ott has discovered 30 contaminants of indoor air. These include volatile organic compounds such as benzene and chloroform, concentrations of pesticides, and the presence of small particles measuring 10 microns or less.

The study points out that for small children, moderate concentrations of house dust is a major source of exposure to lead, cadmium, and other heavy metals. Among the organic pollutants found in homes are polychlorinated biphenyls. Carpets are persistent storage places for toxic compounds. Carpets also house dangerous bacteria and asthma-inducing allergens, such as dust mites, molds, and animal dander. The daily deep vacuuming of carpets and the removal of shoes before entering the house are recommended as effective means of reducing carpet pollution.

Outdoor Disease Vectors

Two deaths have been reported in the northeastern United States from infection by the *Hantavirus.* Nationwide, 106 deaths from this rodent-borne virus have occurred, mainly in the Southwest and New Mexico–Arizona area. In New York, the Hantavirus is carried by the white-footed mouse. Other species of outdoor rodents in the Southwest carry the virus in their feces and urine. Hantavirus causes hemorrhagic fevers in which bleeding occurs from the blood vessels under the skin. The fever is followed by cardiovascular, digestive, kidney, and brain complications. The symptoms of infection are high fever, muscular cramps, headaches, and a violent cough, often resulting in respiratory failure. Research is now being carried out to determine how to protect people from contracting Hantavirus.

Lyme disease was discovered in 1975. It is carried by a tick-borne microbe. The infecting microorganism is carried by one species of tick, *Ixodes scapularis,* which is prevalent in grassy woods. The microbe is the bacterium *Borrelia burgdorferi.* People contract Lyme disease after an infected tick attaches itself to the skin. When the tick bites the person, the bacteria (spirochetes) are released from the salivary glands of the tick and enter the bloodstream of the person, where they multiply. In recent years, antibiotics have been successfully used to cure the disease. A vaccine has been developed to prevent the disease. Research physicians are seeking medications that will reduce the painful symptoms of people who are chronically ill with Lyme disease.

Culex pipiens, the mosquito species bearing the West Nile virus, has surfaced in the New York area. Seven people have died from the viral infection and 60 have been sickened. The West Nile virus kills birds and sickens people by causing *encephalitis.* Symptoms of the disease range from mild, low-grade fever to death-producing brain swelling and fever. The public has been urged to take an active part in the campaign against the mosquito by cleaning up pools (and even small puddles) of stagnant water where mosquitoes breed. Aerial spraying of malathion is now being replaced by use of larvicides such as Vectolex and Altisoids as preventive measures.

Model Questions Related to Ecology

1. The changes that occur in a plant community over a period of a few centuries are referred to as
   (A) evolution
   (B) food chains
   (C) succession
   (D) balance of nature
   (E) symbiosis

2. The last stable community that a particular region can support is called the
   (A) climax
   (B) pioneer
   (C) permanent
   (D) succession
   (E) eutrophication

3. The densest populations of most organisms that live in the ocean are found near the surface. The most probable explanation is that
   (A) the surface is less polluted
   (B) the bottom contains radioactive material
   (C) salt water has more minerals than fresh water
   (D) the light intensity that reaches the ocean decreases with increasing depth
   (E) the largest primary consumers are found near the surface

4. All of the statements about trophic levels and food webs are incorrect EXCEPT
   (A) they describe the general routes of energy flow in ecosystems
   (B) they measure the amounts of energy in ecosystems
   (C) they signify the kind of materials cycled in ecosystems
   (D) they measure the amounts of materials that pass through ecosystems
   (E) they serve as storehouses for the chemical energy initially trapped by plants

5. When present in large numbers, the plant most likely to cause a lake to become filled gradually is the
   (A) lichen
   (B) elodea
   (C) maple tree
   (D) duckweed
   (E) mushroom

6. Growing legumes as part of crop rotation improves the fertility of the soil. Examples of legumes are
   (A) barley and oats
   (B) wheat and corn
   (C) cowpeas and vetch
   (D) rice and wheat
   (E) tapioca and corn

7. Which of the following is the most important role of bacteria and fungi in a community of living organisms?
   (A) Fixation of nitrogen
   (B) Synthesis of antibiotics
   (C) Decomposition of organic material
   (D) Causation of disease
   (E) Alcohol fermentation

In Questions 8–12, choose the term that best fits each symbiotic description.

(A) Commensalism
(B) Mutualism
(C) Parasitism
(D) Saprophytism
(E) Heterotropism

8. Termites eat wood but are unable to digest it without the aid of certain flagellated protozoans that live within the bodies of the termites.

9. Certain young clams attach themselves to the gills of a fish. In a short time, each clam becomes surrounded by a capsule formed by cells of the fish. The clam feeds and grows by absorbing nutrients from the fish's body.

10. Certain luminescent bacteria living in the body of a marine fish obtain nutrition from the fish and, in turn, provide it with a "lantern," a warning signal or recognition device.

11. Beadle and Tatum grew *Neurospora* in a medium containing water, salts, biotin, and several amino acids.

12. The remora is a fish that attaches itself to a shark by suckers and eats the scraps left over by the shark. The shark is apparently unharmed by this relationship.

Base your answers to Questions 13–15 on the diagram below.

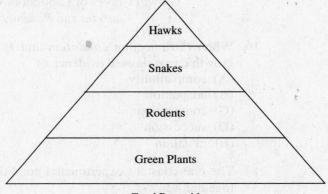

Food Pyramid

13. The greatest amount of energy present in this pyramid is found at the level of the
(A) hawks
(B) snakes
(C) rodents
(D) green plants
(E) decomposers

14. The pyramid implies that, in order to live and grow, 1,000 pounds of snakes would require

(A) less than 1,000 pounds of green plants
(B) 1,000 pounds of rodents
(C) more than 1,000 pounds of rodents
(D) no rodents
(E) less than 1,000 pounds of hawks

15. Which of the following represents the food chain shown in the pyramid?
(A) Green plant, rodent, snake, hawk
(B) Rodent, green plant, snake, hawk
(C) Snake, green plant, rodent, hawk
(D) Hawk, green plant, rodent, snake
(E) Green plant, snake, hawk, rodent

Answer Questions 16–18 on the basis of the graph below.

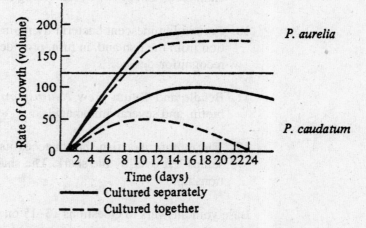

Growth Curves of Laboratory Cultures of *Paramecium
aurelia* and *Paramecium caudatum*

16. When *Paramecium caudatum* and *P. aurelia* were cultured together, the growth curve showed evidence of
(A) compatibility
(B) adaptation
(C) competition
(D) succession
(E) mutation

17. The one constant experimental condition provided when both were grown together was
(A) identical nutrient media
(B) decrease in temperature
(C) introduction of more *Paramecium aurelia* than of *P. caudatum*
(D) variations in light density
(E) decrease in CO_2 concentrations

18. Of the two types of paramecia, *Paramecium aurelia* is probably
(A) larger
(B) stronger
(C) more destructive
(D) better adapted to the nutrient
(E) more predaceous

Answers and Explanations to Model Questions: Ecology

1. C Ecological succession means the orderly replacement of groups of organisms by other organisms. The previously existing organisms alter the environment so drastically that they change the conditions in the environment that enable them to be successful.

Incorrect Choices

A Evolution refers to how species arise from preexisting organisms.

B A food chain is the feeding order or the energy flow in a community.

D The balance of nature refers to the fact that there is a constant recycling of nutritive materials and a population equilibrium.

E Symbiosis expresses the relationships of different kinds of organisms to each other.

2. A The final type of vegetation that can sustain itself without replacement forms a climax community.

Incorrect Choices

B Pioneer organisms are the first to invade an area.

C No community is permanent. Natural and human-made events alter the most stable of climax communities. Fires would be an example of an altering factor.

D Refer to Answer 1.

E Eutrophication is the aging process of a lake.

3. D The depth at which plankton can exist depends on the penetration of sunlight. Sunlight provides the energy for photosynthesis. Since the phytoplankton are abundant near the surface—the region of sunlight penetration—the consumers are also found in this region.

Incorrect Choices

A Pollution does not correlate with the depth of ocean water and cannot be the direct cause for the distribution of organisms.

B The conditions at the ocean bottom are not conducive to abundant life forms. There are a lack of oxygen, colder temperatures, and great pressure.

C The question does not imply that organisms in fresh water do not occupy the surface region.

E All primary consumers will be found in the area of the greatest food supply.

4. A Trophic levels describe the routes of energy flow in ecosystems. Autotrophs (green plants) constitute the first trophic level where the energy potential in the food chain and food web is the greatest. Energy flows from the first to the second trophic level that is constituted by plant-eating animals. The flow of energy is routed to the third trophic level, composed of carnivores. Energy is lost at each successive trophic level.

Incorrect Choices

B Amounts of available energy at each trophic level is variable and measurable in definitive quantities.

C, D, E Not correct choices as indicated in the explanation above.

5. D The duckweed covers the surface of the lake. The blockage of light results in the death and decay of subsurface plants. The lake gradually becomes filled with dead and decaying plants and organisms.

   Incorrect Choices
   A Lichens grow on rock.
   B The elodea is found in shallow water near the edges of lakes and ponds.
   C Maple trees might eventually be the climax community where a lake existed.
   E Mushrooms are the decomposers of a terrestrial environment.

6. C Cowpeas and vetch are legumes. The roots of legumes have swellings called nodules. It is within these nodules that nitrogen-fixing bacteria live. These bacteria convert nitrogen from the air into ammonium ions that are then oxidized by nitrifying bacteria into nitrites and nitrates.

   Incorrect Choices
   A, B, D, E   Incorrect because these choices are not legumes. Tapioca in Choice E is derived from the cassava root.

7. C Without decomposition, there would be no way to recycle what was removed from the environment or ecosystem. The vital materials that sustain a community would be depleted.

   Incorrect Choices
   A Nitrates are restored by both nitrogen fixation and decomposition.
   B No organisms except human beings have been able to utilize antibiotics. Antibiotics must be extracted and purified from the products of bacteria and fungi.
   D In a stable community disease is not a major factor.
   E Although alcohol fermentation is of economic benefit to humans, it is merely the process by which organisms obtain energy.

8. B Both organisms are benefited by the association.

9. C The fish is harmed through the loss of its nutritive material, while the clam derives all the benefits from the association.

10. B Refer to Answer 8.

11. D Saprophytes live on dead matter.

12. A One organism is benefited, but the other is neither harmed nor helped by the association.

13. D Green plants are the producers that can store enough energy to provide the basis upon which a community is built.

   Incorrect Choices
   A The least amount of energy is found at the level of the hawks.
   B, C Energy is lost at each successive level in a food pyramid.
   E Decomposers recycle minerals into the environment from dead matter.

14. **C** Snakes feed on rodents. At each trophic level, energy is lost; therefore the snakes must consume more than their biomass to survive.

   Incorrect Choices
   A Snakes do not feed on plants.
   B The snakes must consume more than their biomass.
   D Rodents are the source of food for the snakes.
   E Hawks feed on snakes.

15. **A** The food pyramid reveals the feeding order in a community.

   Incorrect Choices
   B, C, D, E Each organism in the ascending order of the pyramid feeds upon the organisms at the trophic level below itself.

16. **C** When the two species were grown together, the number of *Paramecium caudatum* decreased as the number of *P. aurelia* increased.

   Incorrect Choices
   A If the two were compatible, both would be able to survive.
   B Since *Paramecium caudatum* did not survive, it was not able to adapt to the environment.
   D Since this was an experiment under conditions that do not naturally occur, no conclusion regarding succession can be formulated.
   E There are no data concerning the role of mutations in the problem.

17. **A** Both were placed in the same nutrient environment, broth.

   Incorrect Choices
   B There are no temperature data.
   C According to the graph, both species were introduced in the same quantities.
   D No data are available for light density.
   E No data are available for CO_2 concentration.

18. **D** The graph indicates that there is some form of competition. Since the nutrients are the limiting factor, one species must be better adapted to the type of food or better adapted in the methods of obtaining the food.

   Incorrect Choices
   A, B, C, E These data are not available from this experiment.

References for Area III: Organisms and Populations

All of the following are from *Scientific American*.

Animals

Bennet-Clark, H. C. "How Cicadas Make Their Noise." May 1998

Blaustein, A., and D. Wake. "The Puzzle of Declining Amphibian Populations." April 1995

Blaustein, A., and R. O'Hara. "Kin Recognition in Tadpoles." January 1986

Fitzgerald, G. "The Reproductive Behavior of the Stickleback." April 1993

Groves, P. "Leafy Sea Dragons." December 1998

Handel, S., and A. Beattie. "Seed Dispersal by Ants." August 1990

Johnsen, S. "Transparent Animals." February 2000

Martini, F. H. "Secrets of the Slime Hag." October 1998

Narins, P. "Frog Communication." August 1995

Safina, C. "The World's Imperiled Fish." November 1995

Tallamy, D. W. "Child Care Among the Insects." January 1999

Thiantafyllou, M., and G. Thiantafyllou. "An Efficient Swimming Machine." March 1995

Wasserman, P. M. "Fertilization in Mammals." December 1988

Wayne, R., and J. Gittleman. "The Problematic Red Wolf." July 1995

Westmorland, F. S. "To Save a Salmon." January 1999

Wilkinson, G. "Food Sharing in Vampire Bats." February 1990

Young, M. W. "The Tick Tock of the Biological Clock." March 2000

Zill, S., and E. Seyfurth. "Exoskeletal Sensors for Walking." July 1996

Ecology

Alley, R. B., and M. L. Bender. "Greenland Ice Cores: Frozen in Time." February 1998

Bell, R. H. V. "Grazing Ecosystem of the Seregenti." July 1971

Berner, R. S., and A. C. Lasaga. "Modeling the Geochemical Carbon Cycle." March 1989

Bongaarts, J. "Can the Growing Human Population Feed Itself?" March 1994

Charlson, R., and T. Wigley. "Sulfate, Aerosol and Climatic Changes." February 1994

Dasgupta, P. "Population, Poverty and the Local Environment." Feburary 1995

Glenn, E. P., J. J. Brown, and J. W. O'Leary. "Irrigating Crops with Seawater." August 1998

Goulding, M. "Flooded Forests of the Amazon." March 1993

Herzog, H., B. Eliasson, and O. Kaarstad. "Capturing Geenhouse Gases." February 2000

Hoagland, W. "Solar Energy." September 1995

Holloway, M. "Nurturing Nature." April 1994

Karl, T. R., and K. Tvenberth. "The Human Impact on Climate." December 1999

Kusler, J., et al. "Wetlands." January 1994

Leffell, D., and D. Brash. "Sunlight and Skin Cancer." July 1996

Martini, F. H. "Secrets of the Slime Hag." October 1998

Packer, C., and A. E. Pusey. "Divided We Fall: Cooperation Among the Lions." May 1997

Robison, B. "Light in the Ocean's Midwaters." July 1995

Rogers, P., S. Whitby, and M. Dando. "Biological Warfare Against Crops." June 1999

Rutzler, K., and I. Feller. "Caribbean Mangrove Swamps." March 1996

Safina, C. "The World's Imperiled Fish." Quarterly Issue, Fall 1998

Schmidt, M. "Working Elephants." January 1996

Whipple, C. "Can Nuclear Waste Be Stored Safely at Yucca Mountain?" June 1996

Zorpette, G. "Chasing the Ghost Bat." June 1999

Human Physiology

Atkinson, M. S., and N. K. Maclaren. "What Causes Diabetes?" July 1990

Axel, R. "The Molecular Logic of Smell." October 1955

Barlow, R. B. "What the Brain Tells the Eye." April 1990

Beardsley, T. "The Machinery of Thought." August 1997

Blaese, R. M. "Gene Therapy for Cancer." June 1997

Brown, W. A. "The Placebo Effect." January 1997

Caldwell, J., and P. Caldwell. "The African AIDS Epidemic." March 1996

Garnick, M., and W. R. Fair. "Combating Prostate Cancer." December 1998

Goldberger, A. L., et al. "Chaos and Fractals in Human Physiology." February 1990

Greene, W. C. "AIDS and the Immune System." September 1993

Greenspan, R. "Understanding the Genetic Construction of Behavior." April 1995

Hoberman, J., and C. Yesalis. "The History of Synthetic Testosterone." February 1995

Hotez, P., and D. Pritchard. "Hookworm Infection." June 1995

Johnson, H., et al. "How Interferons Fight Disease." May 1994

Kantor, F. "Disarming Lyme Disease." September 1994

Lacy, P. "Treating Diabetes With Transplanted Cells." July 1995

LeDoux, J. "Emotion, Memory, and the Brain." June 1994

Levin, T. C., and M. E. Edgerton. "The Throat Singers of Tuva." September 1999

McGinnis, W., and M. Kuziora. "The Molecular Architects of Body Design." February 1994

Nossal, G. J. V. "Life, Death, and the Immune System." September 1993

Novak, M., and A. McMichael. "How HIV Defeats the Immune System." August 1995

O'Brien, S., and M. Dean. "In search of AIDS-Resistance Genes." September 1997

Ott, W. R., and J. W. Roberts. "Everyday Exposure to Toxic Pollutants." February 1998

Paul, W. E. "Infectious Disease and the Immune System." September 1993

Peris, T. T. "The Oldest Old." Jaunuary 1995

Plomin, R., and J. C. DeFries. "The Genetics of Cognitive Abilities and Disabilities." May 1998

Pots, M. "The Unmet Need for Family Planning." January 2000

Prusiner, S. "The Prion Diseases." January 1995

Raichle, M. "Visualizing the Mind." April 1994

Roder, P. M. "The Early Origins of Autism." February 2000

Rose, M. R. "Can Human Aging Be Postponed?" December 1999

Siegal, J. "Narcolepsy." January 2000

Steinman, L. "Autoimmune Disease." September 1993

Streit, W., and C. Colton. "The Brain's Immune System." November 1995

Welsh, M., and A. Smith. "Cystic Fibrosis." December 1995

Weissman, I., and M. Cooper. "How the Immune System Develops." September 1993

White, R. J. "Weightlessness and the Human Body." September 1998

Plants

Cox, P. M. "Water-Pollinated Plants." October 1993

Martinelli, G., and R. Azoury. "The Bromeliads of the Atlantic Forest." March 2000

Meyerowitz, E. "The Genetics of Flower Development." November 1994

Moses, P. B., and N. H. Chua. "Light Switches for Plant Genes." April 1988

Newhouse, J. R. "Chestnut Blight." July 1990

Reganold, J. P., et al. "Sustainable Agriculture." June 1990

Strobel, G. A. "Biological Control of Weeds." July 1991

Taxonomy

Anderson, D. "Red Tides." August 1994

Beland, P. "The Beluga Whales of the St. Lawrence River." May 1996

Blasser, M. "The Bacteria Behind Ulcers." February 1996

Brazaitis, P. M., E. Watanabe, and G. Amato. "The Caiman Trade." March 1998

Carmichael, W. "The Toxins of Cyanobacteria." January 1994

Iovine J. "Genetically Altering Eschericia Coli." (The Amateur Scientist) June 1994

Jacobs, W. "Caulerpa." (tropical alga) December 1994

Johnsen, S. "Transparent Animals." February 2000

LeGueno, B. "Emerging Viruses." October 1995

Martinelli, G., and R. Azoury. "The Bromeliads of the Atlantic Forest." March 2000

Shapiro, J. A. "Bacteria as Multicellular Organisms." June 1988

Vreeland, J. M., Jr. "The Revival of Colored Cotton." April 1999

PART SIX

Test Yourself

Model AP Biology Examination No. 2

Section I Multiple-Choice Questions

120 questions
1 hour and 30 minutes

Directions: Each of the questions or incomplete statements in this section is followed by five suggested answers or completions. Select the ONE that is BEST in each case.

1. To discover whether a trait in an organism is pure or hybrid, the organism should be crossed with a
   (A) pure dominant
   (B) hybrid dominant
   (C) hybrid recessive
   (D) recessive
   (E) litter mate

2. In enzyme synthesis, mRNA functions as
   (A) an activator
   (B) an enzyme
   (C) a source of peptides
   (D) an adaptor
   (E) a template

3. One major characteristic of the thallophytes is the lack of
   (A) cell walls
   (B) vascular tissue
   (C) nuclei
   (D) chloroplasts
   (E) chromosomes

4. Of the following, which statement is INCORRECT about the chromosome number in most seed plants?
   (A) Sepals are diploid.
   (B) Endosperm cells are usually triploid.
   (C) Megaspores are haploid.
   (D) Ovule coats are diploid.
   (E) Microspores are diploid.

5. The oldest xylem in a mature woody stem
   (A) lies directly under the bark
   (B) is just underneath the bark
   (C) is closely applied to the vascular cambium
   (D) loses its function
   (E) conducts nutrients downward

6. Chromosomes do NOT occur in pairs in
   (A) body cells
   (B) somatic cells
   (C) fertilized eggs
   (D) gametes
   (E) zygotes

7. In the flowering process, which organs detect the photoperiod?
   (A) Roots
   (B) Stems
   (C) Leaves
   (D) Stomates
   (E) Flower buds

8. Glucagon, a hormone, is secreted by cells present in the
   (A) liver
   (B) small intestine
   (C) islets of Langerhans
   (D) thyroid
   (E) adrenal cortex

9. The first heterotrophs that evolved on the primitive earth must have been able to obtain energy from
   (A) enzymes present in the environment
   (B) radiant energy of the sun
   (C) inorganic gases in the air
   (D) organic compounds present in the environment
   (E) molecules of ATP in the water environment

10. In Down's syndrome, a portion of chromosome 14 is transferred to chromosome 21. This is an example of
    (A) synapsis
    (B) synapse
    (C) deletion
    (D) translocation
    (E) transcription

11. Differentiation of the ectoderm in a developing bird embryo ultimately provides the organism with the ability to
    (A) reproduce its own kind
    (B) respond to stimuli
    (C) digest corn and other grain
    (D) breathe air
    (E) produce ATP molecules

12. Radioactive isotopes used in trace amounts are useful in diagnosing abnormal conditions. Radioiodine is used in studies of the
    (A) adrenal cortex
    (B) liver
    (C) spleen
    (D) gonads
    (E) thyroid

13. All of the following are characteristics of phylum Chordata EXCEPT
    (A) ventral tubular nerve cord
    (B) notochord at some time in life history
    (C) pharyngeal gill slits at some time in life history
    (D) ventral heart
    (E) endoskeleton

14. In angiosperms, the male gametophyte consists of
    (A) a germinating pollen grain
    (B) a megaspore
    (C) an anther
    (D) an ovule
    (E) a microspore

15. The short segment of the small intestine closest to the stomach in vertebrates is the
    (A) duodenum
    (B) cardiac sphincter
    (C) rumen
    (D) esophagus
    (E) jejunum

Questions 16 and 17

Questions 16 and 17 refer to the diagram below of a seed.

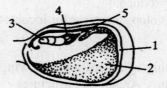

16. Which represents the nutrient-rich structure that provides nourishment to the developing embryo of a flowering plant?
    (A) 1
    (B) 2
    (C) 3
    (D) 4
    (E) 5

17. The embryo of this seed is represented by which of the following structures?
    (A) 1, 2, and 4
    (B) 3, 4, and 5
    (C) 1, 3, and 5
    (D) 1, 4, and 5
    (E) 1 and 2 only

18. The liver produces a chemical substance that prevents blood from clotting inside the blood vessels. The name of this anticoagulant is
    (A) glycogen
    (B) fibrinogen
    (C) porphyrin
    (D) heparin
    (E) heme

19. Of the following, the item that is NOT grouped properly with the other four items is
    (A) centriole
    (B) spindle fiber
    (C) chromosome
    (D) muscle fiber
    (E) centrosome

20. Lichens are able to grow abundantly on bare rock. However, neither the alga nor the fungus partner could inhabit this niche alone. This relationship shows the advantage of
    (A) mutualism
    (B) commensalism
    (C) parasitism
    (D) saprophytism
    (E) synapsis

21. A basic requirement for life is that an organism
    (A) ingest food
    (B) utilize energy
    (C) adapt to other organisms
    (D) reproduce
    (E) develop structures for locomotion

22. The wing of a bird and the wing of an insect are examples of analogous structures. Analogous structures
    (A) have similar evolutionary origins
    (B) show anatomic similarities
    (C) function in the same way
    (D) belong to species of the same phylum
    (E) are derived from the same primary germ layers

23. A compound in cells that functions as an enzyme, a transport molecule, and a receptor site is
    (A) a lipid
    (B) a protein
    (C) a carbohydrate
    (D) a starch
    (E) water

24. The diagram below represents a C_4 plant leaf. In which area do the preliminary reactions that incorporate carbon dioxide into a four-carbon compound occur?
    (A) 1
    (B) 2
    (C) 3
    (D) 4
    (E) 5

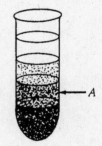

25. A stalk of celery was placed in a beaker of water containing a dye. After a few hours, the celery stalk was cut just beneath the leaves. It was observed that the dye was scattered in the stem. Within which structure in the stem was the dye observed?
    (A) Epidermal cells
    (B) Phloem cells
    (C) Guard cells
    (D) Xylem cells
    (E) Mesophyll cells

26. Cell fractionation results in layers of cellular components separated on the basis of their densities. When the components in section A in the diagram were examined, enzymes required for cellular respiration were identified. Which organelles comprise layer A?
    (A) Nuclei
    (B) Ribosomes and ER
    (C) Mitochondria
    (D) Cellular fluid
    (E) Chloroplasts

27. The mating behavior of a male silk moth is initiated upon the reception by its antennae of a chemical released by a female silk moth. This chemical is known as
    (A) a hormone
    (B) a pheromone
    (C) a neurotransmitter
    (D) an auxin
    (E) a rhodopsin

28. Sugar can be made in the dark if the chloroplasts are provided with a continuous supply of
    (A) chlorophyll
    (B) ATP and NADPH
    (C) ADP and $NADP^+$
    (D) oxygen
    (E) water

29. The size of an animal's egg usually depends on the
    (A) size of the animal
    (B) amount of yolk
    (C) amount of cytoplasm
    (D) number of chromosomes
    (E) rate of development of the embryo

30. Since barbiturates cause paralysis of voluntary muscle, it can be assumed
    that they block impulses traveling
    (A) from the central nervous system to skeletal muscle
    (B) from the autonomic nervous system to involuntary muscle
    (C) from all muscle to the autonomic nervous system
    (D) from smooth muscle to voluntary muscle
    (E) from striated muscle to the central nervous system

31. Vasopressin and oxytocin are produced by the
    (A) posterior pituitary
    (B) anterior pituitary
    (C) adrenal cortex
    (D) adrenal medulla
    (E) adenohypophysis

32. The diagram below represents a biogeochemical cycle.

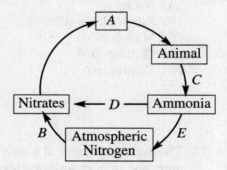

The nitrifying bacteria are represented by the letter
    (A) *A*
    (B) *B*
    (C) *C*
    (D) *D*
    (E) *E*

33. An instinct is best defined as
    (A) a habit
    (B) an operant behavior
    (C) a conditioned reflex
    (D) an operant conditioning
    (E) a series of reflexes

34. In order to break dormancy, most seeds require
    (A) exposure to a cold temperature for a period of time
    (B) exposure to light for a period of time
    (C) planting in soil rich in nutrients
    (D) imbibition
    (E) removal of the seed coat

35. The reduction of $NADP^+$ requires the cooperation of
    (A) photosystem I only in the chloroplasts
    (B) photosystem II only in the chloroplasts
    (C) neither photosystem I nor II
    (D) both photosystems I and II
    (E) RuBP carboxylase and carbon dioxide

36. The gene frequency for eye color in a population is 0.3 for brown-eye alleles and 0.7 for blue-eye alleles. Brown is dominant to blue. What is the frequency for the recessive individuals in the population?
    (A) 0.09
    (B) 0.21
    (C) 0.30
    (D) 0.49
    (E) 0.96

37. The diagram below illustrates phases of a specific life activity being carried out by the cell. Which process is indicated at point *A*?
    (A) Exocytosis
    (B) Endocytosis
    (C) Extracellular digestion
    (D) Intracellular digestion
    (E) Passive transport

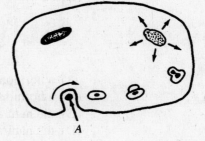

38. Osmosis is a process that
    (A) involves the movement of particles from saturated solutions
    (B) moves water molecules from an area of higher concentration to an area of lower concentration, using energy
    (C) regulates the tonicity on either side of a membrane
    (D) equalizes the concentration of particles by the movement of water molecules
    (E) continues until the medium on each side of the membrane has become hypertonic

39. An example of a tetrapyrrole pigment that is similar to chlorophyll is
    (A) hemoglobin
    (B) anthocyanin
    (C) carotene
    (D) glycine
    (E) cytosine

40. The mitotic spindle, which gives rise to spindle fibers in the polar areas of cells that lack centrioles, is characteristic of
    (A) epithelial cells
    (B) striated muscle cells
    (C) protozoans
    (D) nerve cells
    (E) lily root-tip cells

41. When a cell is placed into distilled water, its cytoplasm decreases in
    (A) size
    (B) salt concentration
    (C) protein content
    (D) volume
    (E) water content

42. The following molecular diagram represents a
    (A) nucleotide of DNA
    (B) nucleotide of tRNA
    (C) nucleotide of mRNA
    (D) nucleoside of DNA
    (E) nucleoside of RNA

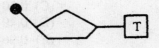

43. A bulb placed on its side in a glass of water will sprout. The curvature at A is caused by

    (A) a hormone stimulating cell growth on the lower side of the stem
    (B) a hormone inhibiting cell growth on the lower side of the stem
    (C) the natural ability of a stem to grow upward
    (D) the natural ability of a stem to grow downward
    (E) the position in which the bulb was placed in the glass of water

44. A unique property of DNA is its ability to
    (A) digest starch
    (B) replicate
    (C) migrate
    (D) form a double helix
    (E) use weak H bonds

45. Fungi differ essentially from green plants in that fungi
    (A) are unicellular plants
    (B) lack cellulose walls
    (C) cannot synthesize proteins
    (D) cannot absorb water
    (E) are unable to make glucose from CO_2 and H_2O

46. The endoplasmic reticulum functions as
    (A) a network that binds cells together
    (B) an ultrastructual framework in the cytoplasm
    (C) a secretory and storage syncytium
    (D) a network of fibers to which the nucleoli are attached
    (E) a control center for protein synthesis

47. The significance of mitosis is that it
    (A) controls cell division
    (B) makes a multicelled organism identical to its parent
    (C) makes the daughter cells identical to the parent cells
    (D) provides opportunity for the replication only of tissue cells
    (E) governs the quantity of cytoplasm in the daughter cells

48. The most primitive system of gathering energy by living organisms is
    (A) photosynthesis
    (B) Krebs cycle
    (C) aerobic respiration
    (D) fermentation
    (E) the cytochrome system

49. Although certain molecules are necessary as respiratory coenzymes, they cannot be synthesized by animal cells. These coenzymes must be taken into the body in the form of
    (A) vitamins
    (B) proteins
    (C) minerals
    (D) carbohydrates
    (E) amino acids

50. During glycolysis, energy is temporarily stored in NADH and then released during the production of
    (A) ATP
    (B) ADP
    (C) pyruvic acid
    (D) citric acid
    (E) lactic acid

51. A plant was placed into a dark box that had a small hole on one side. After several days, it was observed that the stem was bent towards the hole. Which term can be used to describe the plant's reaction?
    (A) Photosynthesis
    (B) Photorespiration
    (C) Phototropism
    (D) Photoperiodism
    (E) Photophosphorylation

52. In plants, the plastids that contain pigments that give fruit, flowers, and autumn leaves their orange and yellow colors are known as
    (A) epiblasts
    (B) leucoplasts
    (C) amyloplasts
    (D) chloroplasts
    (E) chromoplasts

53. The roots of three identical plants were each placed through a hole in covered tubes containing equal amounts of water. The plant in test tube 1 was untreated. The upper surface of the leaves of the plant in test tube 2 were coated with Vaseline. The lower surface of the leaves of the plant in test tube 3 were coated with Vaseline. After 24 hours, the water levels in the tubes were measured. The diagram below indicates the results. Which of the following is a possible explanation for the results?

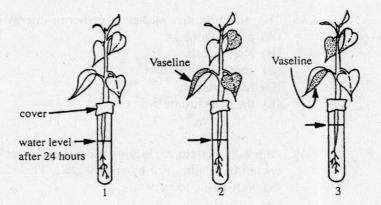

    (A) Vaseline prevents transpiration.
    (B) Less transpiration occurred through the upper leaf surface than through the lower leaf surface.
    (C) More transpiration occurred through the upper leaf surface than through the lower leaf surface.
    (D) Vaseline increases transpiration.
    (E) The greatest amount of transpiration occurred in the untreated plant.

54. It may be possible for mitochondria to exist independently since their DNA molecules contain information for the manufacture of certain
    (A) digestive enzymes
    (B) respiratory enzymes
    (C) carbohydrates
    (D) structural proteins
    (E) emulsifiers

55. Codfish produce many more eggs than will ever survive and develop into adult fish. The significance of this is that the codfish population is able to
    (A) develop a variety of mutational forms
    (B) survive as a species
    (C) survive the grueling trip to the spawning grounds
    (D) increase without limit
    (E) compete with fish predators

56. A primary reason for Mendel's success in his studies of pea plant inheritance was that
    (A) he was the first person to attempt studies of inheritance
    (B) there was plenty of room in his garden
    (C) he studied one trait at a time
    (D) he concentrated on the whole organism
    (E) he was the first to postulate the gene theory

57. In order for pancreatic enzymes to operate in the small intestine, the acidity from the stomach contents must be neutralized. Which substance secreted by the pancreas acts to neutralize the acidity?
    (A) HCl
    (B) KOH
    (C) Na_2CO_3
    (D) NaCl
    (E) H_2CO_3

58. The main structure in the paramecium that aids in excretion is the
    (A) cell membrane
    (B) contractile vacuole
    (C) macronucleus
    (D) oral groove
    (E) food vacuole

59. The process by which genetic material is transferred from one bacterium to another by means of a bacteriophage is known as
    (A) transcription
    (B) translation
    (C) translocation
    (D) transformation
    (E) transduction

60. Which DNA strand represents the complementary base sequence to a portion of a DNA strand indicated below?

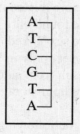

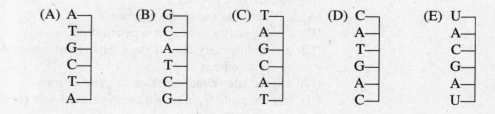

61. The apparatus pictured below was set up to determine the conditions under which an enzyme acts. Each test tube was filled to the three-fourths mark with the liquid indicated. The narrow tubing was charged with boiled egg white. At the end of a 1-day incubation period, the egg white at the ends of the capillary tube in test tube 2 had disappeared. What is the MOST reasonable conclusion that can be drawn from this experiment?
    (A) Hydrolysis of egg white occurs at the end of a long incubation period.
    (B) Egg white can be digested only by the enzyme pepsin.
    (C) Enzymes become activated in an acid medium.
    (D) Egg white must be boiled before enzyme activity can occur.
    (E) Pepsin must be in an acid medium to affect egg white hydrolysis.

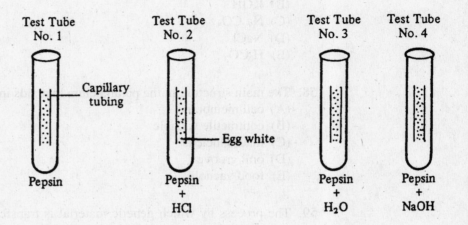

62. The cytoskeleton in cells, which serves mechanical and transport functions, consists of
    (A) cytosol
    (B) cristae
    (C) desmosomes
    (D) filamentous fiber
    (E) glycocalyx

63. Tay-Sachs disease, Fabry's disease, Niemann-Pick disease, and Gaucher's disease are related in that they
    (A) are found only in people of Jewish origin
    (B) are sex-linked diseases
    (C) appear only in Mediterranean peoples
    (D) result from autosomal trisomy
    (E) are lipid-storage disorders

64. All of the following statements concerning the gray crescent in the fertilized frog egg are true EXCEPT
    (A) it bisects the meridian of rotation
    (B) it is the only site of sperm penetration
    (C) it is the primary site of the exchange of materials between the egg and its surroundings
    (D) it is the site of the initiation of gastrulation
    (E) it is the point from which the notochord will grow

65. Powerful hydrolytic enzymes necessary for fertilization are contained in the
    (A) vitelline membrane of the egg
    (B) plasma membrane of the egg
    (C) nucleus of the egg
    (D) acrosome of the sperm
    (E) middle piece of the sperm

66. A codon is located on a molecule of
    (A) DNA
    (B) mRNA
    (C) rRNA
    (D) tRNA
    (E) polypeptide

67. Which means of circulation is characteristic of all organisms?
    (A) Capillary action
    (B) Necrosis
    (C) Phagocytosis
    (D) Diffusion
    (E) Crenation

68. The blastopore appears during the gastrula stage of the frog embryo and becomes the
    (A) lining of the tadpole's oral cavity
    (B) posterior opening of the tadpole's digestive tract
    (C) anterior opening of the tadpole's digestive tract
    (D) ciliated blastocoel
    (E) forerunner of the urinogenital tract

69. If the enzyme hyaluronidase were not produced by a large number of sperm cells, fertilization of the mammalian egg would NOT be successful because
    (A) sperm could not swim against gravity up the oviduct
    (B) the attraction between sperm and egg would not occur
    (C) the cumulus oophorous could not be removed from the egg
    (D) more than one sperm would penetrate the egg
    (E) the fertilization membrane would form at the wrong time

70. Which of the following occurs during the dark reaction of photosynthesis?
    (A) Light energy is absorbed by chlorophyll.
    (B) ATP is produced in large quantities.
    (C) Chlorophyll becomes excited.
    (D) Water is split into oxygen and hydrogen.
    (E) Carbohydrates are synthesized.

71. Which statement concerning the immune system is INCORRECT?
    (A) Plasma cells produce antibodies.
    (B) Lysozyme in saliva and tears attacks bacterial cell walls.
    (C) Humoral immunity depends upon T-cell activity.
    (D) Macrophages are phagocytic.
    (E) Antibodies are specific for antigen.

72. A change affecting the base sequence in an organism's DNA is known as
    (A) replication
    (B) chromosomal mutation
    (C) independent assortment
    (D) gene mutation
    (E) transcription

73. How are C_4 and CAM plants similar in their photosynthetic processes?
    (A) There are no grana in the chloroplast of either type of plants.
    (B) The stomates close during the day in both types of plants.
    (C) Sugar is made without the Calvin cycle in both types of plants.
    (D) Both types of plants use an enzyme different from rubicose in the first step of carbon fixation.
    (E) Sugar is made in the dark in both types of plants.

74. In certain species of bacteria the genetic composition of cells can be changed by adding DNA from other cells. This technique is called
    (A) transformation
    (B) replication
    (C) mutation
    (D) duplication
    (E) translocation

75. Information travels from DNA molecules to proteins by way of intermediary molecules designated as
    (A) proteases
    (B) enzymes
    (C) rRNA
    (D) tRNA
    (E) mRNA

76. The most conclusive evidence of evolution is (are)
    (A) hereditary molecules
    (B) fossil records
    (C) geographic isolation
    (D) diatrophism
    (E) X-ray diffraction

77. Frameshift mutations usually are
    (A) semiactive
    (B) able to mask the normal
    (C) lethal
    (D) capable of reversion
    (E) transitory

78. The membrane of the resting nerve cell is highly permeable to ions of
    (A) Fe^{2+}
    (B) Na^+
    (C) K^+
    (D) H^+
    (E) Cl^-

79. Efferent neurons that discharge adrenaline and noradrenaline into the blood are known as
    (A) glial cells
    (B) terminal branch cells
    (C) Schwann cells
    (D) Negri bodies
    (E) chromaffin tissue

Directions: Each set of lettered choices below refers to the numbered words or statements that follow it. Choose the lettered choice that best fits each word or statement. A choice may be used once, more than once, or not at all.

Questions 80–82

    (A) DNA molecules only
    (B) RNA molecules only
    (C) Both DNA and RNA molecules
    (D) Neither DNA nor RNA molecules
    (E) ATP molecules only

80. Molecules composed of chains of nucleotides.

81. Molecules found in the nucleus of the cell.

82. Molecules that convey genetic information from the nucleus to the ribosomes.

Questions 83 and 84

    (A) Ectoderm
    (B) Endoderm
    (C) Archenteron
    (D) Blastopore
    (E) Mesoderm

83. The region of the vertebrate embryo that gives rise to muscle tissue.

84. The region of the vertebrate embryo that is considered the primary organizer (the region that directs organogenesis).

Questions 85–87

    (A) Mutation
    (B) Linkage
    (C) Hybrid vigor
    (D) Inbreeding
    (E) Codominance

85. A mule, a cross between a horse and a donkey, often is superior to either parent in many ways.

86. Hemophilia and color blindness are carried on the X chromosome.

87. Radiation can cause definite structural changes in the DNA molecules of chromosomes.

Questions 88 and 89

Questions 88 and 89 are based on the chart below. Although the chart is not completely filled in, enough information is provided to answer the questions that follow.

Category	Organism I	Organism II	Organism III	Organism IV
Phylum	Arthropoda			
Class	Hexpoda			
Order	Lepidoptera	Lepidoptera		
Family	Torticidae	Psychidae		Torticidae
Genus	*Archips*	*Solenobia*	*Archips*	*Eulia*
Species	*rosana*	*walshella*	*fevidana*	*pinatubana*

88. Which two organisms are most closely related?
    (A) III and IV
    (B) II and IV
    (C) I and IV
    (D) I and III
    (E) I and II

89. Which organisms belong to the same class?
    (A) I, II, III, and IV
    (B) I, II, and III only
    (C) I and II only
    (D) I and III only
    (E) II and IV only

Directions: Each group of questions that follows is based on a laboratory situation or an experiment. For each, first study the description of the situation. Then select the best answer to each question that follows it.

Questions 90–92

Base your answers to Questions 90–92 on the diagram below, which illustrates an enzymatic reaction, and on your knowledge of biology.

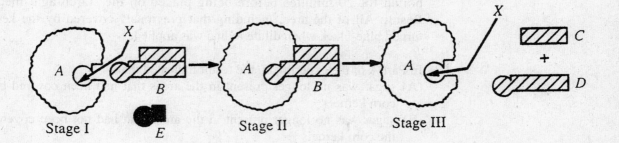

90. The enzyme-substrate complex is best illustrated by
    (A) *X* in stage I of the enzymatic activity
    (B) *A* in stage III of the enzymatic activity
    (C) the union of *E* with *A* in stage I of the enzymatic reaction
    (D) the union of *A* and *B* in stage II of the enzymatic reaction
    (E) molecules *C* and *D* in the enzymatic activity

91. If substance *E* is added to the environment, the enzymatic reaction ceases and products *C* and *D* are not produced. Substance *E* is known as
    (A) a coenzyme
    (B) an active site
    (C) a competitive inhibitor
    (D) a noncompetitive inhibitor
    (E) a cofactor

92. Which of the following best describes the specificity of enzymatic action?
    (A) One gene-one enzyme theory
    (B) Induced-fit hypothesis
    (C) Allosteric regulation
    (D) Feedback mechanism
    (E) Enzyme-substrate complex

Questions 93 and 94

Base your answers on the experiment described below:

A. A number of corn kernels were cut in half and placed with the surface down on a medium of agar containing 1% starch. After several hours, the kernels were removed and dilute iodine was poured over the agar. Result: The areas previously covered by the kernels became brownish in color. The rest of the agar turned blue-black.

B. The same experiment was then repeated, except that the cut kernels were boiled for 20 minutes before being placed on the starch-agar medium. Result: All of the area, including that previously covered by the kernels, turned blue-black when dilute iodine was applied.

93. In Part A of the experiment, the result indicated that
    (A) sugar was no longer present in the areas that had been covered by the corn kernels
    (B) sugar was no longer present in the areas that had not been covered by the corn kernels
    (C) starch was no longer present in the areas that had been covered by the corn kernels
    (D) starch was no longer present in the areas that had not been covered by the corn kernels
    (E) starch was present in the areas that had been covered by the corn kernels

94. A conclusion which may be drawn from this experiment is that uncooked corn kernels may contain
    (A) starch
    (B) sugar
    (C) fat
    (D) iodine
    (E) digestive enzymes

Questions 95–97

In meiosis of a sperm mother cell, chromosomal replication occurs during the first division, resulting in two identical chromatids which are attached to a centromere. In the first division, one homologous chromosome pair (known as a tetrad) separates. In the second division, the chromatids separate, resulting in the formation of four haploid sperm cells. By a similar meiotic division, the mother egg cell forms a functional ovum and three polar bodies.

95. In an organism that has a diploid number of 46 chromosomes, the daughter cells formed as a result of the first division have how many chromosomes?
    (A) 23
    (B) 22
    (C) 92
    (D) 46
    (E) 44

96. According to modern theory, during the second division, the centromeres (kinetochores)
    (A) disappear
    (B) replicate
    (C) disintegrate
    (D) redistribute
    (E) separate

97. After replication, each homologous pair of chromosomes consists of how many chromatids?
    (A) 2
    (B) 4
    (C) 6
    (D) 8
    (E) 10

Questions 98–100

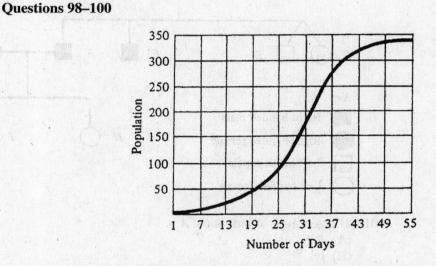

On day 1, three male and three female fruit flies were placed in a flatbottom flask that contained a cornmeal/banana medium. No other flies were added or removed during the course of this experiment. The graph shows the number of living fruit flies in the flask over a period of 55 days.

98. This curve shows that the population of fruit flies in the flask
    (A) increases indefinitely
    (B) increases, then decreases
    (C) increases, then reaches a plateau
    (D) decreases indefinitely
    (E) decreases, then levels off

99. The rate of reproduction is equal to the rate of death on day
    (A) 1
    (B) 7
    (C) 37
    (D) 49
    (E) 25

100. If the volume of the medium in the flask was decreased on day 55, the curve would be expected to
   (A) stay the same
   (B) increase, then level off
   (C) decrease, then level off
   (D) increase indefinitely
   (E) decrease to zero

Questions 101–105

Base your answers on the following pedigree chart, which shows a family history of handedness. Right handedness is dominant over left handedness.

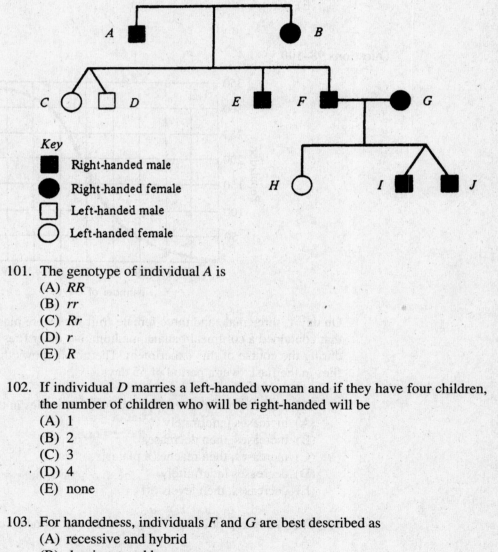

Key
■ Right-handed male
● Right-handed female
□ Left-handed male
○ Left-handed female

101. The genotype of individual *A* is
   (A) *RR*
   (B) *rr*
   (C) *Rr*
   (D) *r*
   (E) *R*

102. If individual *D* marries a left-handed woman and if they have four children, the number of children who will be right-handed will be
   (A) 1
   (B) 2
   (C) 3
   (D) 4
   (E) none

103. For handedness, individuals *F* and *G* are best described as
   (A) recessive and hybrid
   (B) dominant and homozygous
   (C) recessive and homozygous
   (D) dominant and heterozygous
   (E) incompletely dominant

104. Which individuals in the pedigree could possibly be identical twins?
    (A) F and G
    (B) J and I
    (C) C and D
    (D) E and D
    (E) H and I

105. If the trait for handedness were sex-linked, it would have appeared
    (A) in all of the offspring
    (B) most often in the males
    (C) most often in the females
    (D) most often in both males and females
    (E) in the third generation only

Questions 106–110

Each student in a biology laboratory class received two solutions. One solution was distilled water. The other was a salt solution with a concentration of salts slightly greater than that of a living cell. The solutions were labeled X and Y, respectively. The students were instructed to place some freshwater protozoans into each of the solutions and to identify the solutions on the basis of observations. The protozoans in solution X swelled and burst. Those in solution Y shriveled up.

106. These results indicate that
    (A) solution X was salt water
    (B) solution Y was distilled water
    (C) solution Y was salt water
    (D) solution X had other additions
    (E) solution X was tap water

107. The animals in solution X swelled up and burst because
    (A) an antagonistic substance was placed in the solution
    (B) the animals could not adjust to the new environment
    (C) osmotic pressure failed to operate
    (D) a disproportionate amount of water diffused into the animals
    (E) contractile vacuoles in the test organisms lost their function

108. To keep the animals alive in the available solutions X and Y, the laboratory instructor should
    (A) dilute solution Y with solution X and place the animals in this new solution
    (B) boil solution Y and cool, then place the animals in it
    (C) add a few drops of solution X to solution Y, strain, then add the protozoans
    (D) acidify solution X, then add the animals
    (E) evaporate and condense solution X, then add the animals when the solution is cool

109. What would happen to the protozoans if they were placed into pond water?
    (A) They would shrink and disintegrate.
    (B) They would swell and burst.
    (C) They would remain unchanged.
    (D) They would reproduce at an unusual rate.
    (E) Some would shrivel; others would explode.

110. In pond water protozoans excrete water by means of their
    (A) cell membrane
    (B) pellicle
    (C) trichocyst
    (D) contractile vacuole
    (E) pseudopods

Questions 111 and 112

Constituents in Plasma, Filtrate,
and Urine (g/100 ml of fluid)

Constituent	Plasma	Filtrate	Urine
Protein	6.00	0.00	0.00
Glucose	0.10	0.10	0.00
Amino acids	0.05	0.05	0.00
Urea	0.04	0.04	2.00
Uric acid	0.005	0.005	0.05
Salts	0.75	0.75	1.50

111. Which conclusion CANNOT be reached from the data given?
    (A) Filtrate is plasma minus the blood proteins.
    (B) Salts are more concentrated in the urine.
    (C) Calcium salts are more highly concentrated in the urine than in the filtrate
    (D) Protein is present in the plasma in greater amounts than are amino acids.
    (E) Glucose is not present in normal urine.

112. Information that is contradicted by the data given is that
    (A) 0.10 g of glucose is contained in 100 ml of normal blood
    (B) mineral salts are more highly concentrated in the urine than in the plasma
    (C) protein is not filtered into the urinary bladder
    (D) urea is not reabsorbed by the kidney tubules
    (E) uric acid is more highly concentrated in the urine than any other substance

Questions 113–118

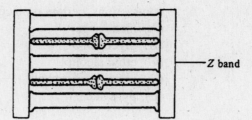

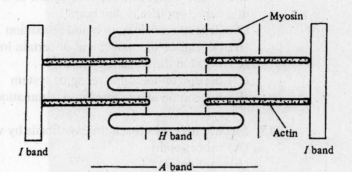

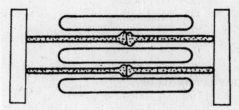

113. The unit of contraction of striated muscle that exists between two Z lines is the
(A) sarcolemma
(B) sarcoplasm
(C) sarcomere
(D) sarcosome
(E) soma

114. Which of the following does NOT occur during muscle contraction?
(A) The Z bands move closer to each other.
(B) The A bands extend from one Z line to the next.
(C) There is a continuous cycle of bridge breaking and reformation.
(D) The overlap of thick and thin filaments increases.
(E) The I bands elongate.

115. Muscle fibrils in living animals are activated to contract by nerve impulses. Such nerve impulses reach the muscle by way of
(A) an afferent nerve fiber
(B) a cranial nerve cell
(C) an associative nerve fiber
(D) an efferent nerve
(E) a sensory nerve

116. If muscle fibers are contracted repeat[...] rest between contractions, certain molecules accum[...]
    (A) lactic acid
    (B) glycogen
    (C) ethyl alcohol
    (D) creatine phosphate
    (E) pyruvate

117. A similarity between the contraction of a [...] of a nerve impulse is that both
    (A) depend on contraction and relaxation
    (B) depend on the movement of certain ior[...]
    (C) occur at the same time rate
    (D) depend on the same receptor system
    (E) depend on above-threshold stimulatior[...]

118. Electric impulses reach the myofibrils by v[...]
    (A) mitochondria
    (B) sarcolemma
    (C) T-system
    (D) glial cells
    (E) neuromuscular junction

Questions 119 and 120

The graph below shows survival rates for five a[...] val curves are calculated, the following assumptions[...]
    (a) All individuals of a given population are t[...]
    (b) No new individuals enter the population.
    (c) No individuals leave the population.

These curves show the relationship of the num[...]ion to units of physiological life span.

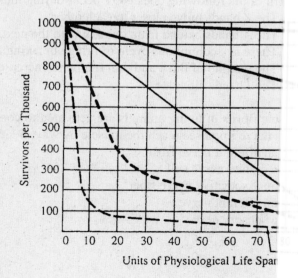

According to the data, it can be assumed that
(A) fruit flies live longer than humans
(B) oysters outlive fruit flies
(C) the population of hydras is steadily declining
(D) the life span of human populations is related to that of oysters
(E) there is a high mortality rate among young oysters

120. The survival curves indicate that
(A) starving fruit flies live out their full life span
(B) human populations are more vulnerable than hydras
(C) human population 2 has a greater rate of survival than human population 1
(D) the hydra has a longer life span than the oyster
(E) fruit flies die directly after pupation

Section II Free-Response Questions #2

Answer all 4 questions
1 hour and 30 minutes

1. Four identical aquatic plants were placed into separate water containers. Each container was exposed to a different light intensity. The gas produced by each plant was collected and measured. The gas was experimentally determined to be oxygen. The results of the experiment are recorded on the graph below.

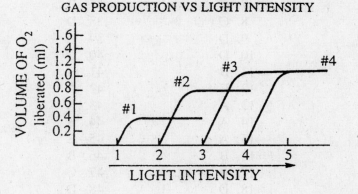

GAS PRODUCTION VS LIGHT INTENSITY

(A) Identify and explain the process that produced the results.
(B) Predict and explain the oxygen production if a fifth plant in a container would have been placed in a higher light intensity than that of the plant in container #4.
(C) Predict and explain the effect of increasing the temperature and carbon dioxide concentration on the rate of oxygen production.

2. Discuss food webs in an ecosystem. Focus your answer on energy flow, conversion, and loss in food chains and on the concepts of trophic levels and pyramids.

3. The various species of finches on the Galapagos Islands (known as Darwin's finches) are living examples of speciation.
   (A) Discuss the evolutionary significance of speciation and the factors that cause this phenomenon.
   (B) Discuss how concepts of genetics give credence to evolutionary theory as related to speciation.

4. In the course of plant evolution, natural selection has favored the dominance of the sporophyte generation.
   (A) Cite evidence to substantiate the above statements.
   (B) Explain why the gametophyte generation continues to exist in the life cycle of plants.

Answers to Model AP Biology Examination No. 2
Section I

1. D	31. A	61. E	91. C
2. E	32. D	62. D	92. B
3. B	33. E	63. E	93. C
4. E	34. D	64. B	94. E
5. D	35. D	65. D	95. A
6. D	36. D	66. B	96. E
7. C	37. B	67. D	97. B
8. C	38. D	68. B	98. C
9. D	39. A	69. C	99. D
10. D	40. E	70. E	100. C
11. B	41. B	71. C	101. C
12. E	42. A	72. D	102. E
13. A	43. A	73. D	103. D
14. A	44. B	74. A	104. B
15. A	45. E	75. E	105. B
16. B	46. B	76. B	106. C
17. B	47. C	77. C	107. D
18. D	48. D	78. C	108. A
19. D	49. A	79. E	109. C
20. A	50. E	80. C	110. D
21. B	51. C	81. C	111. C
22. C	52. E	82. B	112. E
23. B	53. B	83. E	113. C
24. B	54. D	84. D	114. E
25. D	55. B	85. C	115. D
26. C	56. C	86. B	116. A
27. B	57. C	87. A	117. B
28. B	58. A	88. D	118. C
29. B	59. E	89. A	119. E
30. A	60. C	90. D	120. A

Explanations to Model AP Biology Examination No. 2
Section I

1. **D**   Recessive traits will appear only if two alleles for the trait are combined in the gamete. Crossing a recessive with an unknown is known as a back-cross or test-cross.

2. **E**   mRNA carries the triplet code from the DNA strand from which it was replicated. mRNA behaves as a template or pattern for protein synthesis.

3. **B**   The thallophyte division of the plant kingdom contains the most primitive forms of plant life. The species in this group are characterized by a simple body (thallus), lacking true leaves, roots, and stems. Thallophytes lack conducting or vascular tissues.

4. **E**   The microspore mother cell is diploid and divides by meiosis to form four haploid microspores. Each haploid microspore divides once by mitosis resulting in two identical haploid nuclei. These become pollen grains.

5. **D**   Since the oldest xylem is found closest to the central pith or cortex of a plant stem, it is compressed and filled with plant cell exudates such as oils and resins. As the cellular contents disintegrate, the oldest xylem cells in a mature woody cell lose their function.

6. **D**   Gametes are produced by meiosis (a process of nuclear division in which the number of chromosomes is reduced by half), resulting in daughter cells with the haploid or monoploid number of chromosomes.

7. **C**   Leaves contain phytochromes, pigments that respond to the specific wavelengths of light that are visible during the day or night.

8. **C**   Glucagon is a hormone that acts in an opposite manner to insulin. Glucagon is produced by the alpha cells of the islets of Langerhans; insulin, by the beta cells.

9. **D**   Heterotrophs are dependent feeders. This means that they cannot synthesize nutritive material from inorganic compounds but must obtain nutrients from preexisting organic material. The original heterotrophs must have obtained energy from organic compounds that were present in the environment.

10. **D**   Translocation is a chromosomal rearrangement in which a segment of one chromosome fuses with a nonhomologous chromosome.

11. **B**   The ectoderm develops primarily into the nervous and sensory systems of vertebrate organisms. These systems enable the organism to respond to stimuli.

12. **E**   The thyroid gland synthesizes thyroxin, a hormone that contains iodine. Radioactive iodine is easily incorporated into thyroxin molecules and can be followed in the thyroxin synthesis pathway to detect abnormalities of function in the thyroid gland.

13. **A**   A dorsal tubular hollow nerve cord is characteristic of the phylum Chordata. A solid ventral nerve cord is characteristic of the invertebrates.

14. **A**   The pollen grain consists of two haploid nuclei, a sperm nucleus, and a tube cell nucleus. The nuclei were produced by meiosis from cells within the sporangial chamber of the anther.

15. **A**   The duodenum is the first part of the small intestine of vertebrates, extending from the pylorus to the jejunum. It is a region of active digestion and contains the openings of the bile and pancreatic ducts. The

jejunum and ileum are other regions of the small intestine. The cardiac sphincter is a region of the stomach that adjoins the esophagus. The rumen is the large first compartment in the stomach of ruminants, in which cellulose is broken down and food is temporarily stored.

16. B Structure 2 is the endosperm. Endosperm is the nutrient-rich part of the seed that provides nourishment to the embryo.

17. B Structures 3, 4, and 5 represent the embryo of the seed. Structure 3 is the radicle, 4 is the plumule, and 5 is the coleoptile. Structure 1 is the seed coat and is not a part of the embryo.

18. D Heparin is most abundant in the liver and has a great affinity for calcium ions, a mineral necessary in the clotting reaction of the blood.

19. D A muscle fiber is a cell of muscle tissue. The other structures are components of a cell that becomes active during mitosis or meiosis.

20. A Mutualism is the biologic association or vital interrelationship of two organisms that belong to different species. Both organisms are benefited by the relationship and are sometimes unable to survive without it.

21. B All chemical processes of cells require energy to make and break chemical bonds in the syntheses of compounds necessary to metabolism.

22. C Analogous structures are similar in function and often in superficial structure but have different evolutionary origins. The wing of a fly and the wing of a bird are analogous structures.

23. B There are ten thousand different proteins in a human cell. Each has a specific function and structure.

24. B The area designated as 2 consists of the mesophyll cells, which contain chloroplasts. These cells contain phosphoenolpyruvic acid (PEP) to which CO_2 is added to form a four-carbon product.

25. D Xylem cells comprise the vascular tissue, which conducts water up the stem from the roots to the leaves.

26. C Cellular respiration occurs in the mitochondria.

27. B A pheromone is a chemical substance released by an organism which influences the behavior and physiology of another organism of the same species.

28. B The Calvin cycle utilizes ATP and NADPH, which was generated by the light reaction, to reduce CO_2. The reduction of CO_2 ultimately produces sugar. The Calvin cycle is the dark reaction.

29. B The greater the supply of yolk (nutritive material), the larger is the size of the egg that contains it.

30. A The response of voluntary muscle is controlled by motor nerves from the central nervous system.

31. A The posterior pituitary (neurohypophysis) secretes two hormones, oxytocin and antidiuretic hormone (ADH), known also as vasopressin. These hormones are formed in neurons of the hypothalamus and are passed down axons into the posterior pituitary, where they are stored.

32. D The letter *D* represents the nitrifying bacteria. Nitrifying bacteria convert ammonia to nitrates.

33. E An instinct is an innate tendency (a series of reflexes) to a certain action or mode of behavior. Instincts are genetically controlled and are not learned, a fact that separates them from the types of behavior listed in the other choices.

34. D Imbibition is the absorption of water by a seed. Imbibition causes the seed to expand and rupture its coat, and triggers enzymatic activity.

35. D Photosystems I and II cooperate in the noncyclic electron flow in which electrons pass from water to NADP$^+$.

36. D The binomial equation $p^2 + 2pq + q^2 = 1$ is used to determine the frequency of individuals with a specific characteristic in a population; q^2 represents the frequency of the recessive individuals. Here $0.7 \times 0.7 = 0.49$.

37. B Endocytosis is the taking in of macromolecules by pinching in of the cell membrane, indicated at point A.

38. D Osmosis is the diffusion of water through a semipermeable membrane along a concentration gradient.

39. A Hemoglobin is composed of four pyrrole rings in which an iron atom (Fe) is bound in the center. Chlorophyll consists of a tetrapyrrole ring with magnesium bound in the center.

40. E Plant cells do not have centrioles. Spindle fibers emanate from a region known as the polar cap.

41. B The salt concentration of distilled water is extremely low. Therefore, distilled water will flow into a cell where the salt concentration is much higher. The water diffusing into the cell serves to decrease the concentration of salt molecules.

42. A A nucleotide is the basic unit of nucleic acid. A nucleotide consists of a sugar, a base, and a phosphate. A nucleotide of DNA will contain one of four bases: adenine, thymine, guanine, and cytosine.

43. A Hormones stimulate growth in the cells of plants. A hormone accumulates on the underside of the leaf stem at A because of the effect of gravity. The side with the excess hormone grows faster than the other side and curves upward.

44. B Replication is the process by which DNA molecules make exact duplicates of themselves during mitosis. The only biological molecules in which the original molecule serves as a direct model for construction of the new molecule are DNA molecules.

45. E Fungi are nongreen plants that cannot carry on photosynthesis and cannot convert carbon dioxide and water to glucose.

46. B The endoplasmic reticulum is a system of folded membranes that form the ultrastructure of eukaryotic cells. These membranes can be seen only by means of the electron microscope.

47. C Mitosis is a process of cell division through which identical nuclei are formed. The daughter cells thus produced have the same number and types of chromosomes as the parent cell.

48. D Oxygen was absent from the atmosphere of the primitive earth. Therefore, the first organisms that were formed had to rely on anaerobic respiration (fermentation).

49. A Vitamins are coenzymes that act as acceptors of groups or atoms when in loose association with an enzyme.

50. E When pyruvic acid combines with hydrogen in animal cells, lactic acid is produced. This completes anaerobic respiration. First, oxygen is not present so that ATP normally created by the transfer of H to oxygen cannot occur. Second, pyruvic acid functions as a hydrogen acceptor instead of a degradable fuel. All of the potential energy in the pyruvic acid will remain unused.

51. C Phototropism is a growth response of a plant towards light.

52. E Chromoplasts are plant organelles that contain pigments other than chlorophyll.

53. B There are more stomates in the lower surface of a leaf. Therefore, less transpiration occurred when the stomates were sealed by the Vaseline.

54. D Mitochondria contain episomes, discrete genetic elements that can replicate autonomously in bacterial cytoplasm or as an integral part of the chromosome and that transmit genetic information. The episomes contain the codes for the structural proteins of the membranes.

55. B In order to survive as a species, codfish must produce many more eggs than those that are fertilized and developed into independent organisms. Many of the eggs are not fertilized because of destructive or hazardous situations: sperm never reach many of the eggs; great number of eggs are eaten. Also, many immature fish are eaten.

56. C Investigators prior to Mendel tried to explain heredity in terms of many traits, by studying the whole organism. Mendel was unique because he studied pea plant characteristics (for example, height) one at a time.

57. C Pancreatic enzymes are activated in a basic (alkaline) medium. The pH of the stomach contents entering the small intestine is increased by secretion of Na_2CO_3 in pancreatic juice.

58. A Diffusion of carbon dioxide and water vapor takes place across the cell membrane.

59. E Transduction is the transfer of genetic material from bacteria to bacteria by using bacteriophages. Bacteriophages are viruses that attack bacterial cells.

60. C According to a base pair rule, adenine pairs with thymine and guanine pairs with cytosine in a DNA molecule.

61. E Since all of the test tubes contain pepsin, the presence of HCl must be the determining factor.

62. D There are at least three kinds of filamentous fibers in the cytosol: microtubules, microfilaments, and intermediate filaments.

63. E There are many enzymes that catalyze the breakdown of lipids in cells. When the enzyme that catalyzes a particular lipid reaction is inactive or absent, excessive amounts of that lipid begin to accumulate in certain tissues. Lipid-storage disorders in children result in muscular weakness and death.

64. B Sperm may penetrate the frog's egg at any point.

65. D Lysins are hydrolytic enzymes, digestive enzymes contained in the acrosome of the sperm. Sperm are able to penetrate egg cells because of membrane lysins.

66. B A codon is located on messenger RNA. A codon is the triplet code for an amino acid. It is the three-base sequence for a specific amino acid.

67. D Diffusion is the movement of molecules from an area of higher concentration to an area of lower concentration. Diffusion takes place across all cell membranes and within all cells.

68. B In chordates and echinoderms, the blastopore becomes the anal end of the digestive system.

69. C The egg cell is surrounded by a jelly that must be dissolved by hyaluronidase to permit penetration by the sperm.

70. E During the dark reaction of photosynthesis, carbohydrates are synthesized as a result of the fixation of carbon.

71. C T cells are the main agents of cell-mediated immunity. They are involved in the destruction of pathogens that have already entered body cells.

72. D A gene mutation is a change in the sequence of bases in DNA. According to the triplet code, each amino acid is specified by a particular three-base

code. A change in one base may alter one amino acid found in a protein molecule.

73. **D** PEP carboxylase is the enzyme required by both C_4 and CAM plants to produce the first stable four-carbon compound in the first step of the Calvin cycle. RuBP carboxylase (rubicose) is the enzyme required for the first step in the Calvin cycle of C_3 plants.

74. **A** Frederick Griffith (1928) developed a technique of transformation in which pneumococcus bacteria of the virulent strain were changed to non-virulent forms by the incorporation of DNA of the latter.

75. **E** mRNA is produced from the DNA template and therefore carries the triplet code.

76. **B** Fossils are preserved remnants of ancient plant and animal life which show relationships to present-day living organisms.

77. **C** Frameshift mutations are usually lethal because a totally different protein molecule may be produced that has no function in the cell. Frameshift mutations result from the addition or deletion of a single-base pair in the DNA sequence of a gene. The mRNA transcribed from such a faulty gene is translated normally until the point is reached where the addition or deletion occurs. From this point on, the transcription is aberrant and bears no resemblance to the normal molecule.

78. **C** The sodium pump prevents Na^+ ions from crossing the membrane and entering the cell. However, K^+ ions leak out of the cell in response to an ion gradient.

79. **E** Adrenaline and noradrenaline are neurohumors that are synthesized and stored in organs composed of cells that appear brown when stained with chromic acid—hence the name chromaffin.

80. **C** Both DNA and RNA are composed of nucleotides. A nucleotide consists of a sugar, base, and phosporic acid molecule.

81. **C** Both DNA and RNA molecules are found in the nucleus of a cell.

82. **B** RNA molecules, specifically messenger RNA, carry genetic information from the nucleus to the ribosomes.

83. **E** The mesoderm, or middle germ layer, of an embryo gives rise to muscle tissue.

84. **D** The blastopore is the region of the embryo that is considered the primary organizer.

85. **C** A superior characteristic often found in hybrids produced by a cross of two closely related species is known as hybrid vigor.

86. **B** Linkage is the carrying of an allele for a gene on a specific chromosome. The allele for hemophilia and color blindness is located on the X chromosome.

87. **A** A chromosomal mutation is a change in the structure of the DNA of a chromosome.

88. **D** Organisms I and III are most closely related; they belong to the same genus.

89. **A** Organisms I, II, III, and IV belong to the same class. The class is a subdivision of the phylum.

90. **D** An enzyme-substrate complex is a temporary unit formed by an enzyme (A) that is bound by a noncovalent bond to its substrate molecule (B).

91. **C** A competitive inhibitor (E) is a nonsubstrate molecule that binds to an active site because it resembles the substrate molecule.

92. **B** The induced-fit hypothesis states that a substrate, upon entering the active

site, causes the active site to change shape so that the enzyme binds the substrate more tightly.

93. C Starch turns blue-black in the presence of iodine. The absence of starch is denoted by the absence of color change to blue-black.

94. E Enzymes are required to hydrolyze starch to sugar. Amylases are enzymes that digest polysaccharides.

95. A Although the cells at the end of the first meiotic division contain 46 chromosomes, they are not $2n$ cells. These chromosomes are duplicates—one duplicate member for each chromosomal pair. In reality these are haploid cells, each chromosome with a duplicate.

96. E The centromeres separate, freeing the duplicate chromosomes.

97. B Chromatids are strands of a duplicated chromosome held together by a centromere. Each chromosome consists of two chromatids. There are four chromatids in a pair of homologous chromosomes.

98. C After the 49th day, there is no increase or decrease in the population. The graph levels off.

99. D The level graph indicates no population change. This means that the birth and death rates are the same on day 49.

100. C By decreasing the volume, one decreases the carrying capacity of the flask. The number of organisms that can be supported is decreased and then levels off.

101. C Since right-handedness is a dominant trait, the parents must be heterozygous for the trait. Each parent having a recessive gene can produce a left-handed individual.

102. E Left-handedness is determined by two recessive genes. Since both parents are left-handed, none of their children can have dominant genes for right-handedness.

103. D A recessive child can appear if the parents are heterozygous for handedness.

104. B Identical twins are of the same sex.

105. B Sex-linkage occurs on the X chromosome. The Y chromosome of the male does not have counteracting genes.

106. C Solution Y was salt water because it was hypertonic to the cell. The water in the protozoans diffused out through the cell membranes because the relative concentration of water molecules was greater there than in solution Y.

107. D The diffusion of abnormal amounts of water into the cell indicates that the water is hypotonic. The relative concentration of water molecules was greater in solution X than in the cytoplasm of the protozoans.

108. A The purpose of mixing the two solutions is to make an isotonic solution, that is, one where the relative concentration of water molecules is similar to that in the cytoplasm of the protozoans.

109. C Pond water is the natural habitat for protozoans. Because pond water is isotonic for protozoan cells, the animals would remain unchanged.

110. D The contractile vacuole is a homeostatic mechanism in cells of many species of protozoans that functions in the regulation of water, maintaining a functional balance.

111. C The data given do not identify the kinds of salts.

112. E Uric acid is usually only about 0.1 as concentrated in the urine as are other substances, which may be 2 or more times as concentrated.

113. C The sarcomere is the basic structural and functional unit of a muscle fiber.

114. E The *I* bands are pulled toward the center because the actin ro... over the myosin are located in this region. The bands do not elon...

115. D An efferent or motor nerve has a long axon that connects the c... nervous system with the muscle.

116. A Lactic acid is the product of anaerobic respiration. Lack of oxygen to the muscle fibers causes hydrogen to combine with pyruvic acid, converting the latter into lactic acid.

117. B The flow of Na^+ and K^+ ions is required for the transmission of nerve impulses. The flow of Ca^{2+} ions is necessary for muscle contraction.

118. C The tubules of the T-system are indentations of the sarcoplasmic membrane which extends into the cell.

119. E Note that only 10% of the population of young oysters reaches the 20th unit of life span.

120. A Note that almost all of the fruit flies reach the 90th unit of life span before the entire population dies.

Answers to Model AP Biology Examination No. 2
Section II

1. (A) The process that was being investigated by the experiment is *photosynthesis*. In general, photosynthesis is the process by which green plants manufacture carbohydrates in the presence of light. Photosynthesis is not a single process but two processes or stages. These stages are known as the *light reactions* and the *Calvin cycle of dark reactions*.

The light reactions convert solar energy into chemical energy. In the light reactions, ATP and NADPH are produced and oxygen is released. Photolysis, the splitting of water molecules into hydrogen and oxygen, is driven by light energy. Oxygen is a by-product of the light reaction of photosynthesis.

Chlorophyll is a pigment that absorbs light energy. The light energy is in the form of discrete packets called *photons*. The photons eject electrons from chlorophyll by elevating them to an excited state. The ejected electrons travel through an electron transport chain in cyclic and noncyclic pathways. $NADP^+$ is the electron acceptor in the light reaction. NADPH is formed when hydrogen from photolysis and the electrons from the transport chain combine.

The number of photons in light depends upon its intensity. The greater the intensity, the greater the number of photons. The greater the number of available photons, the greater the number of excited electrons and the more water molecules that will be split. However, there reaches a point where all the chlorophyll molecules are excited and the mechanism of photolysis is working at maximum. The rate of NADPH and oxygen production will not increase. Light intensity 4 is that point on the graph.

(B) Since the optimum rate of the light reactions of photosynthesis is reached at light intensity 4, there would be no increase in oxygen liberated at a higher light intensity. The rate of oxygen production in the fifth container would be the same as that in containers 3 and 4.

(C) The increase in temperature and carbon dioxide concentration are factors that affect the final product of the dark reactions. These factors do not influence the light reactions and thus have no effect on oxygen productions.

In the dark reactions, an increase in temperature, up to a point, will speed up the formation of 3-carbon compounds. Increase in temperature speeds up the rate of collisions between substrates and enzymes. Beyond a certain temperature, enzymes are denatured and activity will cease.

An increase in CO_2 concentrations increases the formation of 3-carbon compounds up to a point. Carbon fixation occurs during the Calvin cycle through its Rubisco pathway. The limiting factor is the availability of NADPH produced during the light reactions.

2. In a community, there are many food chains. A food chain is a sequence of organisms through which energy flows. The trophic level is the position that an organism occupies in the food chain.

The food chain is composed of several trophic levels consisting of producers and consumers. Autotrophs, usually green plants, are the producers. They are photosynthetic organisms that convert radiant energy to chemical energy, stored in the bonds of their sugar molecules. Autotrophs form the base of the food chain. Consumers are heterotrophs and must consume food to stay alive. There are several types of consumers. Herbivores feed directly upon the producers. Carnivores feed upon the herbivores or upon other carnivores. The decomposers are agents of decay and recycle minerals through the community. The food chain may be represented by a pyramid.

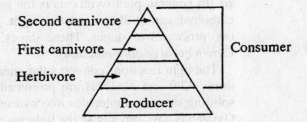

The pyramid also serves to explain the unidirectional flow of energy through the community.

At each trophic level energy is lost. The herbivore does not utilize all of the energy available to it from the plant because the entire plant is not consumed. Some parts of the plant are indigestible and are eliminated. The herbivore also loses energy in the form of heat from cellular respiration. The carnivore faces the same situation but with the additional factor that there is less energy stored than in the body of a herbivore. These facts are consistent with the Second Law of Thermodynamics, which states that as energy is transformed from one form to another, the amount of usable energy decreases.

Because less energy is available at each trophic level, each level can support fewer organisms and less biomass than the preceding level. A pyramid of numbers and a pyramid of biomass resemble the food chain pyramid.

3. (A) According to the theory of natural selection, proposed by Charles Darwin, several conditions encourage speciation. Finches which settled on the Galapagos Islands were the sole bird inhabitants of the islands. Food in unlimited quantities was available to these birds, which reproduced in large numbers. Therefore, large numbers of organisms competed for food. Differences in the birds began to

appear. Some had differently shaped beaks; others varied in size. Some of the phenotypically different types that could not compete for food in the usual way were able to obtain food in other ways. This relieved competition and separated some birds into other niches. If the type of adaptation were not possible, competition for food and natural selection would have eliminated several varieties of finches. Those best adapted to use a specific food source survived in their own niche. Birds occupying different niches became reproductively isolated from each other. The speciation process is caused by adaptive radiation.

(B) Mendelian genetics explains the method in which characteristics are transmitted from parents to offspring. However, the theory of mutation proposed by De Vries accounts for the appearance of variants, the raw material for evolution. The mutations expressed through phenotypes are neutral. However, a change may confer an advantage on those organisms which possess it. The change may allow the organism to secure more food. If this is the case, it survives and gradually reproduces. The variety increases in number until the parental stock becomes extinct. On the other hand, the change may not be advantageous, in which case the organisms perish or fill another niche.

The change in the gene frequencies of a population due to natural selection can also account for speciation. The mathematical representation of this process has been described in the Hardy-Weinberg law. Speciation is a result of evolution, and evolution may be defined in terms of changes in the gene pool of a population.

4. According to the theory of evolution, land plants, like land animals, evolved from aquatic forms. The ancestors of early land plants were most likely green algae. Sexual reproduction in most algae involves the fusion of motile gametes. Only in more complex forms is a heterogamous condition observed.

The first land plants required a very wet environment. This allowed motile sperm to reach the egg cell. A nonmotile egg prompted the changes in method of fertilization. In bryophytes of today, which are the ancestors of early plants, the dominance and independence of the sporophyte generation varies. In the mosses, the sporophyte is parasitic and short lived. In some liverworts the sporophyte is independent. However, the trend seems to be toward the reduction of the sporophyte generation. On land, conditions vary more drastically than in water. Conservation of resources is achieved by the parasitism of one generation on another.

For plants to become totally independent of a watery environment, a type of reproduction comparable to that of internal development and fertilization in animals had to occur. In seed-producing plants, the male gametophyte is protected in a pollen grain and is carried by various means to the site of the female gamete. The gametophyte is protected from drying out.

Evaluation of Model AP Examination No. 2 Results

Below is a breakdown of the test questions by subject area. Place a check mark (✓) in the box below each question answered correctly, and write the total number correct for that area in the space provided at the right. Then determine the percent you answered correctly in that particular subject area. For example, if you answer 4 out of 6 in a subject area, you answered 66% correctly.

Finally, add up the numbers of correct answers for all the areas to get your overall score. You should have a minimum of 90 correct answers. Concentrate your review and test preparation on the areas where your scores were poorest.

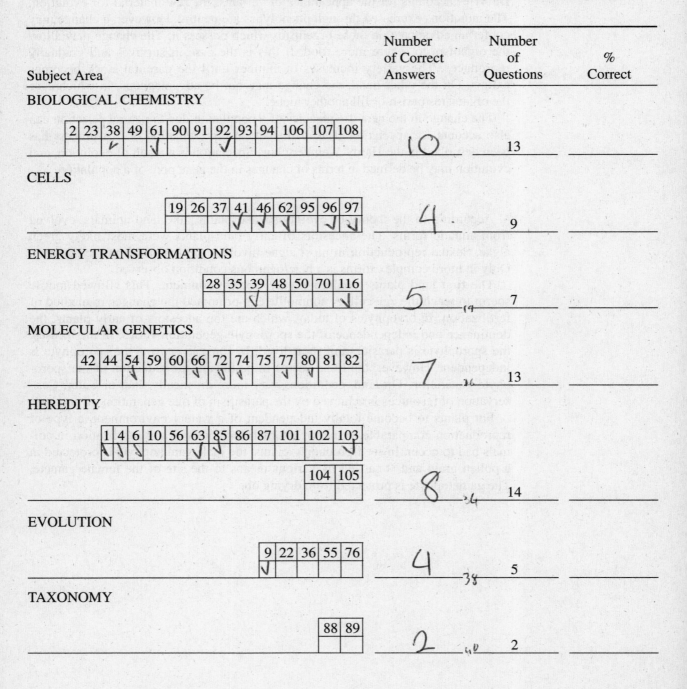

Subject Area		Number of Correct Answers	Number of Questions	% Correct
BIOLOGICAL CHEMISTRY	2 23 38 49 61 90 91 92 93 94 106 107 108	10	13	
CELLS	19 26 37 41 46 62 95 96 97	4	9	
ENERGY TRANSFORMATIONS	28 35 39 48 50 70 116	5	7	
MOLECULAR GENETICS	42 44 54 59 60 66 72 74 75 77 80 81 82	7	13	
HEREDITY	1 4 6 10 56 63 85 86 87 101 102 103 / 104 105	8	14	
EVOLUTION	9 22 36 55 76	4	5	
TAXONOMY	88 89	2	2	

MONERA, PROTISTA, AND FUNGI

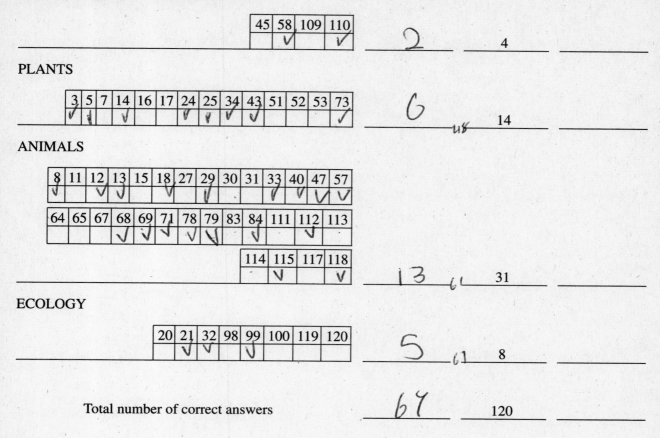

	45	58	109	110
		√		√

2   4

PLANTS

3	5	7	14	16	17	24	25	34	43	51	52	53	73
√	√		√			√	√	√	√				√

6   14

ANIMALS

8	11	12	13	15	18	27	29	30	31	33	40	47	57
√		√	√		√		√			√	√	√	√

64	65	67	68	69	71	78	79	83	84	111	112	113
			√	√	√	√	√		√		√	

	114	115	117	118
		√		√

13   31

ECOLOGY

	20	21	32	98	99	100	119	120
	√	√	√					

5   8

Total number of correct answers   67   120

Model AP Biology Examination No. 3

Section I Multiple-Choice Questions

120 questions
1 hour and 30 minutes

Directions: Each of the questions or incomplete statements in this section is followed by five suggested answers or completions. Select the ONE that is BEST in each case.

1. The action of the contractile vacuole in a paramecium maintains a water balance within the cell. The process of maintaining this balance is known as
   (A) transpiration
   (B) osmosis
   (C) phagocytosis
   (D) homeostasis
   (E) cyclosis

2. Mosses and liverworts are plants that lack many of the terrestrial adaptations of vascular plants. These nonvascular plants, which lack true roots, stems, and leaves, are known as
   (A) angiosperms
   (B) gymnosperms
   (C) bryophytes
   (D) tracheophytes
   (E) monocots

Questions 3 and 4

The diagram below shows the evolution of the vertebrate classes. Answer Questions 3 and 4 on the basis of the chart.

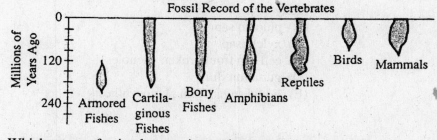

Fossil Record of the Vertebrates

3. Which group of animals came into existence most recently?
   (A) Mammals
   (B) Birds
   (C) Reptiles
   (D) Amphibians
   (E) Armored fish

4. How many vertebrate groups were present on the earth 120 million years ago?
   (A) 1
   (B) 2
   (C) 3
   (D) 4
   (E) 5

5. According to the fossil record, certain species remained unchanged over thousands of years. Then suddenly a new species appeared. The term used to explain this concept of evolution is
   (A) natural selection
   (B) mutualism
   (C) gradualism
   (D) punctuated equilibrium
   (E) genetic drift

6. A basic difference between plants and animals is the ability of plants to
   (A) digest carbohydrates
   (B) fix carbon dioxide
   (C) adapt to environments
   (D) carry out cellular respiration
   (E) resist disease

7. Cyanide bonds easily to the metallic portion of cytochrome molecules. Which of the following is most likely to occur in cyanide poisoning?
   (A) Electron transport to NAD is blocked.
   (B) Electron transfer to FAD is blocked.
   (C) Electron transfer to oxygen is blocked.
   (D) The mitochondrial membrane ruptures.
   (E) The respiratory chain is unaffected.

8. After a student cut a twig from a tree, he observed a droplet of fluid flowing from the cut surface of the branch still attached to the tree. The fluid was most likely
   (A) phloem sap
   (B) xylem sap
   (C) cell sap from broken vacuoles
   (D) guttation fluid
   (E) liquid sugar from photosynthesis

Questions 9 and 10

The diagram below illustrates a phase of protein synthesis. Answer Questions 9 and 10 on the basis of the diagram and your knowledge of biology.

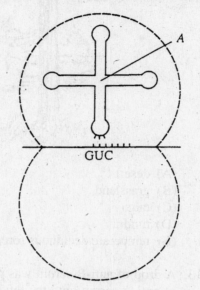

9. Structure *A* represents a molecule of
   (A) DNA
   (B) mRNA
   (C) tRNA
   (D) rRNA
   (E) protein

10. The sequence of bases in structure *A* would most likely be
    (A) CAG
    (B) GUC
    (C) UTC
    (D) CTG
    (E) UUU

11. All of the following are functions of the nephron EXCEPT
    (A) maintenance of pH
    (B) maintenance of electrolyte balance
    (C) control of red blood cell absorption
    (D) regulation of sodium concentration
    (E) secretion of hydrogen ions

12. The biome represented by the diagram below is known as

  (A) desert
  (B) grassland
  (C) taiga
  (D) tundra
  (E) temperate deciduous forest

13. A drop of anti-B serum was placed onto the right side of a slide, and a drop of anti-A serum onto the left side. If the blood was type AB, the blood cells would be clumped on
  (A) the right side only
  (B) the left side only
  (C) both left and right side
  (D) neither left nor right side
  (E) This is not a valid test for blood type.

14. In England, there are two distinct forms of peppered moth. One form is light mottled and the other is black. The table shows the number of each form before, during, and after the Industrial Revolution.

	Before	During	After
light mottled	3,500	3,000	1,000
black	500	1,800	3,200

Which is the most likely explanation for the changes in the population of the moth?
  (A) The soot from the factories increased the mutation rate of the black moth to the light mottled form.
  (B) The light mottled moth adapted to the environmental changes and became black.
  (C) The environmental changes selected against the light mottled moth and for the black moth.
  (D) The light mottled moths migrated into the area and the black moths migrated out of the area.
  (E) The black moth was easily detected and captured by birds while the light mottled moth escaped detection and capture.

15. The sliding filament hypothesis refers to the
    (A) movement of blood
    (B) movement of the amoeba
    (C) transmission at the synapse
    (D) change in membrane permeability
    (E) contraction of muscle

16. When the immature eye of a frog embryo reaches a certain stage in development, a lens begins to form. When such a developing eye tissue is transplanted into the back of another embryo, a lens begins to form in the ectoderm of the host embryo. The term that explains the result of this experiment is
    (A) preformation
    (B) embryonic induction
    (C) regeneration
    (D) differentiation
    (E) parthenogenesis

17. Which compounds are polymers of the molecule shown below?

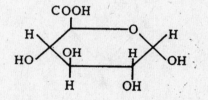

    (A) Glycogen and cellulose
    (B) Cholesterol and estrogen
    (C) Chlorophyll and hemoglobin
    (D) Keratin and protease
    (E) DNA and RNA

18. The location of four genes on chromosome #3 of a fruit fly is based on the following crossover frequencies

e and cu	20%
sp and H	43%
H and e	1%
cu and sp	24%
sp and e	44%

    What is the order of the genes on the chromosome?

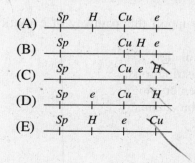

19. When the soil pH is above 7.0, certain plants produce blue flowers. When the soil pH is below 7.0, however, pink flowers are produced. Which statement explains the color change?
    (A) The change in pH serves as a mutagenic agent, which alters the genotype.
    (B) The different colors are produced by chromosomal mutations.
    (C) The environment influences gene expression.
    (D) Mutagenic agents alter the phenotype.
    (E) Polyploidy produces 2n gametes, which can produce flowers of different colors.

Questions 20 and 21

The diagram below represents evolutionary pathways for two hypothetical organisms, (X and Y). Base your answers to Questions 20 and 21 on the diagram and on your knowledge of biology.

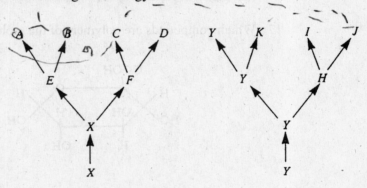

20. Which two organisms have the closest common ancestor?
    (A) A and B
    (B) A and C
    (C) K and J
    (D) C and J
    (E) X and Y

21. Which species was probably the most successful in the environment over time?
    (A) X
    (B) Y
    (C) K
    (D) J
    (E) C

22. The evidence to support this pathway of evolution would best be obtained from the study of
    (A) geographic isolation
    (B) geographic distribution
    (C) comparative embryology
    (D) comparative biochemistry
    (E) paleontology

23. The activation of cyclic AMP by adenyl cyclase occurs through the action of
    (A) steroid hormones on the cell membrane
    (B) protein hormones on the cell membrane
    (C) steroid hormones on the nucleus
    (D) protein hormones on the nucleus
    (E) both steroid and protein hormones on the cell membrane

24. A man hanging from a ledge of a roof eventually loses his grip. All of the following are reasons for losing his grip EXCEPT
    (A) there was a depletion of ATP
    (B) an accumulation of lactic acid impeded the glycolytic pathway
    (C) the man's muscles sustained a condition of tetany
    (D) the man's muscles suffered from oxygen debt
    (E) there was an insufficient store of glycogen in the man's liver

25. Below is a diagram of the heart. Which structure represents the pacemaker?

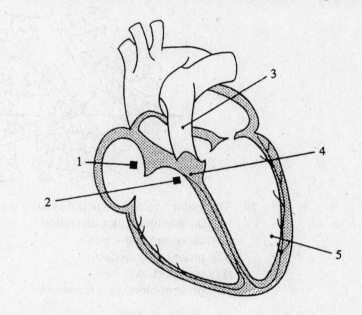

    (A) 1
    (B) 2
    (C) 3
    (D) 4
    (E) 5

26. *Wucheria brancrofti* is a parasitic worm that lives in the lymphatic vessels. Which problem would be caused by the presence of the worms?
    (A) Fat would not be digested.
    (B) Glucose would continue to pass into the lacteals of the villi.
    (C) Complete digestion would not occur in the stomach.
    (D) Cells would not receive enough nutrients.
    (E) Fluids would accumulate in tissue spaces.

27. Biologists have discovered that a certain type of poison interferes with protein synthesis. Which of the following cell structures is most likely to be affected by the poison?
    (A) Cytoplasm
    (B) Centrosome
    (C) Ribosome
    (D) Vacuole
    (E) Mitochondrion

28. A student had a solution of amino acids. In order to identify the amino acids in the solution, she used the process of chromatography. This process separates compounds based mainly on their differences in
    (A) concentration
    (B) color
    (C) heat of activation
    (D) solubility in chromatography solvent and their affinity for the supporting medium
    (E) degree of radioactivity

Questions 29 and 30

Base your answers to Questions 29 and 30 on the diagram below.

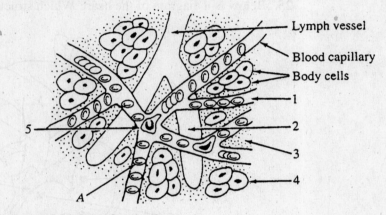

29. The major function of cell *A* is to
    (A) initiate blood clot formation
    (B) carry on phagocytosis
    (C) produce antibodies
    (D) transport oxygen
    (E) prevent blood clot formation

30. All materials passing between body cells and blood must pass through the intercellular fluid (ICF). The ICF is located in area
    (A) 1
    (B) 2
    (C) 3
    (D) 4
    (E) 5

31. In a rare disease condition in humans, the mitochondria are found to have an abnormal structure. This condition is most likely correlated with disturbances in cell
    (A) division
    (B) protein synthesis
    (C) food supply
    (D) energy supply
    (E) digestion

32. Tay-Sachs is a genetic disorder in which there is an absence of a cellular enzyme that acts on lipids. The enzyme is most likely located in the
    (A) peroxisome
    (B) lysosome
    (C) plastid
    (D) chloroplast
    (E) Golgi apparatus

33. The stomates of the leaf of a plant will open when the guard cells
    (A) respond to turgor pressure
    (B) take up H_2O by active transport
    (C) detect CO_2 in the air spaces of the leaf
    (D) accumulate K^+ because of passive transport
    (E) become turgid because of K^+ influx and passive water diffusion

34. The graph below represents the rate of enzyme action on a fixed amount of substrate.

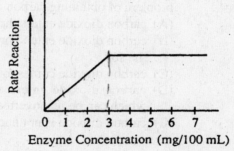

    At which enzyme concentration is all of the substrate reacting with the enzyme?
    (A) 0.5 mg/100 mL
    (B) 1.0 mg/100 mL
    (C) 3.0 mg/100 mL
    (D) 5.0 mg/100 mL
    (E) 7.0 mg/100 mL

35. According to the Linnaean system of classification, the organism *Achillea borealis* is most closely related to
    (A) *Borealis borealis*
    (B) *Borealis lanulosa*
    (C) *Achillea lanulosa*
    (D) *Lanulosa achillea*
    (E) *Borealis achillea*

36. An apple is a fruit. A fruit is most often
    (A) a ripened ovary
    (B) an enlarged ovule
    (C) a fusion of petals and sepals
    (D) a female gametophyte
    (E) a male gametophyte

37. The advantage of asexual reproduction over sexual reproduction is
    (A) small numbers of offspring
    (B) improved offspring
    (C) offspring that differ in appearance from their parents
    (D) offspring with half the chromosome number of the parents
    (E) offspring with the same genetic composition as the parents

38. AIDS is an immune disorder in which there is a reduction in the number of antibody-producing blood cells known as
    (A) red blood cells
    (B) macrophages
    (C) platelets
    (D) lymphocytes
    (E) neutrophils

39. The stomates in plants regulate the passage of gases from the environment to the cells. In some desert plants, the stomates are closed in the daytime and opened at night. The best explanation of how desert plants handle the problem of obtaining carbon dioxide is
    (A) carbon dioxide enters through the spines
    (B) carbon dioxide enters through the root and is carried with the water from the soil
    (C) carbon dioxide comes totally from cellular respiration in the plant cells
    (D) carbon dioxide enters at night and is stored in organic compounds, which can be reconverted to CO_2
    (E) carbon dioxide is not necessary in desert plants

40. A couple whose family history indicates chromosomal genetic disorders seeks genetic counseling. Which procedure would provide them with the genetic information that would help them to plan a family?
    (A) Karyotyping
    (B) Electrophoresis
    (C) Amniocentesis
    (D) Cloning
    (E) Genetic engineering

41. The reabsorption of glucose in the nephron diagramed below occurs as the filtrate passes through region
    (A) 1
    (B) 2
    (C) 3
    (D) 4
    (E) 5

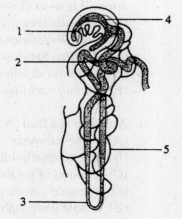

42. A tree dies and falls to the forest floor, where it begins to decay. Termites, worms, spiders, mice, and other organisms are found in or on the tree. Which organism is an example of a primary consumer?
    (A) The tree
    (B) Algae
    (C) Mosses
    (D) Termites
    (E) Spiders

43. The red tide is a phenomenon that occurs when certain organisms multiply very rapidly in water. The organisms responsible for the red tide are known as
    (A) dinoflagellates
    (B) euglena
    (C) diatoms
    (D) kelp
    (E) blue-green algae

44. Darwin noted that the Patagonian hare was similar in appearance and ecological niche to the European rabbit. However, the Patagonian hare is not a rabbit. It is a rodent related to the guinea pig. This example illustrates the principle known as
    (A) allopatric speciation
    (B) adaptive radiation
    (C) divergent evolution
    (D) coevolution
    (E) convergent evolution

45. The process that occurs during synapses and results in the chromosomal changes shown in the diagram below is known as

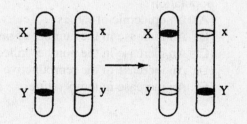

    (A) replication
    (B) crossing-over
    (C) independent assortment
    (D) synaptic transmission
    (E) sex linkage

46. All of the following are the result of active transport EXCEPT
    (A) absorption of glucose by the villi
    (B) sodium absorption by the nephrons
    (C) absorption of fatty acids by the villi
    (D) reabsorption of calcium ions into the sacroplasmic reticulum of muscle cells
    (E) reestablishment of ionic concentrations in nerve cells

47. Below is a diagram of a villus. Structure *A* belongs to the system of the body known as the

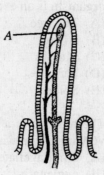

(A) circulatory system
(B) lymphatic system
(C) digestive system
(D) respiratory system
(E) excretory system

48. Which organism would probably be able to grow on a newly formed volcanic island?
(A) Grass
(B) Flowering shrub
(C) Maple tree
(D) Pine tree
(E) Lichen

49. Which would be expected to increase the competition among a squirrel population?
(A) An epidemic of rabies
(B) An increase in the number of predators
(C) An increase in the number killed by cars
(D) An increase in the reproductive rate
(E) An increase in temperature

50. Countercurrent mechanisms play an important role in the process of
(A) coronary circulation
(B) alveoli exchange of gases with the external environment
(C) converting pepsinogen to pepsin in the stomach
(D) reabsorption of filtrate substances in the nephrons of the kidney
(E) peristaltic movements of the small intestine

51. If the cut end of the stem of a white carnation is placed in blue dye, the petals will eventually become blue. This is evidence that
(A) some plants can grow without roots
(B) the cambium layer is present in the stem
(C) xylem tissue is present in the stem
(D) root hairs are adapted for absorption
(E) phloem tissue is present in the stem

Questions 52–54

Base your answers to Questions 52–54 on the diagram below and on your knowledge of biology.

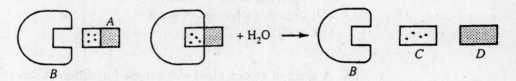

52. Which term describes the process diagramed?
    (A) Hydrolysis
    (B) Respiration
    (C) Photosynthesis
    (D) Photolysis
    (E) Synthesis

53. Compound *B* represents a molecule of
    (A) fat
    (B) protein
    (C) carbohydrate
    (D) phospholipid
    (E) nucleic acid

54. The water in the reaction serves to
    (A) release oxygen into the atmosphere
    (B) acidify the solution
    (C) supply energy
    (D) supply atoms that complete the structures of compounds *C* and *D*
    (E) release carbon dioxide into the atmosphere

55. The diagram below represents a capillary bed. Which statement is INCORRECT?

    (A) The hydrostatic pressure is greater in area *A* than in area *B*.
    (B) The osmotic pressure is greater in area *A* than in area *B*.
    (C) More materials leave the capillary from the tissues in area *A* than in area *B*.
    (D) More materials enter the capillary from the tissues in area *B* than in area *A*.
    (E) More oxygen diffuses into the tissues at area *A* than at area *B*.

56. A response of some plants to exposure to light is a sudden enlargement of the stomates. The importance to the plant of this response is that it makes possible an increased
    (A) intake of carbon dioxide
    (B) intake of water
    (C) intake of oxygen
    (D) discharge of carbon dioxide
    (E) discharge of oxygen

57. According to the theory of organic evolution, the first living things were probably able to obtain energy directly from
    (A) enzymes in the environment
    (B) oxygen in the atmosphere
    (C) carbon dioxide in water
    (D) organic molecules in water
    (E) sunlight

58. A sprig of elodea plant was exposed to different wavelengths of light. If the intensities of the wavelengths were equal, under which wavelength of light would oxygen be produced at the LOWEST rate?
    (A) White
    (B) Green
    (C) Red
    (D) Blue
    (E) Orange

59. The enzymes involved in oxidation-reduction of a substance cannot operate without NAD. The NAD is known as a
    (A) cofactor
    (B) prosthetic group
    (C) coenzyme
    (D) amino acid
    (E) coenocyte

60. The electron transport chain for the photosynthetic reaction is located in the membranous structures called
    (A) stomata
    (B) stroma
    (C) cristae
    (D) mitochondria
    (E) thylakoids

61. In comparison with cellular respiration, fermentation
    (A) is a more efficient catabolic pathway for ATP production
    (B) is a less efficient catabolic pathway for ATP production
    (C) is completely unrelated to cellular respiration
    (D) occurs only in active muscle cells
    (E) is very closely related to bioluminescence

62. The rate at which glycolysis occurs in a cell is stimulated by the concentration of
    (A) ATP
    (B) ADP
    (C) citric acid
    (D) oxygen
    (E) carbon dioxide

63. In the diagrams below, C = consumer, H = herbivore, P = producer. Which diagram represents the biomass in a stable community?

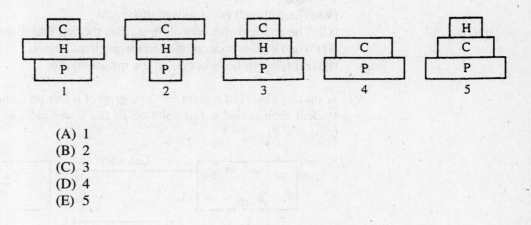

(A) 1
(B) 2
(C) 3
(D) 4
(E) 5

64. Oxidative phosphorylation in chloroplasts, as compared with mitochondria,
    (A) generates ATP by a chemiosmosis mechanism
    (B) extracts high-energy electrons that drop down an electron transport chain from oxidation of food molecules
    (C) utilizes the passage of H^+ ions across ATP synthase to generate ATP
    (D) produces ATP as H^+ diffuses from the intermembrane space to the matrix
    (E) produces ATP as H^+ diffuses across thylakoid membranes into the stroma

65. Which statement is true about the theory of the origin of life on earth?
    (A) It required gaseous oxygen.
    (B) It arose from living systems brought to earth from other planets.
    (C) It has been duplicated in the laboratory.
    (D) It is conceivable in terms of our present knowledge of chemistry and physics.
    (E) It is still continuing.

66. Upon stimulation of a certain brain region in humans, the emotion of anger is observed. The region of the brain that was stimulated is the
    (A) cerebellum
    (B) cerebrum
    (C) medulla
    (D) thalamus
    (E) hypothalamus

67. According to the fluid mosaic model of the plasma membrane, proteins are
    (A) spread out in a continuous layer over both membrane surfaces
    (B) located in the hydrophilic area of the membrane
    (C) embedded in the lipid bilayers
    (D) oriented in a random manner throughout the membrane
    (E) capable of leaving the membrane and dissolving in the solution of the external environment.

68. Which statement concerning prokaryotic cells is correct?
    (A) The plasma membrane consists of proteins embedded in a single lipid layer.
    (B) The cell wall is composed of keratin.
    (C) The endoplasmic reticulum divides the cell into discrete compartments.
    (D) The DNA is concentrated in the nucleoid region.
    (E) Protein synthesis occurs in the mitochondria.

69. A student observed a plant cell in a drop of water on a slide (diagram *A*). The student then added a salt solution to the slide and again observed the cell (diagram *B*).

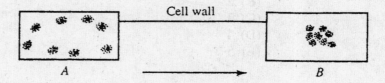

Cell wall

A   *B*

The change in the cell was due to
    (A) osmosis
    (B) facilitated diffusion
    (C) active transport
    (D) cyclosis
    (E) endocytosis

70. Which of the following occurs when the membranes of the alveoli thicken?
    (A) Decrease of oxygen concentration in the blood
    (B) Decrease of carbon dioxide concentration in the blood
    (C) Increase in oxygen concentration in the blood
    (D) Increase in nitrogen concentration in the blood
    (E) Increase in the volume of the blood

71. The clonal selection theory is an important concept in the understanding of how the immune system operates. Which statement concerning the theory is INCORRECT?
    (A) B lymphocytes, upon encounters with antigens, form clones of plasma cells.
    (B) B lymphocytes, upon encounters with antigens, form clones of memory cells.
    (C) Plasma B cells secrete antigen-specific antibodies.
    (D) Macrophages, upon encounters with antigens, secrete specific antibodies.
    (E) Memory cells are long-lived and remain inactive until they encounter an antigen.

72. One of the most important differentiating features between the chlorophyta and cyanophyta is the absence in the cyanophyta of
    (A) cell walls
    (B) cell membranes
    (C) chloroplasts
    (D) cytoplasm
    (E) chlorophyll

73. Black fur is dominant over white fur in guinea pigs. What is the genotype of the parents of a group of offspring that are 50% black and 50% white?
    (A) Both parents are homozygous black.
    (B) One parent is homozygous black and the other is white.
    (C) One parent is heterozygous black and the other is white.
    (D) Both parents are white.
    (E) Both parents are heterozygous black.

74. Termites cannot digest wood, but certain protozoans that live in the gut of the termite digest wood. The termite utilizes some of the digested wood products and provides a safe living area for the protozoan. The relationship between the two organisms is known as
    (A) symbiosis
    (B) commensalism
    (C) mutualism
    (D) parasitism
    (E) saprophytic

Questions 75 and 76

Answer Questions 75 and 76 on the basis of the diagram below. Each circle represents the range for a pair of owls. *X* represents the area where reproduction and nesting takes place.

75. The shaded area represents the area where
    (A) most of the predators of the owls are located
    (B) most of the owls' food is located
    (C) competition is likely to occur between the pairs
    (D) mating occurs
    (E) there is no competition between the two pairs

76. What is the most likely reason why the range of pair *B* is larger than the range of pair *A*?
    (A) Pair *B* is younger.
    (B) Pair *B* can fly faster than pair *A*.
    (C) Pair *B* has more prey near its nest than pair *A*.
    (D) Pair *B* has less food available near its nest than pair *A*.
    (E) Pair *A* does not need as much living space as pair *B*.

77. Certain organs of locomotion, such as the arm of a human, flipper of a whale, foreleg of a horse, and wing of a bat, are often cited as evidence in support for the theory of common ancestry. What type of evidence does this example indicate?
    (A) Comparative anatomy
    (B) Biochemical similarity
    (C) Homologous structures
    (D) Analogous structures
    (E) Geographic isolation

78. Genes in bacteria that are located outside the main chromosome are known as
    (A) plastids
    (B) plasma
    (C) plasmids
    (D) plasmolysis
    (E) plasmalemma

79. Which of the following is (are) NOT associated with nuclear cell division in animals?
    (A) Spindle fibers
    (B) Duplication of chromosomes
    (C) Constriction of dividing cell
    (D) Separation of chromosomes
    (E) Cell plate formation

Questions 80–82

Base your answers to Questions 80–82 on the diagrams below, which represent the developmental stages of a frog.

1   2   3   4   5

80. The cell in diagram 2 is different from the cells in diagram 1 in that cell 2
    (A) can no longer undergo mitosis
    (B) has twice the number of chromosomes as each cell in diagram 1
    (C) has half as many genes as each of the cells in diagram 1
    (D) has half as many chromosomes as each of the cells in diagram 1
    (E) contains only autosomal chromosomes

81. Where would the developmental stages represented by the diagrams take place?
    (A) Amniotic fluid
    (B) Ovary
    (C) Testes
    (D) Uterus
    (E) Watery environment

82. The cell in diagram 2 is formed by the process of
    (A) fertilization
    (B) oogenesis
    (C) mitosis
    (D) spermatogenesis
    (E) cleavage

83. Which process does NOT require the expenditure of ATP?
    (A) Cellular respiration
    (B) Photorespiration
    (C) Fermentation
    (D) Glycolysis
    (E) Photolysis

Questions 84 and 85

Answer Questions 84 and 85 on the basis of the diagram below and your knowledge of biology.

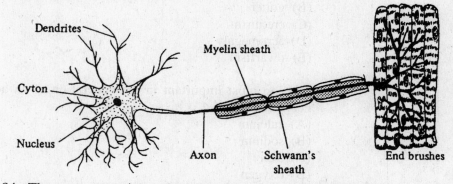

84. The neurotransmitter released at the neuromuscular junction is
    (A) acetylcholine
    (B) epinephrine
    (C) norepinephrine
    (D) serotonin
    (E) adrenaline

85. The type of nerve cell represented by the diagram is a
    (A) motor nerve cell
    (B) sensory nerve cell
    (C) mixed nerve cell
    (D) associative nerve cell
    (E) striated nerve cell

86. During the process of photosynthesis, the oxygen from water is converted into
    (A) water
    (B) NADPH + H$^+$
    (C) ATP
    (D) oxygen gas
    (E) sugars

87. An enzyme that has two binding sites and exists in two or more conformations is known as
    (A) a hydrolytic enzyme
    (B) an allosteric enzyme
    (C) a catalytic enzyme
    (D) a catabolic enzyme
    (E) an anabolic enzyme

88. The process of mRNA transcription occurs in the
    (A) mitochondrion
    (B) nucleus
    (C) cytoplasm
    (D) ribosome
    (E) polysome

89. An experimenter removed the pancreas of a living rat and observed the symptoms that resulted from the operation. The experimenter was trying to produce the symptoms of
(A) diabetes
(B) goiter
(C) cretinism
(D) acromegaly
(E) dwarfism

90. The ion most important in the generation of an action potential across a nerve cell is
(A) calcium
(B) sodium
(C) iron
(D) oxygen
(E) magnesium

91. The diagram below represents different biomes found on a mountainside.

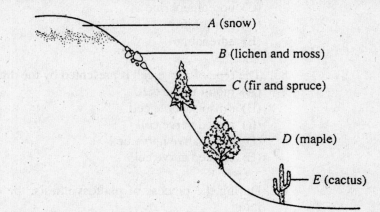

The tundra region would be located on the diagram in zone
(A) A
(B) B
(C) C
(D) D
(E) E

92. Which structure in a human fetus serves as a respiratory organ?
(A) Allantois
(B) Gills
(C) Umbilical cord
(D) Placenta
(E) Yolk sac

93. Which disorder is a result of meiotic nondisjunction?
(A) PKU
(B) Down's syndrome
(C) Hemophilia
(D) Tay-Sachs disorder
(E) Sickle-cell anemia

94. All of the following are characteristics of flowering plants EXCEPT
    (A) they produce autotrophic gametes
    (B) seed dispersal is observed
    (C) they exhibit double fertilization
    (D) some are insect pollinated
    (E) the formation of pollen tubes is essential

Directions: Each set of lettered choices that follows refers to the numbered words or statements immediately after it. Choose the lettered choice that best fits each word or statement. A choice may be used once, more than once, or not at all in each set.

Questions 95–97

(A) Esophagus
(B) Stomach
(C) Small intestine
(D) Liver
(E) Large intestine

95. Gastric juice is produced and secreted by specialized cells in the lining.

96. Chemical digestion is completed, and the products of digestion are absorbed.

97. Peristaltic action transports food from the mouth to the stomach.

Questions 98–102

(A) DNA molecules only
(B) RNA molecules only
(C) Both DNA and RNA molecules
(D) Neither DNA nor RNA molecules
(E) Transfer RNA only

98. May contain the four bases adenine, cytosine, thymine, and guanine.

99. Contain(s) the nitrogenous compound known as urea.

100. Is (Are) present in the nucleus of the cell.

101. Consist(s) of chains of nucleotides.

102. Convey(s) genetic information from the nucleus to the ribosomes.

Directions: Each group of questions that follows concerns a laboratory situation or an experiment. In each case, first study the description of the situation. Then choose the one best answer to the questions that follow.

Questions 103 and 104

In 1940, ranchers introduced cattle into an area. The graph below shows the effect of cattle ranching on the population of two organisms present in the area before the introduction of the cattle. Base your answers to Questions 103 and 104 on the graph and on your knowledge of biology.

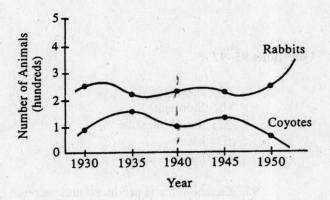

103. The most probable reason for the increase in the rabbit population after 1950 was
    (A) more food became available to the rabbits
    (B) the coyote population declined drastically
    (C) the cattle created a more favorable environment for the rabbits
    (D) the coyotes and cattle competed for the same food
    (E) the coyote population increased

104. If the interrelationship of the rabbits and coyotes was once in balance, what is the most probable explanation for the decline of the coyotes?
    (A) Mutations
    (B) Starvation
    (C) Disease
    (D) Increase in reproductive rate
    (E) Removal by human beings

Questions 105 and 106

Base your answers to Questions 105 and 106 on the following paragraph and on your knowledge of biology.

A student ground 1 gram of fresh liver in a mortar, placed the ground liver into a test tube, and added 1.0 ml of peroxide. The gas that was generated was collected. A glowing splint burst into flames when placed into the gas. The student then repeated the procedure, using 1 gram of boiled liver and 1 gram of liver treated with a strong acid. When peroxide was added to each sample of liver, no gas was generated.

105. The gas that was generated was most likely
    (A) carbon dioxide
    (B) oxygen
    (C) nitrogen
    (D) water vapor
    (E) hydrogen

106. If the substance in the liver that acted on the perioxide was an enzyme, it could
    (A) be recovered from the liver tissue after the reaction ceased
    (B) not be recovered because it was consumed in the reaction
    (C) not be recovered because there is no enzyme in liver that catalyzes the breakdown of peroxide
    (D) not be recovered because grinding would break up the molecule
    (E) be recovered only before the peroxide was added

Questions 107 and 108

Answer Questions 107 and 108 on the basis of the following paragraph and on your knowledge of biology.

Sickle-cell anemia occurs in two forms, major and minor. The major form is usually fatal in childhood, while the minor form is less severe. Normal persons are pure for the normal gene. Persons with the major form are pure for the abnormal gene. Persons with the minor form are hybrid.

107. If both parents have the minor form, what percent of their children are expected to be normal?
    (A) 0%
    (B) 25%
    (C) 50%
    (D) 75%
    (E) 100%

108. Which type of inheritance is illustrated by the paragraph?
    (A) Complete dominance
    (B) Incomplete dominance
    (C) Codominance
    (D) Sex linkage
    (E) Multiple genes

Questions 109–111

Base your answers to Questions 109–111 on the experiment described below and on your knowledge of biology.

In an experiment two potted plants were subjected to the same conditions. The leaves of both plants were exposed to air containing radioactive carbon ($^{14}CO_2$). The roots were immersed in a nutrient solution containing radioactive phosphorus (^{32}P). Part of the vascular system of one plant was ringed or tied off. At the end of 6 hours, the plants were observed and the following sketches made:

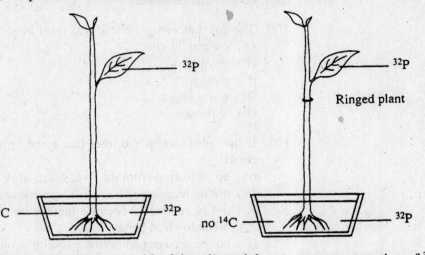

109. The tissue in the stem and leaf that showed the greatest concentration of ^{32}P was the
    (A) sclerenchyma
    (B) parenchyma
    (C) xylem
    (D) phloem
    (E) cambium

110. The fact that no ^{14}C was found in the roots of the tied or ringed plant stem is attributed to
    (A) the interference of transport in the xylem
    (B) the interference of transport in the phloem
    (C) the interference of transport in both the xylem and phloem
    (D) the lack of photosynthetic activity in the leaf
    (E) the lack of absorption by the roots

111. In dicots, it is easy to tie off one kind of vascular tissue because each type of tissue is separated by a layer of cells known as
    (A) cambium
    (B) spongy cells
    (C) parenchyma
    (D) pith
    (E) sclerenchyma

Questions 112–114

Base your answers to Questions 112–114 on the paragraph and diagram below and on your knowledge of biology.

A student conducted an experiment to determine the conditions that affect the flowering of chrysanthemums. A group of the plants that was subjected to 14 hours of darkness flowered (*A* in the diagram below), while a group subjected to less than 14 hours did not flower (*B* in the diagram). In a second part of the experiment, the student interrupted the 14-hour dark period with a flash of red light. No flowering was observed (*C* in the diagram). In a third series of experiments, the plants were exposed to 14 hours of darkness, but their dark period was interrupted first by red light (R) and then by infrared (F). Flowering occurred (*D* in the diagram).

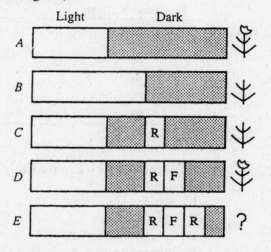

112. The photoperiod of these plants may be considered to be
    (A) long day, short night
    (B) short day, long night
    (C) long day, long night
    (D) short day, short night
    (E) indeterminate

113. Which of the following can be concluded from the experimental data?
    (A) Red light alone inhibits flowering.   ✓
    (B) Red light alone is required for flowering.
    (C) Infrared light inhibits flowering.
    (D) Infrared light is required for flowering.
    (E) Exposure to infrared light reduces the length of the dark period required for flowering.

114. In the experiment marked *E* in the diagram, what result would be expected?
    (A) No flowering will occur.
    (B) Flowering will occur.
    (C) Not enough information is provided to venture an educated guess.
    (D) Flowering will occur only 50% of the times.
    (E) The plants will die from overexertion.

Questions 115 and 116

Answer Questions 115 and 116 on the basis of the experiment described in the paragraph that follows.

A dog was placed in a special room free of stimuli. On repeated trials a tone was sounded for 5 seconds. The dog was given food approximately 2 seconds after the tone. Trials, 1, 5, 15, 20, 25, and 30 were test trials. The tone was sounded for 30 seconds, but no food was given.

The following data were collected:

Test Trial Number	Number of Drops of Saliva Secreted	Number of Seconds Between Tone and Salivation
1	0	0
5	28	15
15	22	8
20	58	3
25	60	1
30	55	3

115. The experimenter was most likely studying
    (A) habit
    (B) instinct
    (C) simple reflex
    (D) conditioned reflex
    (E) imprinting

116. The experiment is most closely associated with the work of
    (A) Pavlov
    (B) Darwin
    (C) Mendel
    (D) Watson
    (E) Lorenz

Questions 117 and 118

Questions 117 and 118 refer to an experiment with 50 newborn rats to determine the importance of two nutrients, A and B, in their diets. The dashed-line curve shows the normal growth rate of rats based on previous experiments. The solid-line curve shows the growth rate of the 50 newborn rats, which were fed a normal diet containing nutrients A and B from birth to point X. At point X, the rats were deprived of both nutrients. At point Y, nutrient A was again added to the diet. At point Z, nutrient B was added and nutrient A was continued.

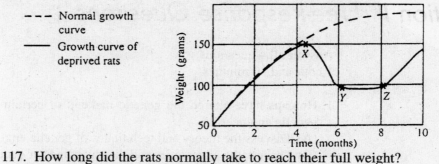

117. How long did the rats normally take to reach their full weight?
(A) 2 months
(B) 5 months
(C) 6 months
(D) 7 months
(E) 8 months

118. If the experiment had continued as described except that at point Z nutrient B had not been returned to the diet of the 50 rats, it is reasonable to conclude that these rats would most likely have
(A) lived for 4 months and then died
(B) remained about half the size of normally developed rats
(C) continued to gain weight, but at a slower rate than the normal rats
(D) become sexually immature adults
(E) continuously lost weight

Questions 119 and 120

The ability to roll the tongue into a U shape is considered a dominant trait. Individuals who cannot do this are considered to have a recessive genotype.

Students in a biology class carried out an experiment to study the trait in their community. The following data were collected from adults in the community:

	Tongue Rollers	Not Tongue Rollers
Number	640	360
Percent	64	36

119. The frequency of the gene for tongue rolling in the population of this community was approximately
(A) 0.64
(B) 0.36
(C) 0.8
(D) 0.6
(E) 0.4

120. Approximately what proportion of the people in this community would be heterozygous for tongue rolling?
(A) 35%
(B) 16%
(C) 24%
(D) 48%
(E) 80%

Free-Response Questions #3

Answer all 4 questions
1 hour and 30 minutes

1. Humans have altered the genetic makeup of certain microorganisms through genetic engineering.
   (A) Discuss the theory and techniques of genetic engineering.
   (B) Discuss the advantages and disadvantages of genetic engineering.

2. There are several types of nucleic acids in a cell.
   (A) Discuss the structure and function of each nucleic acid in the synthesis of a polypeptide.
   (B) Discuss the role of transcription and translation in the synthesis of a polypeptide.

3. One of the factors that enable an organism to survive is the ability of the body to respond to the presence of foreign matter.
   (A) Discuss the difference between the humoral and cell-mediated immune response.
   (B) Discuss the function of B-cell lymphocytes and T-cell lymphocytes in the immune response.
   (C) Discuss the formation, biochemical nature, and function of antibodies.

4. In an experiment, a student worked with two species of flour beetles, *Tribolium confusum* and *Tribolium castaneum*. Four pairs of *T. confusum* w ̇ e placed into a container of flour. Four pairs of *T. castaneum* were place ̇ ̇o a second container of flour. Four pairs of *T. confusum* and four pair ̇ ̇ *castaneum* were placed into a third container of flour. Each container h ̇ ̇ me amount of flour and was kept under the same environmental c ̇ ̇ The table below indicates the results of the experiment for a nine mc ̇ ̇

	Population of Organism Per Contai ̇		
Months	*T. confusum* Container #1	*T. castaneum* Container #2	*T. co* ̇ ̇ *astaneum* ̇3
1	8	8	8
2	18	16	̇9
3	32	46	̇0
4	48	58	̇0
5	70	88	̇5
6	85	109	̇0
7	40	60	̇0
8	35	43	̇5
9	20	15	5
10			
11			

(A) Make a prediction as to the population of each container for the tenth and eleventh months.
(B) Explain the factors operating in containers #1 and #2 to produce the observed results.
(C) Discuss the possible causes for the observed results in container #3.

Answers to Model AP Biology Examination No. 3
Section I

1. D	31. D	61. B	91. B
2. C	32. B	62. B	92. D
3. B	33. E	63. C	93. B
4. E	34. C	64. E	94. A
5. D	35. C	65. D	95. B
6. B	36. A	66. E	96. C
7. C	37. E	67. C	97. A
8. B	38. D	68. D	98. A
9. C	39. D	69. A	99. D
10. A	40. A	70. A	100. C
11. C	41. B	71. D	101. C
12. C	42. D	72. C	102. B
13. C	43. A	73. C	103. B
14. C	44. E	74. C	104. E
15. E	45. B	75. C	105. B
16. B	46. C	76. D	106. A
17. A	47. B	77. C	107. B
18. B	48. E	78. C	108. C
19. C	49. D	79. E	109. C
20. A	50. D	80. B	110. B
21. B	51. C	81. E	111. A
22. E	52. A	82. A	112. B
23. B	53. B	83. E	113. A
24. E	54. D	84. A	114. A
25. A	55. B	85. A	115. D
26. E	56. A	86. D	116. A
27. C	57. D	87. B	117. E
28. D	58. B	88. B	118. B
29. D	59. C	89. A	119. E
30. C	60. E	90. B	120. D

Explanations to Model AP Biology Examination No. 3
Section I

1. D Homeostasis is the process by which a water b͟a͟l͟a͟n͟c͟e is maintained in a
      paramecium. Water balance is maintained by the ͟c͟o͟n͟t͟r͟a͟c͟t͟i͟l͟e vacuole.

2. C Bryophytes are nonvascular plants. Bryophytes ͟l͟a͟c͟k roots, st͟e͟m͟s, and
      leaves.

3. B The birds appeared on the earth a little more tha͟n ͟1͟5͟0 ͟m͟i͟l͟l͟i͟o͟n ͟y͟e͟a͟r͟s ago.

4. E Five vertebrate groups were present on the earth ͟2͟0͟0 ͟m͟i͟l͟l͟i͟o͟n ͟y͟e͟a͟r͟s ago.

5. D Punctuated equilibrium is the term used to expla͟i͟n ͟t͟h͟e ͟s͟u͟d͟d͟e͟n appearance
      of a new species. Niles Eldredge proposed this ͟t͟h͟e͟o͟r͟y ͟b͟a͟s͟e͟d ͟o͟n obser-
      vations that organisms tend to remain unchang͟e͟d ͟f͟o͟r ͟m͟i͟l͟l͟i͟o͟n͟s ͟o͟f years
      and then to be suddenly replaced by a different s͟p͟e͟c͟i͟e͟s.

6. B The basic difference between plants and animal͟s ͟i͟s ͟t͟h͟e ͟a͟b͟i͟l͟i͟t͟y ͟o͟f plants to
      fix carbon dioxide. Carbon fixation is a part of th͟e ͟p͟h͟o͟t͟o͟s͟y͟n͟t͟h͟e͟t͟i͟c reaction.

7. C In cyanide poisoning, the electron transfer to ͟o͟x͟y͟g͟e͟n ͟i͟s ͟b͟l͟o͟c͟k͟e͟d. The
      cytochromes make up the part of the respirato͟r͟y ͟c͟h͟a͟i͟n ͟t͟h͟a͟t ͟t͟r͟ansfers
      electrons to oxygen. Cytochromes contain iron. Th͟e ͟c͟y͟a͟n͟i͟d͟e ͟b͟i͟n͟d͟s ͟t͟h͟e iron.

8. B Xylem is conducting tissue that transports mate͟r͟i͟a͟l͟s ͟f͟r͟o͟m ͟r͟o͟o͟t͟s to the
      leaves of plants. The only upward flow of mat͟e͟r͟i͟a͟l͟s ͟o͟c͟c͟u͟r͟s ͟t͟hrough
      the xylem.

9. C Structure *A* represents a molecule of tRNA, whi͟c͟h ͟i͟s ͟a ͟c͟l͟o͟v͟e͟r-shaped
      molecule composed of nucleotides. tRNA transp͟o͟r͟t͟s ͟a͟n ͟a͟m͟i͟n͟o acid
      to its appropriate position along an mRNA stran͟d.

10. A The sequence of bases in structure *A* would ͟m͟a͟t͟c͟h ͟C͟C͟A ͟a͟n͟d G. The
       region of structure *A* that carries a specific trip͟l͟e͟t ͟i͟s ͟k͟n͟o͟w͟n as the
       anticodon. An anticodon follows the base pairi͟n͟g ͟o͟f ͟n͟u͟c͟l͟e͟i͟c acids.
       Cytosine pairs with guanine; adenine pairs with u͟r͟a͟c͟i͟l.

11. C All of the choices are functions of the nephron e͟x͟c͟e͟p͟t ͟f͟o͟r ͟r͟e͟d blood
       cell absorption. Nephrons are filtering units in ͟t͟h͟e ͟k͟i͟d͟n͟e͟y͟s. ͟K͟idneys
       are excretory organs that control pH, elect͟r͟o͟l͟y͟t͟e ͟a͟n͟d ͟s͟odium
       concentration, and secretion of hydrogen ions.

12. C The biome represented by the diagram is known ͟a͟s ͟t͟a͟i͟g͟a. ͟T͟a͟i͟g͟a is a
       region dominated by spruce and fir trees (conif͟e͟r͟s). ͟M͟o͟o͟s͟e ͟a͟r͟e ͟c͟ommon
       in the taiga.

13. C If the blood was type AB, the blood cells would ͟a͟g͟g͟l͟u͟t͟i͟n͟a͟t͟e ͟o͟n ͟bo͟th the
       left and right side of the slide. AB blood type co͟n͟t͟a͟i͟n͟s ͟a͟n͟t͟i͟g͟e͟n͟s ͟t͟h͟a͟t react
       with the antibodies of both anti-A and anti-B ser͟a.

14. C Environmental changes selected against the lig͟h͟t ͟m͟o͟t͟t͟l͟e͟d ͟m͟o͟t͟h ͟a͟nd for
       the black moth. The Industrial Revolution pollut͟e͟d ͟t͟h͟e ͟e͟n͟v͟i͟r͟o͟n͟m͟e͟n͟t. The
       black moths blended into the dark gray envir͟o͟n͟m͟e͟n͟t ͟w͟h͟i͟l͟e ͟t͟h͟e light
       mottled moths became very visible and easy ͟p͟r͟e͟y ͟a͟g͟a͟i͟n͟s͟t ͟t͟h͟e dark
       environment.

15. E The sliding filament hypothesis refers to the ͟c͟o͟n͟t͟r͟a͟c͟t͟i͟o͟n ͟o͟f ͟m͟uscle.
       According to the hypothesis, two muscle cell pr͟o͟t͟e͟i͟n͟s, ͟a͟c͟t͟i͟n ͟a͟n͟d ͟m͟yosin,
       slide over each other, shortening the muscle.

16. **B** Embryonic induction best explains the reason why an eye from transplanted eye tissue develops in the back region of the host embryo. Embryonic induction states that specific cells influence and direct tissues in a region of the embryo.

17. **A** The molecule represented by the diagram is a glucose molecule. A polymer is a compound composed of repeating units. Starches are polymers of glucose, and glycogen and cellulose are starches.

18. **B** $\underset{Sp \qquad\qquad Cu\; H \quad e}{\rule{8cm}{0.4pt}}$ represents the order of the genes on chromosome #3 of the fruit fly. The percentage of crossing-over can be used to determine the distance between genes and thus their positions in relationship to each other.

19. **C** Environmental factors turn genes "on" and "off." Thus, soil pH influences gene expression.

20. **A** The diagram represents the evolutionary history of two organisms. Each history is expressed as a branching tree. Of the choices given, organisms A and B have the closest common ancestor, they branch from organism E.

21. **B** The organism that can adapt to the environment survives. Organism Y has continued to exist since it first appeared. Therefore, organism Y has been the most successful in the environment over time.

22. **E** The evidence to support this evolutionary pathway is best obtained from the study of paleontology. Paleontology is the study of fossils.

23. **B** The activation of cyclic AMP by adenyl cyclase occurs through the action of protein hormones on the cell membrane. The cell membrane contains receptor sites for protein hormones. The bond between the hormone and the receptor site activates an enzyme which initiates a series of reactions in which glycogen is broken down.

24. **E** An insufficient store of glycogen in the liver is NOT a reason for a man hanging from a ledge to lose his grip. Other factors operated before the glycogen reserves could have been used up. These factors included a depletion in ATP, an accumulation of lactic acid, muscular tetany, and oxygen debt.

25. **A** Structure 1 represents the pacemaker. Structure 1 is the S-A node, which regulates the rhythmic beating of the chambers of the heart.

26. **E** Fluids would accumulate in tissue spaces if a parasitic worm lived in the lymphatic vessels. The lymphatic vessels collect the tissue fluid and transport it back into the bloodstream.

27. **C** The ribosome of the cell is most likely to be affected by the poison. Protein synthesis occurs in the ribosomes.

28. **D** In chromatography, closely related substances are separated from each other on the basis of their solubility in the solvent and their affinity for the supporting medium (silica-gel, filter paper, and so on).

29. **D** Cell A is a red blood cell. Red blood cells transport oxygen.

30. **C** The gap between cells (area 3) is known as the intercellular space and is filled with the intercellular fluid.

31. **D** The condition in which mitochondria have an abnormal structure is most likely correlated with disturbances in cell energy supply. The mitochondria are organelles in which cellular respiration occurs.

32. **B** The lysosome contains over 40 hydrolytic enzymes. In Tay-Sachs disorder, the enzyme that acts on a lipid is absent.

33. **E** Guard cells become turgid when they take in water by osmosis. An osmotic gradient is created when K^+ ions are drawn into the guard cells.

34. **C** At a concentration of 3.0 mg/100 mL, all of th
the enzyme. The rate of enzyme activity increa
enzyme increases. The rate of activity levels of
molecules are engaged by the enzyme.

35. **C** *Achillea borealis* is most closely related to *Ach
to Linnaean classification, every organism h
name. Here the two organisms belong to the
different species.

36. **A** A fruit is a ripened ovary of a flower.

37. **E** The advantage of asexual reproduction ove
offspring with the same genetic makeup as the p
possess the same characteristics that made th
particular environment.

38. **D** The cells involved in antibody production in
lymphocytes. Antibodies are chemicals that dest

39. **D** In desert plants, carbon dioxide enters at night
compounds which can be reconverted to carb
pathway of carbon dioxide fixation in desert pl
cycle or the Hatch-Slack pathway.

40. **A** Karyotyping is a procedure by which speci
chromosomes are photographed and arranged i
used to detect abnormalities in the number and sl

41. **B** The reabsorption of most of the glucose occu
through region 2, which represents the proximal
nephron of the kidney.

42. **D** The termites are primary consumers. Primary c
Herbivores eat plants.

43. **A** Dinoflagellates are the organisms responsibl
dinoflagellates make up part of the phytoplankton
explosive increase of these organisms.

44. **E** Convergent evolution is the development of total
the same manner because of adaptation to similar

45. **B** Crossing-over is the exchange of genetic mater
chromosomes that occurs during synapsis in the p

46. **C** Fatty acids enter the villi by diffusion, which
transport. The absorption of glucose in the villi, t
by the nephrons, the reabsorption of calcium ions
reestablishment of ionic concentrations in nerv
transport.

47. **B** Structure *A* belongs to the lymphatic system and
closed-ended tubes that absorb fats from the diges

48. **E** Lichens would probably be able to grow on a
island. Lichens are able to live on bare rock and d

49. **D** An increase in the reproductive rate would increa
squirrel population. The greater the number of orga
food and living space, the greater is the competition

50. **D** A countercurrent mechanism plays an important
the kidney. The flow of fluids in opposite directi
constitutes the countercurrent mechanism, which
and concentration of materials in the nephrons
filtrate.

51. **C**   The fact that the petals of a carnation turn blue when the cut end is placed into blue dye is evidence that xylem tissue is present in the stem. Xylem tissue transports water and minerals up the stem to the leaves.

52. **A**   Hydrolysis is the process by which a molecule is split through the action of an enzyme and through the addition of water.

53. **B**   Compound *B* is an enzyme. An enzyme is a protein.

54. **D**   The water in the reaction serves to supply atoms that complete the structures of compounds *C* and *D*. The water molecule provides the electrons and ions to the cleaved molecules.

55. **B**   The osmotic pressure is greater in area *A* than in area *B* is an incorrect statement. Proteins in the blood control osmotic pressure. Because the protein composition of the blood is the same throughout the system, the osmotic pressure is constant.

56. **A**   The importance of the sudden enlargement of stomates upon exposure to light in plants is that it makes possible an increased intake of carbon dioxide. The stomates are openings in leaves which allow for the exchange of gas with the atmosphere. In the presence of sunlight, the plant requires carbon dioxide for photosynthesis.

57. **D**   According to the theory of organic evolution, the first living things were probably able to obtain energy directly from organic molecules in water. It has been proposed that organic molecules, which were produced through chemical reactions in the atmosphere, were anaerobically broken down by the first living things.

58. **B**   When a sprig of an elodea plant was exposed to different wavelengths of light, the lowest rate of oxygen would be produced under the green wavelength of light. Oxygen is a product of photosynthesis. Green wavelengths of light are least effective in photosynthesis.

59. **C**   A coenzyme is a small nonprotein organic molecule that is loosely attached to an enzyme. NAD is a coenzyme. Vitamins are coenzymes, and NAD is a vitamin B derivative.

60. **E**   The electron transport chain for photosynthesis reactions is located in the thylakoids. Thylakoids are stacks of membranes within the chloroplasts.

61. **B**   Fermentation is less efficient in ATP production than is cellular respiration. Fermentation is the process by which glucose is broken down anaerobically, and glucose is converted to pyruvate. Approximately 4 ATPs are produced through fermentation; 36 ATPs through cellular respiration.

62. **B**   The control of the rate of glycolysis depends upon the activity of the allosteric enzyme, phosphofructokinase. The enzyme is activated by ADP and inhibited by ATP.

63. **C**   Diagram 3 represents the biomass in a stable community, where there are fewer numbers of organisms at each level of the food chain. There are more producers than herbivores, and more herbivores than carnivores.

64. **E**   Thylakoids are membranous sacs located within the chloroplasts. The stroma is the space outside of the thylakoids. Oxidative phosphorylation, the production of ATP, occurs in the chloroplasts.

65. **D**   The statement which is true about the origin of life on earth is that it is conceivable in terms of our present knowledge of chemistry and physics.

66. **E**   The region of the brain that is stimulated when a human expresses the emotion of anger is the hypothalamus. The hypothalamus is the center that controls emotions, temperature, thirst, and pain.

67. C According to the fluid mosaic model of the
    embedded in the lipid bilayers.

68. D A prokaryotic cell lacks an organized nucleus.
    prokaryotic cell consists of a chromosome locate
    the cell, the nucleoid.

69. A Osmosis is the movement of water down a conc
    there is less free water outside of the cell, wate
    The cell shrinks.

70. A When the membranes of the alveoli, air sacs in th
    a decrease of oxygen concentration in the blo
    occurs in the alveoli. The thickened membranes
    oxygen into the blood.

71. D Macrophages, upon an encounter with an an
    antibodies is an incorrect statement concernir
    theory. Macrophages are phagocytic cells that er
    not produce antibodies.

72. C One of the most important differentiating featur
    phyta and cyanophyta is the absence of chloropl
    The cyanophyta are prokaryotic cells, which
    organelles. Blue-green algae are chlorophyta.

73. C According to the Punnet square, one parent must
    and the other be white for the offspring to be 50%

	B	b
b	Bb	bb
b	Bb	bb

74. C The relationship between the termite and pr
    mutualism. Mutualism is a relationship between tv
    both organisms are benefitted by the association.

75. C The shaded area, which represents the territory sha
    the area where competition is likely to occur betwe

76. D The most likely reason why the range of pair B is
    A is that pair B has less food available near its nest
    extended its range to increase its chance of capturin

77. C Homologous structures are similar in form and p
    The arm of a human, the flipper of a whale, the fore
    wing of a bat are homologous structures.

78. C Plasmids are genes in bacteria that are locat
    chromosome. Plasmids contain genes that can b
    bacterial cells.

79. E Cell plate formation is not associated with nuclear c

80. B The cell in diagram 2 is different from the cells in d
    has twice the number of chromosomes as each cell
    in diagram 2 is a fertilized egg cell or zygote. A z
    two gametes fuse. Gametes contain half of the ch
    the zygote.

81. E The developmental stages of a frog represented by t
    in a watery environment. The embryo of a frog deve
    of the female in the external environment.

82. A Fertilization is the process by which the cell in diagram 2 is formed. Fertilization is the union of the nuclei of two gametes.

83. E Photolysis is part of the photosynthetic process by which water is broken down into hydrogen and oxygen. Sunlight is the source of energy for the process.

84. A The neurotransmitter released at the junction between the nerve and the muscle is acetylcholine, the neurotransmitter at the motor end plate in the central nervous system. The motor end plate is the region where the motor neuron synapses with a muscle fiber.

85. A The type of nerve cell represented by the diagram is a motor neuron. A motor neuron transmits impulses to an effector (muscle or gland).

86. D The oxygen from water is converted to oxygen gas. Oxygen is released into the atmosphere during photosynthesis.

87. B An enzyme with two binding sites is known as an allosteric enzyme. The two sites are the active site and the allosteric site. The allosteric site binds with a product that inhibits the enzyme action.

88. B The process of mRNA transcription occurs in the nucleus. Transcription is the process by which an mRNA strand is made from a DNA template. mRNA carries the genetic code from the nucleus to the ribosome.

89. A The experimenter was trying to produce the symptoms of diabetes when she removed the pancreas from a living rat. The pancreas produces insulin, which regulates the passage of sugar from the blood into the cells.

90. B The ion most important to nerve cell activity is sodium. The wavelike movement of sodium across a cell membrane is known as an impulse. Nerve cells function to conduct impulses.

91. B Letter *B* represents the biome known as the tundra, which is delineated by a climax community of lichens and mosses.

92. D The placenta in the human fetus serves as a respiratory organ. The placenta is a vascularly rich, spongy area within the uterus. Materials are exchanged between the mother and the fetus in this area.

93. B Down's syndrome is a result of meiotic nondisjunction, that is, the failure of chromosomes to separate from each other during meiosis. The individual receives an extra chromosome. One of the results is mental retardation.

94. A All of the choices are characteristics of flowering plants except they produce autotrophic gametes. The egg is housed in the ovule, and the sperm in the pollen grain. The gametes are not autotrophic and cannot produce their own food.

95. B Gastric juice is produced and secreted by specialized cells in the stomach lining.

96. C Chemical digestion is completed and the products of digestion are absorbed in the small intestine.

97. A Peristaltic action in the esophagus transports food from the mouth to the stomach.

98. A Only DNA molecules may contain the four bases adenine, cytosine, thymine, and guanine. RNA molecules contain uracil bases instead of thymine bases.

99. D Neither DNA nor RNA molecules contain the nitrogenous base urea. Urea is a metabolic waste.

100. C Both DNA and RNA molecules are present in the nucleus of a cell. DNA is the chromosomal material, and RNA is synthesized in the nucleus.

101. C Both DNA and RNA molecules consist of (
      Nucleotides are the building blocks of nucleic aci

102. B Only RNA molecules carry genetic information
      ribosomes. RNA molecules are transcribed from l

103. B The most probable reason for the increase in the
      1950 was that the coyote population decli
      relationship indicated by the graph is one of pre
      prey upon the rabbits, keeping their population in

104. E The most probable explanation for the decline of
      by humans. Since the decline in coyotes (
      introduction and since cattle and coyotes do not
      food, the individuals that introduced the cattle mu
      the coyote population. Because coyotes are
      perceived as dangerous to the cattle and were hunt

105. B The gas that was generated was most likely oxy
      glowing splint to burst into flames. Oxygen is als
      decomposition of peroxide.

106. A If the substance in the liver that acted on the pero:
      could be recovered from the liver tissue after t
      Enzymes are not consumed in the reactions that th

107. B If both parents have the minor form, 25% of their
      to be normal.

	S	s
S	SS	Ss
s	Ss	ss

108. C Codominance is the type of inheritance represen
      Codominance is a condition in which an individual
      traits of both parents.

109. C The tissue in the stem and leaf that showed the gre
      ^{32}P was the xylem. Xylem is vascular tissue in p
      water and minerals from the roots in the soil to the l

110. B The fact that no ^{14}C was found in the root of the
      attributed to the interference of transport in the phlc
      the leaf for photosynthesis. The products of
      transported from the leaves to the roots through
      known as phloem.

111. A In dicots it is easy to tie off one kind of vascular tiss
      of tissue is separated by a layer of cells known as ca
      undifferentiated plant tissue.

112. B The photoperiod of the plants is considered to be s
      The plants must have long periods of darkness to flo

113. A From the experiment it may be concluded that rec
      flowering. No flowers were produced by plants wh
      interrupted by a flash of red light.

114. A In the experiment marked E no flowering would b
      The red light would interrupt the dark period of the p

115. D The experimenter was most likely studying a co
      conditioned reflex is a learned response to a stimulus

116. A The experiment is most closely associated with the v
      formulated the concept of the conditioned reflex.

117. E In 8 months rats normally reached their full weight.
118. B The rats would most likely have remained about half the
        developed rats if nutrient *B* had not been returned at poin
        would continue to weigh less than 100 grams.
119. E According to the Hardy-Weinberg principle, $p + q = 1$; q represents the
        recessive allele in the gene pool. Since 36% of the people have the
        recessive allele, $\sqrt{0.36} = 0.6$. Since $p + 0.6 = 1$, $p = 0.4$.
120. D According to the Hardy-Weinberg principle, $p^2 + 2pq + q^2 = 1$; $2pq$
        represents the heterozygous individual. $2pq = 2(0.6)(0.4) = 0.48 = 48\%$.

Answers to Model AP Biology Examination No. 3
Section II

1. (A) The technique used in genetic engineering is called *recombinant technology*. This includes gene splicing and gene cloning.

In *gene splicing*, a specific section of DNA is removed from a eukaryotic cell and is inserted into a plasmid. A *plasmid* is a small circular DNA molecule outside of the main chromosome.

Bacterial cells are stripped of their cell walls. The stripped cells are subjected to differential centrifuging, which separates the plasmids from the other cell parts.

The plasmids are then cleaved (opened up) by restriction enzymes. The enzymes work on one particular point of the plasmid DNA.

DNA fragments from the eukaryotic cells are added to the cleaved plasmid. Ligase, an enzyme, sutures or sews up the plasmid DNA with the eukaryotic DNA. Successfully spliced plasmids are separated by centrifuging. They are heavier than the pure or incomplete plasmids.

The new recombinant DNA plasmids are then *cloned* and are added to the media containing susceptible bacterial cells or protoplasts. The protoplasts incorporate the plasmids into their cells. The plasmids reproduce as the protoplasts reproduce and colonies of cells are formed.

(B) The advantanges of genetic engineering are as follows.
1. Vital chemicals for the benefit of humans can be produced by bacterial organisms.
2. Vital chemicals can be produced cheaply. Interferon, insulin, hoof and mouth antigen, and human growth hormone are produced through genetic engineering. Previously, the growth hormone had to be extracted from cadavers at great expense.

The disadvantages of genetic engineering are as follows.
1. An "uncontrolled microbe" that has no known treatment could be produced.
2. Microbes that produce dangerous substances such as toxins could be made for germ warfare.

2. There are two basic types of nucleic acids, DNA (deoxyribose nucleic acid) and RNA (ribose nucleic acid). RNA is further divided into mRNA (messenger), tRNA (transfer), and rRNA (ribosomal).

DNA is a double-stranded molecule made up of the poly...
nucleotide consists of a sugar, base, and phosphoric acid...
bases (triplet code) in a DNA sequence signifies a specific...
DNA contains the codes for the arrangement of amino aci...
polypeptide is a chain of amino acids. Proteins are polypept...

mRNA is a single strand of nucleotides composed of th...
same bases as DNA with the exception of the base uracil,...
of thymine. mRNA is copied from the DNA molecule....
process of copying the codes from the DNA molecule. DN...
The DNA becomes the template for the synthesis of mRN...
the complementary base sequence for the arrangement...
polypeptide. The three-base code in mRNA is called a codo...
the codons for the polypeptide to the ribosomes.

tRNA is a short chain of nucleotides shaped like a cl...
tRNA bonds to a specific amino acid. One side chain l...
ribosome. The anticodon loop is the complementary t...
recognizes a codon on mRNA. tRNA functions to transpor...
specific site on the mRNA molecule. For every amino acic...
tRNA molecule.

rRNA is ribosomal RNA. The ribosome is composed of s...
chained RNA combined with proteins. The ribosome is th...
protein synthesis occurs. rRNA is synthesized from a...
molecule that makes up the nucleolus.

Translation is the conversion of the genetic code i...
Translation includes initiation, elongation, and termination....
mRNA combines with the ribosome. Elongation is the...
polypeptide chain. As the ribosome moves down the mRNA...
the appropriate amino acid into the ribosome. Eventually the...
portion of the mRNA that contains a stop code. The synthes...
is terminated.

3. The humoral immune response involves the product...
Antibodies are produced and secreted by B lymphocytes.

When B lymphocytes encounter an antigen (foreign b...
undergo mitosis, and differentiate, forming clones of plasm...
cells. The plasma cells secrete thousands of molecules of antil...

Antibodies are plasma proteins known as immunoglobin...
molecule made up of four polypeptide chains. Two chain...
heavy chains, and two chains as light chains. The chains a...
disulfide linkages. There are two regions on each chain, v...
constant region. The constant regions of the light and heavy c...
the variable regions are different. It is the variable region tha...
antigen. The union between the antibody and antigen is kno...
antibody complex. Two antigens can be bound to the same...
The binding of antibody molecules to antigens induces phag...
engulfing of molecules by white blood cells. In addition, th...
complex activates circulating plasma proteins known a...
complement causes the rupture of the antigens.

Memory cells store the information for the production of...
invasion by the same antigen occurs, the memory cells give ri...

This response, called a secondary immune response, occurs more rapidly than the initial response to the antigen.

The cell-mediated immune response involves the T lymphocytes. T lymphocytes produce antibodies. Unlike the antibodies produced by B lymphocytes, the antibodies produced by T lymphocytes are integrated into cell membranes of the lymphocyte. The antibodies on the cell surface recognize antigens and the T lymphocytes surround the antigen. The T lymphocytes usually attack cells that have been affected by viruses, parasites, or tissue grafts.

Both B lymphocytes and T lymphocytes originate in the bone marrow, where the B lymphocytes develop. The T lymphocytes, however, migrate to and develop in the thymus.

4. (A) In each of the containers, the population will continue to decline until no organisms remain. In container #1 and #2, the principles of intraspecific competition are operating. In container #3, both the principles of interspecific and intraspecific competition are operating.

(B) In container #1 and #2, intraspecific competition is operating. Intraspecific competition is a competition among individuals of the same species for basic resources. When the organisms were introduced into the containers, there were unlimited resources (food and space). The organisms steadily reproduced. By the sixth month, the carrying capacity of the environment had been exceeded. The *carrying capacity* is the ability of the environment to support a limited number of individuals. The organisms competed for the limited food supply. Overcrowding, disease caused by overcrowding conditions, and the accumulation of toxic wastes also contributed to the population's decline.

These populations exhibit a J-shaped growth curve. The *J-shaped curve* is characteristic of organisms that crash periodically. However, under laboratory conditions there is no opportunity for the population to recover from resource deprivation.

(C) In container #3, two types of competition were exhibited, intraspecific and interspecific competition. *Interspecific competition* is the competition of two different species for the same resources. According to Gause's Competition Exclusion Principle, no two organisms may occupy the same niche. The niche is the position on the food chain. Both species of *Tribolium* were competing for the same resources. The species that is better adapted to the environment will survive while the other will become extinct. The adaptation of *T. castaneum* over *T. confusum* is not quite clear. Predation may be involved. Predation is the hunting and eating of one organism by another organism.

Since *T. castaneum* lives on flour without predation as shown in container #2, intraspecific competition for the already reduced resources led to its demise. Its population also followed a J-shaped curve in container #3.

Evaluation of Model AP Examination No. 3

Below is a breakdown of the test questions by subject
mark (✓) in the box below each question answered correctl
number correct for that area in the space provided at the righ
percent you answered correctly in that particular subject area
answer 4 out of 6 in a subject area, you answered 66% corre

Finally, add up the numbers of correct answers for all t
overall score. You should have a minimum of 90 correct a
your review and test preparation on the areas where your sco

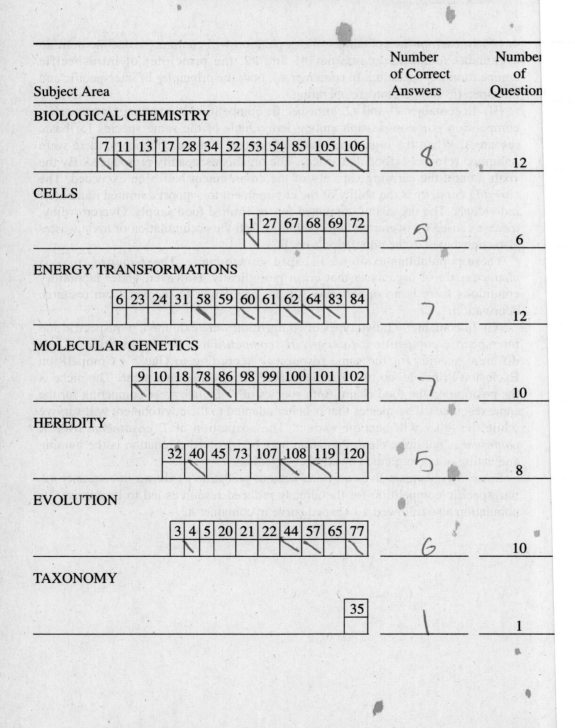

Subject Area	Number of Correct Answers	Number of Questions
BIOLOGICAL CHEMISTRY		
7 11 13 17 28 34 52 53 54 85 105 106	8	12
CELLS		
1 27 67 68 69 72	5	6
ENERGY TRANSFORMATIONS		
6 23 24 31 58 59 60 61 62 64 83 84	7	12
MOLECULAR GENETICS		
9 10 18 78 86 98 99 100 101 102	7	10
HEREDITY		
32 40 45 73 107 108 119 120	5	8
EVOLUTION		
3 4 5 20 21 22 44 57 65 77	6	10
TAXONOMY		
35	1	1

STA, FUNGI

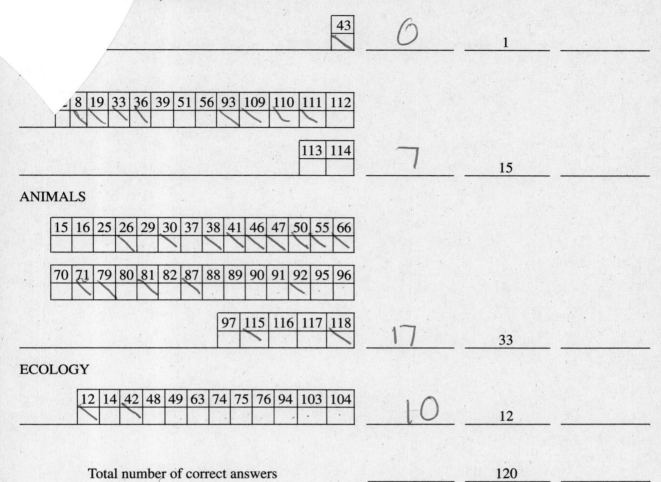

	43

O   1 _____

8	19	33	36	39	51	56	93	109	110	111	112

113	114

7   15 _____

ANIMALS

15	16	25	26	29	30	37	38	41	46	47	50	55	66

70	71	79	80	81	82	87	88	89	90	91	92	95	96

97	115	116	117	118

17   33 _____

ECOLOGY

12	14	42	48	49	63	74	75	76	94	103	104

10   12 _____

Total number of correct answers   120 _____

Plants —re-read!
Animals

Model AP Biology Examination No. 4

Section 1 Multiple-Choice Questions

120 questions
1 hour and 30 minutes

Directions: Each of the questions or incomplete statements in this section is followed by five suggested answers or completions. Select the ONE that is BEST in each case.

1. Flagella in protozoans perform the same function as
   (A) contractile vacuoles in protozoans
   (B) roots in plants
   (C) tentacles in Hydra
   (D) tracheal tubes in insects
   (E) wings in birds

2. Polymerization of macromolecules in a cell is usually accomplished by the process of
   (A) anaerobic respiration
   (B) dehydration synthesis
   (C) hydrolysis
   (D) absorption
   (E) photosynthesis

3. In an experiment, hamsters were fed a diet containing a controlled dose of a mutagenic agent. The mutagenic agent caused chromosomal damage. The hamsters were also fed varying amounts of β-carotene. The graph below indicates the results of the experiment.

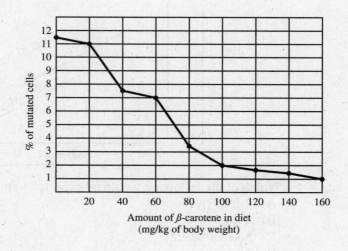

Which statement accurately describes the result of the experiment?

(A) The greatest effect of β-carotene on the percentage of mutated cells occurs at high dosages.

(B) The effect of a mutagenic substance cannot be counteracted by β-carotene.

(C) β-carotene protects hamsters from the effect of radiation.

(D) The presence of β-carotene prevents base substitution.

(E) β-carotene causes chromosomal damage.

4. Which of the following structures are common to both prokaryotic and eukaryotic cells?
(A) Plasmids
(B) Endoplasmic reticulum
(C) Mitochondria
(D) Ribosomes
(E) Vacuoles

5. Fungi can absorb food from their environment as a direct result of
(A) intracellular digestion, only
(B) extracellular digestion, only
(C) extracellular and intracellular digestion
(D) action of vacuoles
(E) sporangia

6. According to the heterotroph hypothesis of the origin of life, which change in the environment MOST LIKELY contributed to the evolution of aerobic organisms?
(A) The appearance of photosynthetic organisms
(B) A decrease in the availability of solar energy
(C) An increase in atmospheric hydrogen concentration
(D) A decrease in atmospheric oxygen concentration
(E) The increase in the availability of inorganic nutrients

7. Organisms move in a variety of ways. Each method of locomotion requires the use of energy. The following graph shows the energy cost of different locomotive methods for different-sized animals.

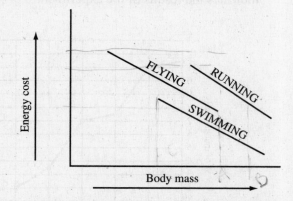

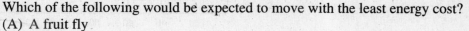

Which of the following would be expected to move with the least energy cost?
(A) A fruit fly
(B) A dog
(C) A killer whale
(D) A cheetah
(E) A hummingbird

8. In order to make apples picked in the fall available in the winter, the apples are stored in a way that will retard ripening. The action of which plant hormone must be inhibited?
(A) Abscisic acid
(B) Auxin
(C) Cytokinin
(D) Gibberellin
(E) Ethylene

9. The hypothesis that the eukaryotic cell evolved from the association of prokaryotic cells living inside larger prokaryotes is known as the
(A) Jacob-Monod model
(B) endosymbiotic model
(C) induced-fit model
(D) fluid mosaic model
(E) autogenous model

10. Which compound is LEAST likely to contain the following molecule?

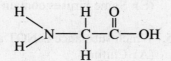

(A) Actin
(B) Keratin
(C) Hemoglobin
(D) Cellulose
(E) Amylase

11. The diagram below represents a pair of homologous chromosomes before and after synapsis. Which type of mutation is indicated after synapsis?

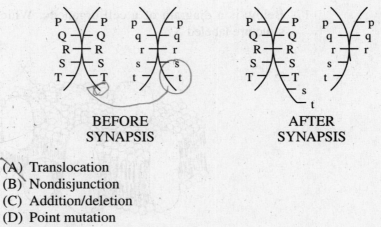

BEFORE
SYNAPSIS

AFTER
SYNAPSIS

(A) Translocation
(B) Nondisjunction
(C) Addition/deletion
(D) Point mutation
(E) Inversion

12. An organelle that is found in both plant and animal cells is the
    (A) centriole
    (B) chloroplast
    (C) mitochondrion
    (D) cell wall
    (E) contractile vacuole

13. What is the order of genes on a chrosome whose frequencies are A-B, 10 map units; A-C, 30 map units; A-D, 25 map units; B-C, 20 map units; B-D, 35 map units?
    (A) A-C-D-B
    (B) D-A-B-C
    (C) C-D-B-A
    (D) A-C-D-B
    (E) B-A-C-D

14. Which statement concerning viruses is NOT true?
    (A) Some viruses contain both single-stranded DNA and RNA.
    (B) Some viruses may contain single-stranded RNA.
    (C) Some viruses contain an enzyme-reverse transcriptase.
    (D) Some viruses may contain double-stranded DNA.
    (E) Some viruses contain double-stranded RNA.

15. Which substance is NOT a polymer of glucose?
    (A) Chitin
    (B) Cellulose
    (C) Glycogen
    (D) Cholesterol
    (E) Starch

16. Which of the following is MOST LIKELY the result of deforestation?
    (A) Soil erosion
    (B) Silting of rivers and streams
    (C) Increased flooding
    (D) Decreased atmospheric carbon dioxide utilization
    (E) all of the above

17. Below is a diagram of a cell organelle. Which is NOT a function of the structure labeled X?

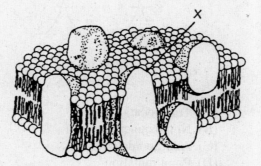

(A) Acts as a binding site for elements of the cytoskeleton.
(B) Forms channels for the passage of material across the membrane.
(C) Can function as enzymes for facilitated and active transport.
(D) Are receptor sites for specific substances such as hormones.
(E) Acts as a binding site for phospholipids giving the organelle a rigid quality.

18. What is the main difference between an ectotherm and an endotherm?
    (A) An ectotherm obtains body heat from the environment; an endotherm generates and retains its own body heat.
    (B) An ectotherm always has a higher body temperature than an endotherm.
    (C) Endotherms employ behavioral modification to control the range of body temperature.
    (D) An ectotherm loses heat to its environment and an endotherm absorbs heat from its environment.
    (E) Ectotherms live in water; endotherms live on land.

19. The graph below shows the number of pushups that an individual completed in each of four 2-minute trials during a 15-minute period. Which statement BEST explains the test results?

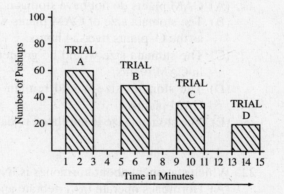

    (A) There was an insufficiency of stored glycogen in the muscle tissue.
    (B) Carbohydrates were changed into alcohol, which caused muscle fatigue.
    (C) Insufficient oxygen reaching the muscle cells resulted in lactic acid build-up.
    (D) Carbon dioxide was retained by the muscle tissue.
    (E) There was a genetic defect in the transport system of the individual.

20. The inner membrane of chloroplasts is similar to the inner membrane of mitochondria in that both
    (A) form compartments
    (B) contain attached ATP synthase
    (C) contain electron transport chains
    (D) have channels for the movement of H^+
    (E) all of the above

21. The graph below represents data obtained from an experiment that tracked stomata size of C_3 plants over 24 hours. How would the stomata size of CAM plants compare to the stomata size of C_3 plants over a 24-hour period?

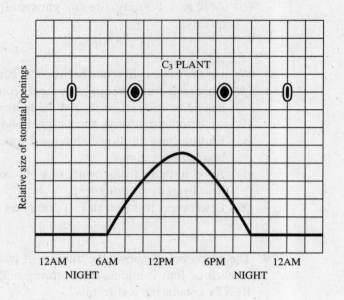

(A) CAM plants do not have stomata.
(B) The stomata size of CAM plants would reflect the same size distribution as the C_3 plants over 24 hours.
(C) The stomata size would be greater during the night than during the day in CAM plants.
(D) The stomata size would remain constant throughout the 24 hours in CAM plants.
(E) The stomata size is unrelated to day or night periods in CAM plants.

22. Which statement about hormones is INCORRECT?
(A) Hormones operate in vertebrate and invertebrate organisms.
(B) Hormones are secreted by special cells of glands in the endocrine system.
(C) Hormones travel to a target organ by way of the circulatory system.
(D) Hormones cause a rapid response of the target organ.
(E) Hormones may be regulated through a feedback system.

Questions 23 and 24

Base your answer to Questions 23 and 24 on the diagrams below and on your knowledge of biology.

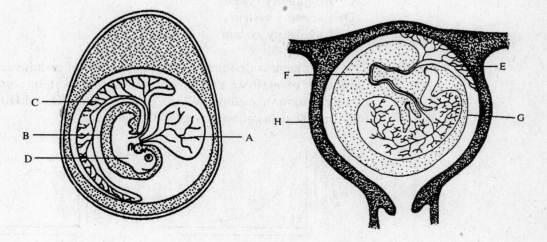

23. The removal of nitrogenous wastes from the bird and human embryo occurs through structures
    (A) A and G
    (B) B and G
    (C) C and E
    (D) C and H
    (E) B and F

24. The embryonic structures that have a similar function in the bird and human embryo are
    (A) A and F
    (B) C and H
    (C) A and E
    (D) B and G
    (E) B and F

25. The essential difference between the humoral and cell-mediated immune system is that
    (A) antibodies from the humoral system act against antigens in the plasma and antibodies from the cell-mediated system operate against antigens inside of body cells
    (B) cells from the humoral system are concentrated in the lymphatic system and cells from the cell-mediated system are concentrated in the liver
    (C) the humoral immune system recognizes thousands of antigens and the cell-mediated immune system recognizes only five antigens
    (D) humoral immunity is a function of lymphocytes and cell-mediated immunity is a function of neutrophils
    (E) cells of the humoral immune system are produced in bone marrow and cells of the cell-mediated immune system are produced in the thymus

26. In vertebrates, which system does NOT have a direct opening to the external environment?
    (A) Reproductive system
    (B) Digestive system
    (C) Respiratory system
    (D) Excretory system
    (E) Circulatory system

27. An experiment is designed to measure the rate of maltase activity. Measured amounts of maltose are slowly added to a system containing a fixed concentration of maltase. Which graph would MOST LIKELY result from plotting the data collected from the experiment.

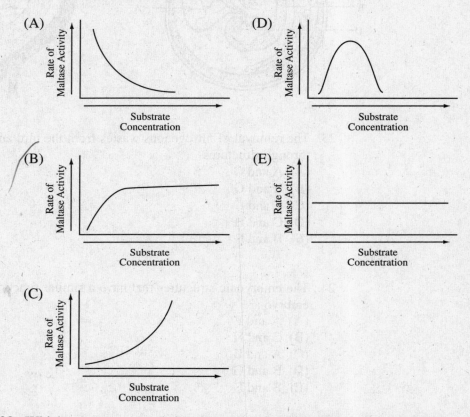

28. Which statement concerning the mode of action of hormones is TRUE?
    (A) Steroid hormones cannot pass through the plasma membrane.
    (B) Peptide hormones operate at the nuclear membrane level.
    (C) Steroid hormones have specific binding sites on the plasma membrane.
    (D) Peptide hormones affect the transcription of specific genes.
    (E) Peptide hormones operate at the level of the cell membrane via synthesis of cyclic AMP.

29. Which osmoregulatory structure is CORRECTLY paired with its organism?
    (A) Flame cell system—earthworm
    (B) Malpighian tubule—protozoan
    (C) Nephron—dog
    (D) Tracheal tubes—horse
    (E) Alveoli—fish

30. A kidney can be transplanted from one identical twin to another without being rejected because
   (A) antibodies in the plasma do not react to antigens in the cells of the transplanted kidney
   (B) the twins have the same major histocompatibility-complex genes (MHC) so that the transplanted kidney does not stimulate a cell-mediated response
   (C) the twins were exposed to each other's histiocompatibility-complex as fetuses and as a result no cell-mediated response is stimulated
   (D) macrophages cannot destroy the transplanted organ because it is too large
   (E) B lymphocytes do not react to cell surface antigens of individuals with the same MHC genes

31. Which organelle contains enzymes that are effective in catalysis deamination, detoxification of alcohol, and degradation of hydrogen peroxide?
   (A) Leucoplast
   (B) Lysosome
   (C) Centrosome
   (D) Tonoplast
   (E) Peroxisome

32. Geranium leaves grow in positions that permit the optimum use of light as a result of
   (A) photosynthesis
   (B) phototropism
   (C) photoperiodism
   (D) photophosphorylation
   (E) photorespiration

Questions 33 and 34

Base your answers to Questions 33 and 34 on the diagram and your knowledge of biology.

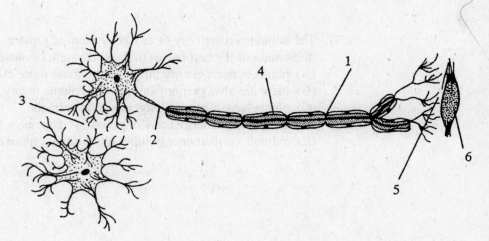

33. Which ion is essential for the complete passage of an impulse from region 5 to 6?
    (A) Sodium
    (B) Potassium
    (C) Chlorine
    (D) Calcium
    (E) Iron

34. A substance that interferes with neurotransmission may operate at regions
    (A) 1, 2
    (B) 2, 3
    (C) 3, 6
    (D) 4, 5
    (E) 3, 5

35. Salmon return to the stream in which they hatched to spawn. This phenomenon is known as
    (A) imprinting
    (B) habituation
    (C) associative learning
    (D) conditioning
    (E) reasoning

36. Which of the following is INCORRECTLY paired with its characteristics?
    (A) Platyhelminthes-bilateral, acoelomates with protosome developmental pattern
    (B) Echinodermata-bilateral, coelomate with deuterostome developmental pattern
    (C) Annelida-bilateral, pseudocoelomate with protostome developmental pattern
    (D) Chordata-bilateral, coelomate with deuterostome developmental pattern
    (E) Nematoda-bilateral, pseudocoelomate with protostome developmental pattern

37. The annual productivity of any ecosystem is greater than the annual increase in biomass of the herbivores in the ecosystem because
    (A) plants convert energy input into biomass more efficiently than animals
    (B) there are always more animals than plants in any ecosystem
    (C) plants have a greater longevity than animals
    (D) during each energy transformation, some energy is lost
    (E) animals convert energy input into biomass more efficiently than plants

Questions 38 and 39

Base your answers to Questions 38 and 39 on the pedigree chart below. The pedigree traces the inheritance of a metabolic disorder. Affected individuals are indicated by the darkened circles and squares. Squares represent males; circles females.

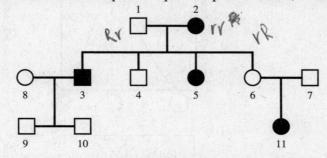

38. What is the pattern of inheritance for this metabolic disorder?
    (A) A dominant allele
    (B) Incomplete dominance
    (C) A recessive allele
    (D) Sex-linkage
    (E) Multiple alleles

39. What is the genotype of individual number 6?
    (A) Homozygous dominant
    (B) Homozygous recessive
    (C) Heterozygous dominant
    (D) Heterozygous recessive
    (E) Codominant

40. Which of the following is NOT a vertebrate?
    (A) Amphioxus
    (B) Lamprey eel
    (C) Shark
    (D) Snake
    (E) Bird

41. The amount of DNA in a diploid cell in the G_1 phase of a cell cycle is designated as Y. How should the amount of DNA in the anaphase stage of Meiosis II be designated?
    (A) 4Y
    (B) 2Y
    (C) Y
    (D) 0.75Y
    (E) 0.5Y

42. Below are graphs describing the fates of a hypothetical population of organisms in which there is variation in color. The arrows represent selective pressures. Which graph represents a stabilizing mode of selection?

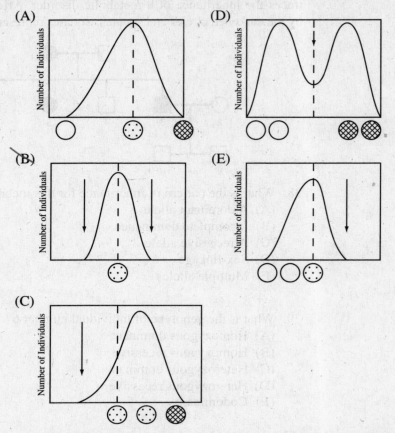

43. The essential difference between fermentation and cellular respiration is that
    (A) fermentation occurs in the mitochondria and cellular respiration occurs in the cytoplasm of a cell
    (B) oxygen is required for fermentation but not for cellular respiration
    (C) ATP is produced through the chemiosmotic process during fermentation and cellular respiration
    (D) an electron transport system operates in the process of cellular respiration but not in the process of fermentation
    (E) alcohol is a product of cellular respiration and lactic acid is a product of fermentation

Questions 44 and 45

Base your answers to Questions 44 and 45 on the diagrams below and on your knowledge of biology.

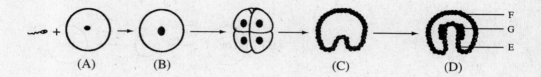

44. The stage described by diagram D is called
    (A) cleavage
    (B) differentiation
    (C) blastulation
    (D) organogenesis
    (E) gastrulation

    or C

45. Which pair of tissues is derived from the primary germ layer labeled G?
    (A) Bone and cardiac muscle
    (B) Nervous tissue and the lining of the digestive system
    (C) Blood and skin
    (D) Lens of eye and skin
    (E) Lining of reproductive tract and pancreas

46. Which statement concerning the M phase of the cell cycle is TRUE?
    (A) The cell increases in size during the M phase.
    (B) The M phase is usually short in duration.
    (C) Mitosis and cytokinesis occur during the M phase.
    (D) Chromosomes duplicate during the M phase.
    (E) Protein synthesis occurs during the M phase.

47. The condition in which hair appears on the outer edge of the earlobe is known as hairy pinna. The gene for hairy pinna is located only on the Y chromosome. No corresponding gene is found on the X chromosome. Which statement about hairy pinna is CORRECT?
    (A) Hairy pinna will usually not occur in normal females.
    (B) The gene for hairy pinna is an autosomal dominant gene.
    (C) Half as many women as men will carry the gene for hairy pinna.
    (D) Males with a normal X chromosome and the gene for hairy pinna will not have hairy earlobes.
    (E) The gene for hairy pinna is an autosomal recessive gene.

48. Which diagram BEST represents spermatogenesis?

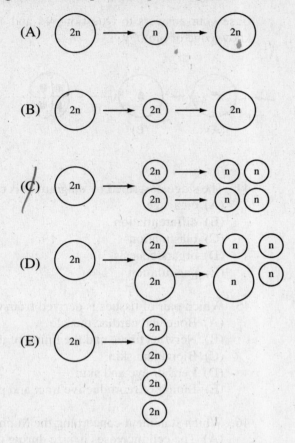

49. The survival of organisms depends upon the exchange of vital substances between the internal and external environment. One method of control is through countercurrent exchange. All of the following are examples of countercurrent exchange EXCEPT
    (A) the absorption of oxygen from the movement of water over the lamella of fish gills
    (B) the reabsorption of minerals and water in the kidneys of mammals
    (C) the retention of heat by the feet of ducks living in cold water
    (D) the absorption of nutrients across the villi in the human intestine
    (E) the cooling of blood in the nasal passages of a dog by panting

Questions 50–52

In Questions 50 to 52, choose the term that best applies to the relationships described in each case.

(A) Parasitism   (D) Competition
(B) Predation   (E) Commensalism
(C) Mutualism

C

50. Certain mites carry out most of their life activities within the flower of a particular plant species. Hummingbirds that pollinate these flowers transport the mites from old to new flowers.

51. A dove eats the fruit of a particular plant species. The undigested seeds pass out of the dove's body with its excrement and germinates away from the parent plant.

52. In certain areas of the Southwest, the population of gila woodpeckers is decreasing and the population of starlings is increasing. Both birds nest in the cavities of giant cacti.

53. In an experiment, a U-tube was divided in half by a membrane impermeable to starch. A 10% starch solution was placed into the right side of the tube. A 6% starch solution of equal amount was placed into the left side.

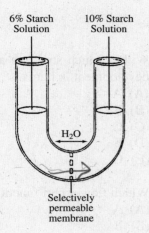

6% Starch Solution   10% Starch Solution

H₂O

Selectively permeable membrane

Which of the following would MOST LIKELY occur?
(A) Water would move across the membrane from left to right until equilibrium was reached.
(B) Water would move across the membrane from right to left until equilibrium was reached.
(C) Water would not move in either direction across the membrane.
(D) Starch would move across the membrane from left to right.
(E) Starch would move across the membrane from right to left.

54. Usually, the passage of a complete set of chromosomes from a parent cell to each daughter cell is accomplished by the process of
(A) gametogenesis
(B) reduction-division
(C) nondisjunction
(D) meiotic cell division
(E) mitotic cell division

Questions 55–57

Base your answers to Questions 55 to 57 on the diagram below and on your knowledge of biology.

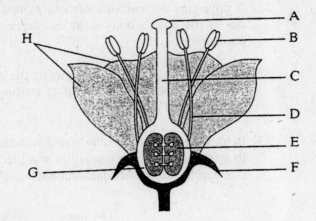

55. Within which structure are the growth stages of a plant similar to prenatal development in humans?
    (A) A
    (B) B
    (C) C
    (D) E
    (E) G

56. Which flower part functions in a manner similar to the human testes?
    (A) A
    (B) B
    (C) C
    (D) D
    (E) F

57. Within which structure does a process occur that is similar to that which occurs in the human oviduct?
    (A) A
    (B) B
    (C) C
    (D) E
    (E) G

58. Which term is CORRECTLY defined?
    (A) Introns are noncoding segments of DNA interspersed between the coding segments.
    (B) Exons are the noncoding regions of messenger RNA.
    (C) A genome is one half of an organism's genetic material that contains nontranscribable genes.
    (D) Transposons are segments of DNA that are always located in one particular region of a chromosome.
    (E) Nucleosomes are the building blocks of DNA.

Etnies

59. In the diagram of a twig, the structure labeled X functions to

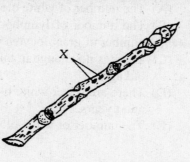

(A) increase the surface area for the absorption of water and minerals
(B) regulate the upward movement of sugar from the mesophyll into sieve tubes
(C) stimulate meristematic activity
(D) enable solar energy to penetrate into the mesophyll
(E) allow gas exchange to occur between the plant cells and the external environment

60. Anything that interferes with the functioning of microtubules interrupts cellular division because
(A) DNA replication requires the action of microtubules
(B) the dissolution of the nuclear membrane requires the action of microtubules
(C) microtubules are required to align chromosomes in a cell in preparation for their distribution to each daughter cell
(D) multinucleated cells will be produced
(E) microtubules are required to form the cell plate in animal cells

Questions 61 and 62

Base your answers to Questions 61 and 62 on the two graphs below and on your knowledge of biology. The graphs indicate the normal number and distribution of white blood cells in 1mm³ of human blood for people of different age groups. Granulocytes consist of neutrophils, basophils, and eosinophils.

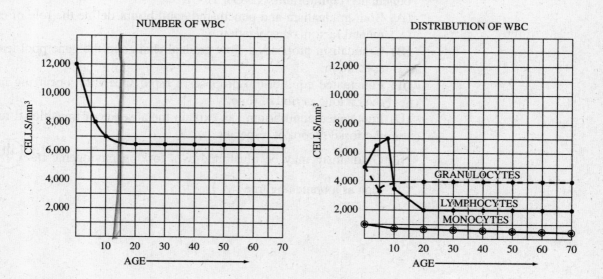

61. Which statement is CORRECT?
    (A) The number of white blood cells do not vary after the age of 10 years.
    (B) The number of lymphocytes and monocytes remain constant while the number of granulocytes vary with age.
    (C) There is no change in number or distribution of blood cells from birth to old age.
    (D) There are fewer white blood cells per mm³ in early childhood than in the adult years.
    (E) The number of white blood cells remains constant from birth to old age.

62. Which would MOST LIKELY occur during a severe bacterial infection?
    (A) The number of white blood cells would decrease.
    (B) The number of granulocytes would decrease and the number of white blood cells would increase.
    (C) The number of granulocytes would increase and the number of white blood cells would increase.
    (D) The number of monocytes would increase and the number of lymphocytes would decrease.
    (E) The number of white blood cells and their distribution throughout the circulatory system would remain unchanged.

63. One difference between a bean seed and a corn kernel is that
    (A) a bean seed has a seed coat but a corn kernel does not have a seed coat
    (B) a bean seed has two cotyledons; the corn kernel has one cotyledon
    (C) the corn kernel contains stored food; the bean seed does not contain stored food
    (D) the corn kernel has a plumule; the bean seed does not have a plumule
    (E) the bean embryo has two apical meristems; the corn kernel has one apical meristem

64. Which statement concerning the two theories of evolution, gradualism and punctuated equilibrium, is NOT TRUE?
    (A) Both gradualism and punctuated equilibrium define the role of environmental factors on speciation.
    (B) Gradualism proposes a slow gradual change in the gene pool leading to speciation.
    (C) Punctuated equilibrium proposes a rapid speciation occurring in spurts over a long period of time.
    (D) Punctuated equilibrium can explain the absence of transitional fossils in the fossil record of a species.
    (E) Gradualism may be illustrated as a forking phylogenic tree rather than as a branching tree

Questions 65–68

Below is a diagram showing the relationships that exist in an arid ecosystem. Base your answers to Questions 65 through 68 on the diagram and your knowledge of biology.

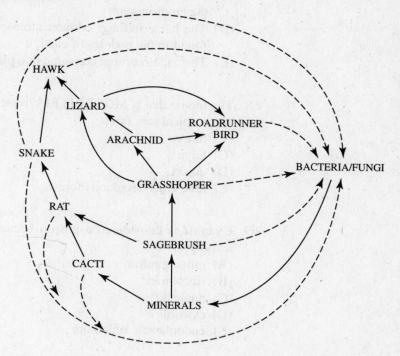

65. Which organisms are CORRECTLY paired to demonstrate a secondary consumer nutritional relationship?
    (A) snake → rat
        arachnid → grasshopper
    (B) cacti → rat
        sagebrush → grasshopper
    (C) grasshopper → arachnid
        grasshopper → bacteria/fungi
    (D) grasshopper → roadrunner
        sagebrush → rat
    (E) snake → hawk
        roadrunner → bacteria/fungi

66. Between which two organisms would there MOST LIKELY be the greatest competition?
    (A) rat and snake
    (B) lizard and arachnid
    (C) cacti and sagebrush
    (D) grasshopper and bacteria/fungi
    (E) arachnid and roadrunner

67. In the diagram, which statement CORRECTLY describes the role of bacteria/fungi?
    (A) The bacteria/fungi convert radiant energy into chemical energy.
    (B) The bacteria/fungi directly provide a source of nutrition for animals.
    (C) The bacteria/fungi are saprophytic agents restoring inorganic material to the environment.
    (D) The bacteria/fungi convert atmospheric nitrogen into minerals and are found in the nodules of cacti.
    (E) The bacteria/fungi consume live plants and animals.

68. The biome that is MOST LIKELY represented by the diagram is the
    (A) tropical rain forest
    (B) tundra
    (C) taiga
    (D) desert
    (E) temperate deciduous forest

69. Cyanide is considered a poison because it interferes with the cytochrome activity of
    (A) mitochondria
    (B) ribosomes
    (C) nucleus
    (D) cytosol
    (E) endoplasmic reticulum

70. An adaptation in land plants that reduces transpiration is the
    (A) mesophyll
    (B) rhizoid
    (C) stomate
    (D) cuticle
    (E) pith

71. Which statement concerning the illustrated organelle is NOT true?

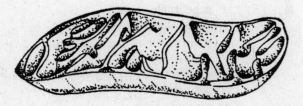

    (A) The electron transport chain and Krebs cycle enzymes are located within the organelle.
    (B) The inner membrane is the site for the chemiosmotic production of ATP.
    (C) NAD^+ is an electron carrier molecule in the electron transport chain of the organelles.
    (D) Oxygen is the final electron acceptor within the organelle.
    (E) ATP is formed via an electron transport chain and chemiosmotic synthesis within the organelle.

72. The hummingbird's beak has adapted to a tubular flower while the tubular flower adapted to accommodate the bird's beak. This reciprocal adaptation is known as
    (A) divergent evolution
    (B) convergent evolution
    (C) adaptive radiation
    (D) punctuated evolution
    (E) coevolution

73. All of the statements concerning the genetics of prokaryotes and eukaryotes are true EXCEPT
    (A) the regulator in prokaryotes directs the production of protein that blocks RNA polymerase
    (B) extrachromosomal factors in both prokaryotes and eukaryotes can turn genes "on and off"
    (C) reverse transcriptase reverses the direction of messenger RNA transcription from the DNA template in prokaryotic cells
    (D) extrachromosomal inheritance is observed in both prokaryotes and eukaryotes
    (E) nucleosome organization of chromosomal material occurs in eukaryotes

Questions 74–76

For each statement in Questions 74 through 76, select the system that is involved.

    (A) Respiratory system   (D) Nervous system
    (B) Endocrine system   (E) Immune system
    (C) Excretory system

74. Homeostatic control is maintained through the secretion of chemical substances that travel to target organs by way of the bloodstream.

75. The exchange of gases between the external and internal environment occurs across moist membranes.

76. The kidney responds to the amount of oxygen reaching the tissues by producing erythropoietin, which stimulates red blood cell formation.

77. In the process of oxidative phosphorylation, the final electron acceptor in the electron transport chain is
    (A) carbon dioxide
    (B) oxygen
    (C) water
    (D) NAD$^+$
    (E) ATP

78. Two adaptations of bryophytes that made it possible to inhabit terrestrial habitats are
    (A) vascular tissue and a waxy cuticle
    (B) true roots and stems and a waxy cuticle
    (C) vascular tissue and protected reproductive cells
    (D) a waxy cuticle and protected reproductive cells
    (E) true roots and stems and protected reproductive cells

79. In the evolution of flowering plants, major trends have been toward
    (A) increase in the number of flower parts, gametophytic dominance, and ovaries above the petals and sepals
    (B) reduction in the number of flower parts, sporophytic dominance, and ovaries below the petals and sepals
    (C) fusion of flower parts, ovaries above petals and sepals, and radial flower symmetry
    (D) increase in the number of flower parts, ovaries below petals and sepals, and bilateral flower symmetry
    (E) fusion of flower parts, gametophytic dominance, and ovaries below petals and sepals

80. The color of the flower of a particular variety of hydrangea produces blue flowers when the soil pH is above 7.0. When the soil pH is below 7.0, pink flowers are produced. Which statement BEST explains the color changes?
    (A) The pH of the soil acts as a mutagenic agent.
    (B) Environmental factors influence gene expression.
    (C) Multiple alleles are activated by pH.
    (D) Chromosomal mutations produce different color effects.
    (E) Epistasis influences gene expression.

Questions 81–83

Below is a diagram representing an enzymatic reaction. Answer Questions 81 through 83 based on the diagram and on your knowledge of biology.

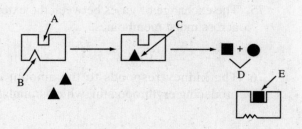

81. Which letter represents the enzyme-substrate complex?
    (A) A
    (B) B
    (C) C
    (D) D
    (E) E

82. Which letter represents an allosteric site?
    (A) A
    (B) B
    (C) E
    (D) F
    (E) G

83. Molecule E acts as a
    (A) cofactor
    (B) noncompetitive inhibitor
    (C) competitive inhibitor
    (D) coenzyme
    (E) substrate

Based on the diagram and your knowledge of biology, answer the following:

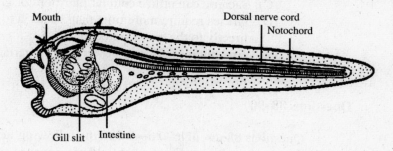

84. The sea squirt is a member of the same phylum as the
    (A) squid and clam
    (B) crayfish and scorpion
    (C) dog and human
    (D) starfish and sea urchin
    (E) earthworm and leech

85. The greatest amount of oxygen will be lost from human blood as it travels through the
    (A) capillaries surrounding the alveoli
    (B) left atrium of the heart
    (C) arteries
    (D) capillaries of the body tissues
    (E) veins

86. The advantage of having a digestive tract with a mouth and an anus is that it allows
    (A) different parts of the gut to perform different digestive functions
    (B) an animal without teeth to have a means of grinding food
    (C) an animal to eat large quantities of food at one time
    (D) an animal to eat large pieces of food
    (E) an animal to expel undigested waste easily

87. The graphs show the growth patterns of bacteria and viruses. Which statement BEST explains the difference in the growth patterns?

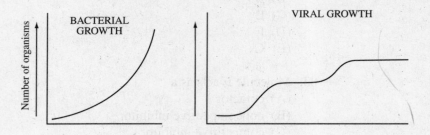

(A) Viral reproduction occurs only inside living host cells while bacterial reproduction occurs extracellularly.

(B) Sexual reproduction in bacteria is faster than asexual reproduction in viruses.

(C) Bacteria can utilize cellular nutrition for growth better than viruses.

(D) Viruses require more time than bacteria to transfer their genetic material directly to the host's mitochondria.

(E) Viral binary fission is slower than bacterial binary fission.

Questions 88–90

Questions 88 through 90 deal with the following facts that are used to support the theory of evolution. From what field of science is each fact derived?

(A) Anatomy   (D) Embryology
(B) Biochemistry   (E) Paleontology
(C) Biogeography

88. The blood proteins of 12 species of quenon monkeys inhabiting eastern Africa were analyzed and an evolutionary tree for the monkeys was established.

89. The gill pouches in fish develop into gills while those of a human become the Eustachian tubes.

90. X rays from a well-preserved mammoth skeleton indicated many similarities to the modern elephant.

91. A man with blood type A marries a woman with blood type AB. What is the chance that they will have a child with blood type O?
(A) 100%
(B) 75%
(C) 50%
(D) 25%
(E) 0%

92. The diagram below represents a phylogenic tree of the evolution of the even-toed ungulate.

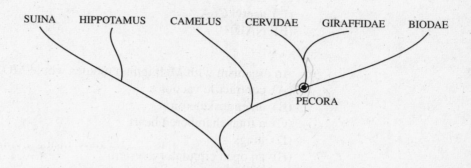

The most likely explanation for the branching pattern seen in the circled region is that
(A) environmental changes caused extinction
(B) inbreeding led to speciation
(C) no speciation occurred
(D) speciation was influenced by environmental changes
(E) only the best adapted organisms survive from generation to generation

Questions 93 and 94

Base your answers to Questions 93 and 94 on the chart below. The chart shows the data collected from a study of the gene pool of a large population of meadow mice over a 60-year period. Dark-colored fur is dominant over cream-colored fur.

	1900	1910	1920	1930	1940	1950	1960	1970
frequency of allele D (dark fur)	0.01	0.19	0.36	0.51	0.64	0.75	0.84	0.90
frequency of allele d (cream colored)	0.99	0.81	0.64	0.49	0.36	0.25	0.16	0.10

93. In 1970 what was the frequency of the individuals with cream-colored fur in the population?
(A) 1.0%
(B) 10%
(C) 18%
(D) 81%
(E) 90%

94. The best explanation for the data shown in the chart is that
(A) the allele for dark fur was needed by the species and therefore suddenly appeared in the gene pool
(B) organisms carrying the dark-colored allele have a reduced survival rate
(C) environmental factors favored individuals with the dark-colored fur allele
(D) environmental factors favored individuals with the cream-colored fur allele
(E) genetic drift was the main factor influencing the distribution of genes in this gene pool

95. Which of the following is NOT a product of the Krebs cycle?
    (A) ATP
    (B) CO_2
    (C) $FADH_2$
    (D) acetyl CoA
    (E) NADH

96. An organism with Malpighian tubules would MOST LIKELY possess
    (A) contractile vacuoles
    (B) an endoskeleton
    (C) a four-chambered heart
    (D) lungs
    (E) an open circulatory system

97. If 23 white, 26 brown, and 53 tan-colored offspring resulted from the mating of two tan-colored birds, what percentage of the offspring from the union of a tan-colored bird with a white-colored bird will be white?
    (A) 100%
    (B) 75%
    (C) 50%
    (D) 25%
    (E) 0%

Questions 98–100

Answer Questions 98 to 100 on the basis of the following diagram of a plant organelle in which a biochemical reaction takes place.

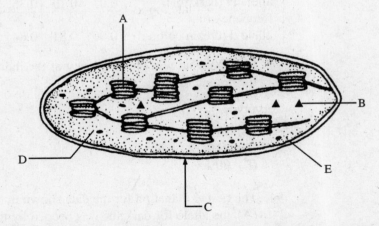

98. The electron transport chains required for the process of photophosphorylation are located in the structures labeled
    (A) A
    (B) C
    (C) D
    (D) E
    (E) B

99. The formation of ATP through cyclic and noncyclic photophosphorylation occurs in the structure labeled
   (A) A
   (B) B
   (C) C
   (D) D
   (E) E

100. Molecule B is a product resulting from a reaction occurring in structure A. Molecule B is most likely
   (A) ATP
   (B) NADPH
   (C) ATP synthase
   (D) chlorophyll
   (E) PGAL

101. The calcium concentration in the cytoplasm of the muscle cell is regulated by the cell's
   (A) nucleus
   (B) lysosome
   (C) Golgi apparatus
   (D) sarcoplasm reticulum
   (E) mitochondrion

102. The villi in the small intestine of humans and the intestinal folds of the earthworm function to
   (A) increase the surface area for the absorption of digestive products
   (B) remove metabolic wastes
   (C) direct the passage of food in one direction through the digestive system
   (D) secrete hormones regulating the reproductive process
   (E) circulate blood throughout the organisms

Questions 103 and 104

Base your answers to Questions 103 and 104 on the diagram below of a human karyotype and on your knowledge of biology.

103. Which genetic disorder would MOST LIKELY be diagnosed from the karyotype?
(A) PKU
(B) Hemophilia
(C) Tay-Sachs
(D) Down's syndrome
(E) Diabetes

104. Which of the following procedures was MOST LIKELY utilized to obtain the genetic material for a karyotype of an unborn individual?
(A) Genetic engineering
(B) Gene therapy
(C) Cyclic photophosphorylation
(D) Biopsy
(E) Amniocentesis

105. In which diagram of the human heart below do the arrows CORRECTLY describe the normal path of blood?

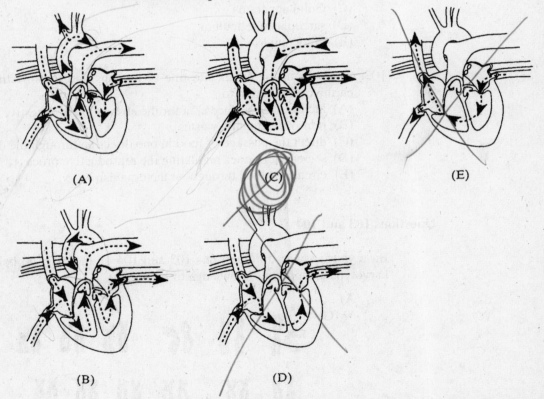

(A)   (C)   (E)

(B)   (D)

106. A chemical substance released by cells infected with a virus that prevents viral penetration into uninfected cells is known as
(A) interleukin-1
(B) interneuron
(C) interferon
(D) antihistamine
(E) antibody

Questions 107–109

The diagram below represents a function of the nucleic acid DNA. Answer Questions 107 through 109 based on the diagram and on your knowledge of biology.

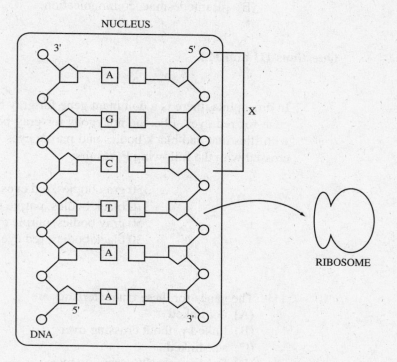

107. The production of messenger RNA from DNA is known as
    (A) replication
    (B) recombination
    (C) translation
    (D) transcription
    (E) translocation

108. What is the MOST LIKELY nucleotide sequence of the messenger RNA?

(A) A
    G
    C
    T
    A
    A

(B) U
    C
    G
    A
    U
    U

(C) T
    C
    G
    A
    T
    T

(D) T
    C
    G
    U
    T
    T

(E) T
    U
    G
    A
    T
    T

109. Which sequence represents an anticodon of section X in the diagram?

(A) A
    G
    C

(B) U
    C
    G

(C) T
    G
    C

(D) T
    C
    G

(E) U
    U
    U

110. Actin is a protein involved in all of the following EXCEPT
    (A) muscle contraction
    (B) cytoplasmic streaming
    (C) animal cell division
    (D) pseudopodia formation
    (E) plasmodesmata communication

Questions 111 and 112

In drosophila, there is a dominant gene for gray body color and another dominant gene for red eyes. Flies homozygous for gray bodies and red eyes were crossed with flies that had black bodies and purple eyes. The F_1 offspring were then test-crossed with the following F_2 results:

FF

> 350 gray bodies, red eyes
> 370 black bodies, purple eyes
> 50 gray bodies, purple eyes
> 30 black bodies, red eyes

FF

111. The genes for these characteristics are
    (A) not linked
    (B) linked without crossing over
    (C) sex-linked
    (D) linked with 10% crossing over
    (E) linked with 90% crossing over

112. Which of the following is TRUE about the 50 gray, purple-eyed flies?
    (A) There is no linkage.
    (B) The linked alleles are Gr and gr.
    (C) The linked alleles are the same as in the F_1 generation.
    (D) The linked alleles cannot be determined without a test cross.
    (E) None of the above

Questions 113 and 114

Base your answers to Questions 113 and 114 on the diagram and on your knowledge of biology.

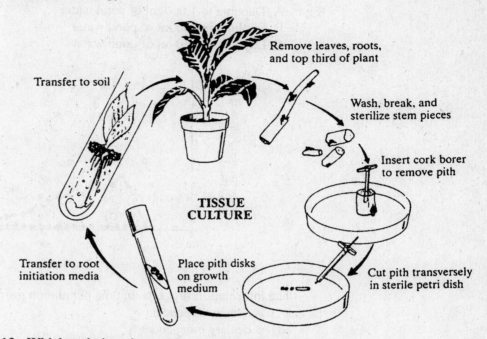

113. Which technique is represented in the diagram?
    (A) Genetic engineering
    (B) Karyotyping
    (C) Hybridization
    (D) Amniocentesis
    (E) Cloning

114. Which statement concerning the process is TRUE?
    (A) The test tube plant will never grow.
    (B) The test tube plant is genetically identical to the parent plant.
    (C) The plant disk cells grow by the process of meiosis.
    (D) The test tube plant contains half of the genetic material of the parent cells.
    (E) The test tube plant will not produce chlorophyll.

115. A group of 100 daphnia was placed into each of three culture jars of three different sizes. The graph below shows the average number of offspring produced per female each day in each jar.

Key: A Daphnia in 1,000 ml of pond water
       B Daphnia in 500 ml of pond water
       C Daphnia in 250 ml of pond water

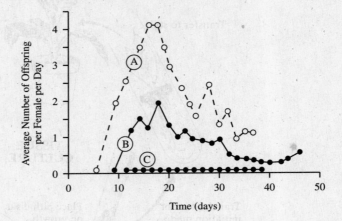

The information suggests that the population growth rate
(A) is density dependent
(B) is density independent
(C) depends on a predator-prey relationship
(D) increases exponentially after 20 days
(E) levels off when the carrying capacity is reached

116. One of the major evolutionary advancements that made terrestrial life possible was the development of
(A) live birth
(B) amniotic eggs
(C) endothermy
(D) fins
(E) hard shells

117. In an experiment, an equal number of winged and wingless fruit flies were added to each of two culture vials containing the same amount of nutritive material. A sticky piece of flypaper was placed into one culture vial. After a week, only wingless fruit flies survived. Which statement would be CLOSELY related to Darwin's theory of evolution?
(A) The type of food provided in the experiment acted as a selective agent against the winged flies.
(B) Mutations were responsible for the survival of the wingless flies in the experiment.
(C) In the natural environment wingless flies are better adapted than winged flies.
(D) The flypaper acts as a selective agent against the winged flies.
(E) Wingless flies out-competed the winged flies for the natural resources.

118. Wolves and bobcats may be said to occupy the same trophic level because they
   (A) both use their food with about 10% efficiency
   (B) are both secondary consumers
   (C) both live in the same habitat
   (D) are both large mammals
   (E) both eat a wide range of dietary items

Questions 119 and 120

In an experiment, six test tubes each containing the same amount of catalase and peroxide were maintained at different temperatures. The amount of bubbles generated by the reactions in the tubes were recorded and graphed. Answer Questions 119 and 120 based on the experiment.

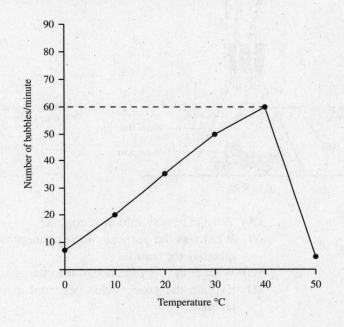

119. What is the origin of the bubbles measured in the experiment?
   (A) Hydrogen and oxygen gases resulted from the breakdown of water.
   (B) Oxygen was produced from the degradation of catalase.
   (C) Carbon dioxide was produced by the degradation of catalase.
   (D) Hydrogen was generated by the degradation of hydrogen peroxide.
   (E) Oxygen was generated by the degradation of hydrogen peroxide.

120. The best explanation for the sharp decline in the rate of reaction at 50° C is that the
   (A) activation energy was lowered
   (B) substrate was destroyed
   (C) molecular configuration of catalase was altered
   (D) molecular collisions between enzyme and substrate declined
   (E) catalase was consumed in the reaction

Section II Free-Response Questions #4

Answer all 4 questions
1 hour and 30 minutes

1. Below is a diagram of a volumeter into which a grasshopper was placed. A small tube containing a slightly basic phenopthalein solution was covered with mesh and placed in the volumeter with the grasshopper. The temperature was maintained at 25° C. At the end of an hour, the results were noted and recorded.

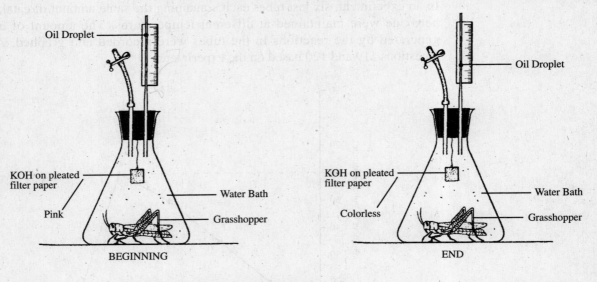

BEGINNING   END

(A) What results should be recorded?
(B) What was the purpose of the phenopthalein? How can titration be used to quantify the results?
(C) What is the significance of the change in position of the oil droplet?
(D) Explain the process that occurred in the grasshopper that produced the results.

2. Certain drugs and toxins interfere with the transmission of impulses between a nerve cell and a muscle.
(A) Describe the normal transmission from nerve to skeletal muscle.
(B) Propose several ways that the drugs may interfere with the normal transmission.

3. Steroid hormones affect the expression of specific genes in target cells.
(A) Discuss the nature of a gene.
(B) Describe the role of a gene in protein synthesis.
(C) Describe how a steroid hormone affects the expression of a gene.

4. Compare the structural and chemical similarities and differences between chemiosmosis in chloroplasts and mitochondria.

Answers to Model AP Biology Examination No. 4
Section I

1. E		31. E		61. A		91. E
2. B		32. B		62. C		92. D
3. A		33. B		63. B		93. A
4. D		34. E		64. E		94. C
5. B		35. A		65. A		95. D
6. A		36. B		66. B		96. E
7. C		37. D		67. C		97. C
8. E		38. C		68. D		98. A
9. B		39. C		69. A		99. C
10. D		40. A		70. D		100. B
11. C		41. E		71. C		101. D
12. C		42. B		72. E		102. A
13. B		43. D		73. C		103. D
14. A		44. E		74. B		104. E
15. D		45. A		75. A		105. A
16. E		46. C		76. B		106. C
17. E		47. A		77. B		107. D
18. A		48. C		78. D		108. B
19. C		49. D		79. B		109. A
20. E		50. E		80. B		110. E
21. C		51. C		81. C		111. D
22. D		52. D		82. A		112. B
23. C		53. A		83. B		113. E
24. D		54. E		84. C		114. B
25. A		55. D		85. D		115. A
26. E		56. E		86. A		116. B
27. B		57. D		87. A		117. D
28. E		58. A		88. B		118. B
29. C		59. E		89. D		119. E
30. B		60. C		90. A		120. C

Explanations to Model AP Biology Examination No. 4
Section I

1. E Flagella in protozoans and wings in birds are locomotive structures.
2. B Dehydration synthesis is the process by which simple molecules are converted into complex molecules by the removal of water and through the action of enzymes.
3. A The lowest percentage of mutated cells occurred at 160 mg/kg body weight of β-carotene.
4. D Ribosomes are present in both prokaryotic and eukaryotic cells. Ribosomes are the sites of protein synthesis.
5. B Extracellular digestion is the digestion of material outside of the cell. Fungi secrete digestive enzymes into the substrate. The nutrients are absorbed from the substrate.
6. A Oxygen was absent in the early earth's atmosphere. The first life forms were anaerobic heterotrophs. Oxygen became available with the appearance of photosynthetic organisms. Aerobic organisms require oxygen for cellular respiration.
7. C Swimming costs less in terms of energy expenditure for all sizes of organisms. A killer whale moves by swimming.
8. E Ethylene initiates or hastens changes that cause fruit ripening.
9. B The endosymbiotic model proposes that early prokaryotes were engulfed by larger prokaryotes and existed with the larger cell in a symbiotic relationship.
10. D Cellulose is a polymer of glucose. The molecule in question is an amino acid.
11. C Addition is the adding of a segment of one chromosome to another homologous chromosome. Deletion is a loss of a chromosome segment.
12. C Mitochondria are found in both plant and animal cells. ATP is produced in the mitochondria.
13. B B is located between A and C. D is farthest from A and B.
14. A Viruses contain either RNA or DNA, not both.
15. D Cholesterol is a steroid. Steroids are lipids.
16. E Roots of trees anchor and hold the soil in place. Loss of soil results in soil erosion, silting, and area flooding. Carbon dioxide utilization decreases since photosynthesis decreases with loss of plants.
17. E Structure X represents a protein molecule in a cell membrane. Protein molecules are not binding sites for phospholipids.
18. A An ectotherm is an organism that cannot regulate its internal body temperature. Ectotherms depend upon environmental heat and behavioral modifications to control body temperature. Endotherms maintain a constant body temperature through their own metabolic processes.
19. C Muscle cells under strenuous activity switch to anaerobic respiration due to a lack of oxygen. Lactic acid is an end product of anaerobic respiration.
20. E All of the statements refer to the similarity of the inner membrane of mitochondria and chloroplasts.
21. C The stomata size would be greater during the night than in the day for CAM plants. Closure of stomata during the day prevents water loss in arid regions.

22. **D** The effect of a hormone is slow due to the mode of transport and cellular activation.

23. **C** The allantois is the excretory organ of the avian embryo and the placenta is the excretory organ of the human embryo.

24. **D** Structures B and G represent the amnionic sac. The fluid in the amnionic sac protects the embryo and keeps the cells moist.

25. **A** The humoral immune system produces antibodies that react with antigens in the plasma. The cell-mediated immune system produces antibodies that act against cells infected with antigens.

26. **E** The circulatory system in vertebrates is a closed system. In a closed system, the fluid remains in the blood vessels and continuously circulates throughout the body. There is no opening to the external environment.

27. **B** Graph B would most likely result from plotting the experimental data. The limiting factor in the experiment is the concentration of enzymes. The maximum rate of reaction is reached when all enzyme molecules are engaged. The rate will remain constant as long as there is sufficient substrate to keep all the enyzme molecules engaged.

28. **E** Peptide hormones cannot pass through the cell membrane. The peptide hormone binds to a specific receptor initiating cyclic AMP synthesis. cAMP triggers a response in the target cells.

29. **C** The nephron is the filtering unit of the kidney in dogs and other mammals.

30. **B** The major histocompatibility complex (MHC) consists of protein molecules embedded in the plasma membranes. There are 20 genes with 50 alleles that code for the protein molecules. Therefore, no two people except identical twins have the same membrane proteins. Membrane proteins stimulate cell-mediated immune responses.

31. **E** The peroxisome is a single-membraned microbody that is formed from the endomembrane system and produces nonhydrolytic enzymes. The enzymes have many functions in plant and animal cells.

32. **B** Phototrophism is the growth response of a plant toward light.

33. **D** Depolarization of presynaptic membranes cause calcium ions to rush into the neuron, which initiates the reactions for the release of neurotransmitters.

34. **E** Region 3 is a synapse and region 5 is a neuromuscular junction. Transmission of impulses require neurotransmitters to conduct impulses across those regions.

35. **A** The response of salmon is known as olfactory imprinting. Imprinting is learned behavior related to innate behavior that is acquired during a critical period.

36. **B** Echinodermata-bilateral, coleomate with deuterostome developmental pattern.

37. **D** Less energy is available at each trophic level because energy is lost by organisms through respiration and incomplete digestion of food sources. Therefore, fewer herbivores can be supported by the vegetative material.

38. **C** Since affected and unaffected individuals result from the marriage of a normal parent #1 to an affected parent #2, the affected offspring receive one gene for the disorder from each parent. Parent #1 is heterozygous. The gene for the disorder must be recessive.

	A	a
a	Aa	aa
a	Aa	aa

39. **C** Both #6 and #7 parents are normal. Since their offspring is affected with the metabolic disorder, both parents contributed a gene for the disorder. Parent #6 is therefore heterozygous.

40. **A** Amphioxus does not have a backbone. It is an invertebrate chordate.

41. **E** Y = 2n (diploid chromosome number. The chromosomal material duplicates prior to Meiosis I (2Y). The cells resulting from Meiosis I contain one half of the doubled genetic material: $\frac{1}{2} \times 2Y = Y$. Meiosis II reduces the genetic material of gametic cells by one half: $\frac{1}{2} \times Y = 0.5Y$.

42. **B** Stabilizing selection operates on the extremes. Stabilizing selection reduces phenotypic variations but maintains the status quo in the gene pool.

43. **D** An electron transport system operates in the process of cellular respiration but not in fermentation. An electron transport system consists of a series of molecules through which an electron is passed from one molecule to another. Oxygen is the final molecule in the transport chain.

44. **E** Gastrulation is the invagination of cells from the hollow ball stage producing a 3-cell layered embryo.

45. **A** The mesoderm is represented by the germ layer labeled G. Bone and cardiac muscle are derived from the mesoderm.

46. **C** Mitosis (nuclear division) and cytokinesis (cytoplasmic division) occur during the M phase.

47. **A** The gene for hairy pinna is located on the Y chromosome. Only males have the Y chromosomes. Females have two X chromosomes.

48. **C** Spermatogenesis is the process by which four haploid (n) gametes of equal size are produced.

49. **D** The absorption of material across the villi in the human intestinal tract occurs by passive and active transport.

50. **E** Commensalism is an association between two organisms in which one organism benefits and the other is neither harmed nor benefitted by the association.

51. **C** Mutualism is an association between two organisms in which both are benefitted from the association.

52. **D** Competition is the struggle between different species to obtain and use the same natural resources.

53. **A** Water moves along a concentration gradient from an area of high concentration to an area of lower concentration. There is less free water on the right side than on the left side of the U-tube.

54. **E** Mitosis is the process of nuclear division in which each nucleus receives the same chromosomal material.

55. **D** Structure E is the ovule inside which seed development occurs. The seed contains the plant embryo.

56. **B** Structure B is the anther. Anthers and human testes produce sperm.

57. **D** Fertilization occurs in the ovule of flowers and in the human oviduct.

58. **A** Introns are noncoding segments of DNA that are interspersed between the coding segments of DNA.

59. **E** Structure X represents a lenticel. Lenticels are openings in the stem of a plant that allow the exchange of gas between plant cells and the external environment to occur.

60. **C** Microtubules move chromosomes around a cell during the process of mitosis ensuring the equal distribution of the chromosomes to the daughter cells.

61. A According to the graphs, the number and distribution of white blood cells remain unchanged from ten years of age.

62. C In severe bacterial infection, the number of granulocytes increase. Neutrophils are the specific granulocytes that increase in number. Neutrophils phagocytize bacteria. The increase in neutrophils increases the total number of white blood cells.

63. B Flowering plants are divided into two groups, monocots and dicots. A bean seed is a dicot; it has two cotyledons. A corn kernel is a monocot; it has one cotyledon. Cotyledons provide nutritive material to the developing plant embryo.

64. E Gradualism may be illustrated as a tree with many branches. Punctuated equilibrium may be represented by a forking tree.

65. A Secondary consumers are carnivores that feed upon the flesh of other animals. Snakes eat lizards and arachnids eat grasshoppers.

66. B Both lizards and arachnids eat grasshoppers. Competition is the struggle between different species for the same resources.

67. C Saprophytic agents gain nourishment from dead organic matter. They are the organisms of decay. Bacteria and fungi are saprophytes.

68. D The desert is a biome characterized by low precipitation, high evaporation rate, and a cactus climax community.

69. A Cytochromes are iron-containing proteins in the electron chain. The electron transport chain operates in the mitochondria in the production of ATP.

70. D The cuticle is a waxy waterproofing coat that prevents water loss from the leaf and stem.

71. C NAD^+ is a coenzyme that helps in the transfer of electrons during the redox reactions in the Krebs cycle.

72. E Coevolution is the evolution of two different species interacting with and influencing the adaptations of each other.

73. C Reverse transcriptase is an enzyme of RNA viruses that uses RNA as a template for DNA synthesis.

74. B The endocrine system produces hormones that control the activity of target organs.

75. A The respiratory system functions to remove carbon dioxide from an organism and to deliver oxygen to the cells of the organism.

76. B Erythropoietin is a hormone that is produced chiefly by the kidney. Hormones are produced and secreted by endocrine organs or clusters of endocrine cells within exocrine organs.

77. B Oxygen is the final electron acceptor in the electron transport chain in the process of oxidative phosphorylation or cellular respiration.

78. D A waxy cuticle prevents water loss. Protected reproductive cells increases the chances for survival in adverse environmental conditions.

79. B The evolutionary trends in flowering plants is toward a reduced number of flower parts as a result of a fusion of parts, a change from radial to bilateral flower symmetry, and inferior ovaries (ovaries below the petals and sepals).

80. B The color of hydrangea flowers is due to the effect of pH on gene action. The production of pigment is controlled by a gene but the type of pigment depends upon the chemical substrate.

81. C An enzyme-substrate complex is a temporary union of an enzyme and its substrate.

82. A An allosteric site is a receptor site on some part of the enzyme away from the active site.

83. B Letter E represents a noncompetitive inhibitor. A noncompetitive inhibitor alters the configuration of the active site so that the enzyme is no longer functional.

84. C The sea squirt belongs to the phylum Chordata. Dogs and humans also belong to Chordata.

85. D Capillaries are single-celled blood vessels through which gases are exchanged with the cells of the body.

86. A A one-way digestive tract contains specialized areas that have a specific function to perform. This makes the digestion of materials more efficient.

87. A Viral reproduction occurs in waves or steps because viral reproduction involves invasion of the host cell, incorporation of viral genetic material into the host's genome, viral manufacture, and viral release from the host cell.

88. B Comparative biochemistry or molecular biology investigates the similarities in the chemistry of living things.

89. D Comparative embryology studies the similarities in the development of living things.

90. A Comparative anatomy studies the similarities and differences in the structures of living things.

91. E The genotype for AB blood type is $I^A I^B$. The genotype for A blood type is $I^A I^A$ or $I^A i$. The genotype for O type blood is ii. There is no chance of having a child with O type blood since the child must receive one recessive i allele from each parent.

92. D Speciation occurs when environmental factors change the composition of the gene pool.

93. A By using the equation $p^2 + 2pq + q^2 = 1$, the frequency of a particular genotype in a population may be determined. q^2 = the homozygous recessive condition. Therefore, $(0.10)(0.01) \times 100 = 1.0\%$.

94. C Environmental factors operate to favor the reproduction of individuals with specific characteristics. Therefore, these individuals contribute more offspring in the successive generations, changing the frequency of genes in the gene pool.

95. D Acetyl CoA is a product of the conversion of pyruvate, which was formed in the cytoplasm.

96. E Insects possess Malpighian tubules. Insects also have an open circulatory system.

97. C Tan-colored birds are hybrid. They have one allele for white color and one allele for brown color. The tan birds are incompletely dominant. White birds are homozygous for color. Therefore, a cross between BW and WW birds results in 50% white-colored offspring and 50% tan-colored offspring.

	W	*B*
W	*WW*	*WB*
W	*WW*	*WB*

98. A The electron transport chains are located in the structure labeled A. Structure A represents the thylakoid in the chloroplast.

99. C The formation of ATP occurs in the structure labeled D. Structure D represents the stroma.

100. B Molecule B represents NADH, which is produced through the electron transport chain.

101. D The sarcoplasm reticulum is a specialized endoplasmic reticulum that is an intracellular storehouse for calcium.

102. A Villi in the human small intestine like intestinal folds of the earthworm increase the surface area for the absorption of nutrients.

103. D Karyotypes are used to screen for abnormalities in chromosome structure and number. Down's syndrome is a result of the presence of an extra chromosome.

104. E Aminiocentesis is the process by which fetal cells are drawn along with amniotic fluid from the amniotic sac. Chromosomes may be observed for abnormalities in structure and number from a preparation of the fetal cells.

105. A Blood flows from the vena cava into the right atrium. From the right atrium the blood flows into the right ventricle and then to the lungs by way of the pulmonary artery. Blood returns from the lungs by way of the pulmonary veins into the left atrium. Blood flows from the left atrium to the left ventricle and out of the heart to all parts of the body by way of the aorta.

106. C Interferon is a chemical messenger released from cells infected with a virus. Interferon coats uninfected cells limiting further viral infection.

107. D Transcription is the production of mRNA from a DNA template.

108. B The nucleotide sequence for messenger RNA is complementary to the DNA neucleotides.

109. A The anticodon is a 3-base nucleotide sequence of a tRNA (transfer) molecule. The tRNA sequence is complementary to the messenger RNA codon.

110. E Plasmodesmata are channels in the cell wall of plants that serve to bind cells together.

111. D In a simple dihybrid test cross, a 1:1:1:1 ratio is expected since the alleles for each gene are located on different chromosomes. Since this ratio was not observed, linkage is suspected. If the genes are linked, the organisms in the minority are produced because of crossing over. There is a total of 800 organisms, 80 resulting from crossing over.

$$\frac{80}{800} = 0.10 \times 100 = 10\%$$

112. B The genotype of the gray, purple-eyed flies is Ggrr. The gray-purple flies receive a chromosome with the gr linkage from test-crossed parents. The second chromosome must have the Gr linkage.

113. E Cloning is the technique shown in the diagram. It is the process by which an entire individual develops from a few parent cells.

114. B The test tube plant is a clone. A clone is an individual produced from the cells of a parent plant.

115. A The ability of Daphnia to produce offspring is affected by population density. The reproductive rate decreases as the population increases.

116. B The amniotic egg is a fluid-filled shelled egg that provides all the materials that an organism requires in order to live on land.

117. D Because the winged flies could fly, they came into contact with the flypaper. In this situation, *having wings* was a disadvantage to the flies.

118. B Both wolves and bobcats are carnivores. Carnivores consume the flesh of other animals.

119. E Catalase catalyzes the breakdown of hydrogen peroxide into water and oxygen.

120. C Heat denatures enzymes. Heat altered the molecular configuration of catalase, thus changing the shape of the active site.

Answers to Model AP Biology Examination No. 4
Section II

1. (A) The following results should be recorded in the experiment:

O_2 consumption	CO_2 production

(B) Phenolpthalein is an indicator. It is pink in a weak basic solution. As the pH of the basic solution decreases, phenolpthalein becomes colorless. The grasshopper produced CO_2. Carbon dioxide combined with the water in the phenolpthalein solution forming carbonic acid. The phenolpthalein solution becomes colorless in an acid.

The amount of CO_2 produced is determined through the titration of the phenolpthalein solution. A 0.04% NaOH solution was added to the colorless phenolpthalein from a burette until an end point was reached (pink color). The number of milliliters required to reach the end point is multiplied by 10 to obtain the number of micromoles of CO_2 produced by the grasshopper.

A micromole is a basic unit of gas measurement. One milliliter = 10 micromoles.

(C) The glass tube entered the chamber of the volumeter and the volumeter chamber contained air. The oil droplet in the glass tube prevented air from entering or leaving the volumeter once the chamber was sealed. At the beginning of the experiment, there was equal air pressure on each side of the oil droplet. As the grasshopper consumed O_2, the air pressure in the volumeter decreased and the oil droplet was drawn towards the volumeter. The difference between the position of the droplet measured by the calibrated milliliter scale at the beginning and end of the experiment was the amount of O_2 consumed by the grasshopper. The amount of oxygen consumed is multiplied by 10 to convert milliliters into micromoles.

(D) Cellular respiration is the process that occurred in the grasshopper that produced the results in the experiment. Cellular respiration is the process by which ATP is synthesized from the complete breakdown of glucose in the presence of oxygen. The process involves two phases, glycolysis and the Krebs cycle. Both phases provide electrons for the electron transport system that operates in the mitochondria. The electron transport system is coupled with ATP synthesis through a mechanism known as chemiosmosis. O_2 is the final electron acceptor in the transport system. The grasshopper consumed O_2. CO_2 is a waste product of cellular respiration. It was released by the grasshopper.

2. (A) The site at which an axon terminal contacts a muscle cell is known as a neuromuscular junction or motor end plate. A gap between the nerve cell and muscle is called a synapse.

An impulse traveling down an axon causes a permeability change in the presynaptic membrane of a nerve terminal. Ca^{++} ions flood into the nerve terminal from the synapse. The Ca^{++} causes synaptic vesicles to fuse with the presynaptic membrane. When the vesicles fuse to the membrane, a neurotransmitter is released by exocytosis. In vertebrates, the neurotransmitter is acetylcholine.

The neurotransmitter flows across the synapse to the postsynaptic membrane. The union of the neurotransmitter and special receptors in the postsynaptic membrane cause a change in the permeability of the postsynaptic membrane to Na^+, K^+, and Cl^-. A wave of depolarization travels along the plasma membrane of the muscle.

Skeletal muscle is made up of many muscles or fibers surrounded by a plasma membrane. Each fiber is composed of two types of myofilaments: actin—a thin filament, and myosin—a thick filament.

When a muscle cell or fiber contracts, the filaments slide past each other longitudinally. The sliding filament hypothesis accounts for the shortening of the muscle cell.

The sliding of the filaments depends on the interaction of actin and myosin. Myosin consists of an elongated tail and a pair of globular heads. The heads function as ATPase.

ATP is bound to the myosin heads. When ATP is hydrolyzed, the myosin heads change shape. The new configuration binds to an active site on actin forming cross bridges. As the myosin heads relax and return to their previous configuration, the actin is pulled toward the center of the cell.

When the muscle fiber is at rest, the active sites on actin are blocked by a regulatory protein, tropomyosin, which wraps around the actin. Other protein molecules form a troponin complex. The troponin complex has three binding sites, one for actin, one for tropomyosin, and one for Ca^{++}. When Ca^{++} ions are present, the troponin complex changes shape affecting the position of troponin. The active sites of actin are exposed and a cross-bridged formation occurs.

Ca^{++} ions are retained inside the sarcoplasmic reticulum by ion pumps. When the permeability of the sarcoplasmic reticulum changes, Ca^{++} ions are released and contraction occurs. The sarcoplasmic membrane lies near a series of tubules, transverse tubules, formed by the infolding of the sarcoplasmic membrane.

A wave of depolarization spreads from the sarcoplasmic membrane down the transverse tubules. The depolarization affects the permeability of the sarcoplasmic reticulum. Ca^{++} ions are released.

(B) Drugs may interfere with normal synaptic transmission between nerve and muscle cells in several ways:

1. They prevent or encourage the release and/or formation of a neurotransmitter;
2. They bind to the neurotransmitter;
3. They block the receptors at the postsynaptic membrane;
4. They prevent the removal of the neurotransmitter from the postsynaptic membrane or they mimic the neurotransmitters.

Curare is a drug that structurally mimics acetlycholine. It binds to the postsynaptic receptor sites but does not open ion gates. It cannot be removed from the receptors because acetylcholinesterase is unable to destroy it.

The toxins of botulism operate at the presynaptic membrane. They may cause a constant release of neurotransmitters resulting in muscular paralysis.

The rabies virus blocks the postsynaptic receptors.

3. (A) A gene is that region of a DNA molecule that contains information for the production of a protein. A DNA molecule consists of a double strand of nucleotides. A nucleotide is composed of a sugar (deoxyribose), a phosphate, and a nitrogenous base. There are four kinds of nucleotides based upon the presence of one of four bases: guanine, cytosine, adenine, and thymine.

The nucleotide strands are held together by weak hydrogen bonds between the nitrogenous bases. There is specific base pairing observed between DNA nucleotide strands: adenine pairs with thymine, guanine pairs with cytosine. Because of the base pairing, the nucleotide strands are antiparallel to each other. The strands are also arranged in a helical pattern. Thus, a DNA molecule is referred to as a double helix.

(B) Proteins are composed of different kinds of amino acids arranged in a specific sequence. The sequence of amino acids in a protein is encoded in a triplet code of DNA. Each amino acid is specified by a triplet base code in DNA; therefore, a gene contains the triplet codes for a particular protein. The assembly of amino acids into the proper sequence occurs in the ribosomes. The information for the protein synthesis is carried from DNA in the nucleus to the ribosome by way of messenger RNA.

Messenger RNA is synthesied from a DNA template or pattern. The process is called transcription. RNA polymerase II, an enzyme, pries the two nucleotide strands apart and binds RNA nucleotides together to form mRNA. RNA polymerase II first binds to a region of the gene known as the promoter. The promoter consists of a series of nucleotides that is recognized by the RNA polymerase II. The promoter region also includes an initiation site from which transcription begins.

RNA polymerase II cannot recognize and bind to a promoter site without the assistance of transcription factors. Transcription factors are proteins that attach to the promoter site. The RNA polymerase II recognizes the complex between protein and DNA as its binding site.

There are over 100 transcription factors. These factors can be divided into three groups of domains according to the way in which they bind to DNA.

(C) Steroid hormones pass through the cell membrane of a target cell, enter the nucleus, and affect transcription. The steroid hormone binds to a nuclear receptor protein forming a complex. This complex can then bind to an acceptor protein that recognizes a specific region of DNA. The acceptor protein acts as a transcription factor that is recognized by RNA polymerase II.

4. Both mitochondria and chloroplasts are double-membraned organelles and both contain an outer and inner membrane. The inner membrane of a mitochonrion is folded. The folds are known as cristae. The space inside the inner membrane is known as a matrix and the space between the two membranes is called the intermembrane space.

Thylakoids are membranous sacs located inside the inner membrane of chloroplasts. The thyalakoids are arranged in stacks called grana. The stroma is the space between the thylakoids and the inner membrane.

Both organelles contain electron transport chains. Electron transport chains are located within the membranes of the cristae and thylakoids. The electron carriers in the chain convey electrons from their excited state down to their ground state. This is accomplished through a series of redox reactions. The final electron acceptor in the mitochondria is oxygen. The final electron acceptor in the chloroplasts is chlorophyll.

Both mitochondria and chloroplasts produce ATP through a mechanism known as chemiosmosis. As electrons move through the transport chain, H^+ is pumped across the membranes of the cristae and thylakoids by specific carrier molecules. An H^+ gradient known as a proton-motive force is established. H^+ can only diffuse across the membranes through channels in ATP synthase. ATP synthase is a protein complex embedded in the membranes of cristae and thylakoids. The passage of H^+ through the ATP synthase channels drives the phosphorylation of ADP.

There are several differences in the chemiosmosis mechanism in mitochondria and chloroplasts. In the mitochondria, the electrons are supplied by the oxidation of food through the glycolytic process and the Krebs cycle. The process of ATP production is called oxidative phosphorylation. The mitochondria transfer chemical energy from food to ATP.

In the chloroplast, chlorophyll supplies electrons. Chlorophyll is embedded in the membranes of the thylakoids. Light energy ejects electrons from chlorophyll. The process of ATP synthesis is called photophosphorylation. Chloroplasts change light energy into chemical energy.

In the mitochondria, the catalytic heads of ATP synthase are on the matrix side of the inner membrane. The H^+ ions are pumped into the intermembrane that serves as a proton reservoir. ATP is produced in the matrix.

In the chloroplasts, the catalytic heads of ATP synthase are on the stroma side. H^+ ions are pumped into the thylakoid compartments. The thylakoid compartments serve as proton reservoirs. ATP is produced in the stroma. The ATP is used as the energy source for sugar production during the Calvin cycle.

Evaluation of Model AP Examination No. 4 Results

Below is a breakdown of the test questions by subject area. Place a check mark (✓) in the box below each question answered correctly, and write the total number correct for that area in the space provided at the right. Then determine the percent you answered correctly in that particular subject area. For example, if you answer 4 out of 6 in a subject area, you answered 66% correctly.

Finally, add up the numbers of correct answers for all the areas to get your overall score. You should have a minimum of 90 correct answers. Concentrate your review and test preparation on the areas where your scores were poorest.

Subject Area	Number of Correct Answers	Number of Questions	% Correct

BIOLOGICAL CHEMISTRY

2	10	15	27	28	53	81	82	83	110	119	120

Number of Questions: 12

CELLS

1	4	12	17	31	101

Number of Questions: 6

ENERGY TRANSFORMATIONS

19	20	43	69	71	77	95	98	99	100

Number of Questions: 10

MOLECULAR GENETICS

11	58	73	107	108	109	113	114

Number of Questions: 8

HEREDITY

13	38	39	41	47	54	60	80	91

93	94	97	103	104	111	112	117

Number of Questions: 17

EVOLUTION

6	9	42	64	72	79	88	89	90	116

Number of Questions: 10

TAXONOMY

36	40	84	92

Number of Questions: 4

MONERA, PROTISTA, FUNGI

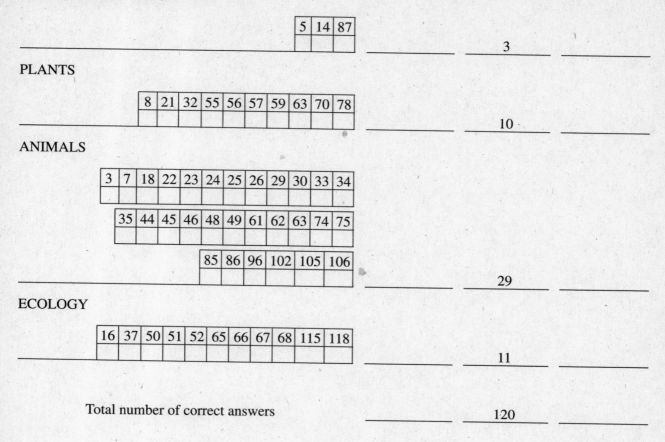

5	14	87

3

PLANTS

8	21	32	55	56	57	59	63	70	78

10

ANIMALS

3	7	18	22	23	24	25	26	29	30	33	34

35	44	45	46	48	49	61	62	63	74	75

85	86	96	102	105	106

29

ECOLOGY

16	37	50	51	52	65	66	67	68	115	118

11

Total number of correct answers   120

Appendices

APPENDIX A

Metric—English Equivalents

1. Linear Measures

1 millimeter	= 0.039 inch	= 1,000 micrometers
1 centimeter	= 0.394 inch	= 10 millimeters
1 decimeter	= 3.937 inches	= 10 centimeters
1 meter	= 39.37 inches	= 10 decimeters
1 kilometer	= 3,281 feet	= 1,000 meters

1 inch	= 25.4 millimeters	= 2.54 centimeters
1 foot	= 12 inches	= 30.48 centimeters
1 yard	= 36 inches	= 0.91 meter
1 rod	= 16½ feet	= 5.03 meters
1 mile	= 5,280 feet	= 1.61 kilometers

To Convert	Multiply by	To Obtain
inches	2.54	centimeters
feet	30.00	centimeters
centimeters	0.39	inches
millimeters	0.039	inches
meters	3.2808	feet
feet	0.3048	meters

2. Weights

1 milligram	= 0.001 gram
1 centigram	= 0.01 gram
1 decigram	= 0.1 gram
1 gram	= 0.001 kilogram
1 kilogram	= 1,000 grams (2.204 pounds)

1 ounce = 28.35 grams
1 pound = 453.59 grams
1 ton   = 907.18 kilograms

To Convert	Multiply by	To Obtain
ounces	28.3	grams
pounds	453.6	grams
pounds	0.45	kilograms
grams	0.035	ounces
kilograms	2.2	pounds

3. Liquid Measures

1 milliliter = 0.03 fluid ounce
1 centiliter = $\frac{1}{3}$ fluid ounce
1 deciliter = $3\frac{1}{3}$ fluid ounces
1 liter = 1.06 quarts

$\frac{1}{8}$ cup = 1 fluid ounce = 30 milliliters
1 cup = 8 ounces = 240 millilters
1 pint = 16 ounces = 0.47 liter
1 quart = 32 ounces = 0.95 liter
1 gallon = 128 ounces = 3.78 liters

To Convert	Multiply by	To Obtain
fluid ounces	30.00	milliliters
quarts	0.95	liters
milliliters	0.03	fluid ounces
liters	1.06	quarts

APPENDIX B

Commonly Used Prefixes and Suffixes

A major step in preparing for the examination in Advanced Placement Biology is to learn the vocabulary required by the subject matter. Like all scientific disciplines, biology has technical terms that have been derived from Latin (L) and Greek (Gr) roots. If you will learn this list of commonly used prefixes and suffixes, you can increase your reading comprehension of scientific materials and improve your understanding of the concepts to which these technical terms are related. A hyphen (-) following the derivative indicates a prefix. If the - precedes the derivative, the entry is a suffix. Note that some derivatives can be both prefixes and suffixes.

a-, an-   without, not; as achromatic (Gr)

ab-, abs-   off, from, separated; as abscission (L)

ad-   attached to; as adhere (L)

-ad   to, toward; as dorsad (L)

amphi-   on both sides; as amphiaster (Gr)

ana-   up, back, anew; as anaphase (Gr)

ante-   before; as anterior (L)

anti-   against, opposed to, antagonistic to; as antibody (Gr)

arch-, archi-, archaeo-   primitive, first; as Archaeopteryx (Gr)

bi-   twice; as biennial (L)

bi-, bio-   life; as biome (Gr)

-blast, blasto-   an embryonic layer or cell; as osteoblast (Gr)

cata-, same as **kata-**   against or down; as catabolism

-cle, -clus, -cule   little, small; as molecule (L)

cyst-   bladder; as cystolith (Gr)

-cyte, cyto-   cell; as cytotoxic (Gr)

de-   from, away, down; as dehydration (L)

derm-, -derm   cell layer, skin; as dermis (Gr)

di-   twice; as dichotomous (Gr)

dia-   between, through; as diapause (Gr)

diplo-   double, twice; as diploid (Gr)

dis-   parting from, depriving; as distal (L)

ect-, ecto-   outside, external; as ectoderm (Gr)

en-, em-   before, in; as enteric (Gr)

endo-   within, inside; as endosperm (Gr)

epi-   over, outside, anterior; as epiglottis (Gr)

ex-   from, without; as exergonic (L)

-genesis   production, reproduction; as biogenesis (Gr)

-gonium   a mother cell or structure; as oedogonium, sporogonium (Gr)

gono-   offspring, reproductive; as gonophore (Gr)

gyn-   female reproductive organ; as gyne (Gr)

hetero-   different; as heterozygous (Gr)

holo-, hol-   entire; as holoblastic (Gr)

homo-   same, alike; as homozygous (Gr)

hyper- above, over, beyond; as hypertonic (Gr)

hypo- less than, under, below; as hypotonic (Gr)

-id member of a class, itself, individual, daughter or son; as arachnid (Gr)

in- in, into, toward; as induction (L)

in- not, against; as indehiscent (L)

inter- among, between; as interphase (L)

intra- within; as intracellular (L)

iso- equal, the same; as isogamete (Gr)

-ite a part of, a division; as somite (Gr)

-itis inflammation of; as tonsilitis (Gr)

kata-, same as **cata-** down against; as katabolic (Gr)

macr-, **macro-** long, large; as macrophage (Gr)

mega- great, large; as megaspore (Gr)

mero-, **mere-** part; as merogamete (Gr)

meso- middle; as mesoderm (Gr)

met-, **meta-** posterior, after, outside; as metaphase (Gr)

micr-, **micro-** small; as microsome (Gr)

mon-, **mono-** one, alone, single; as monoploid (Gr)

-oid, **-eidos** in the form of: as haploid (Gr)

-ole, **-ola**, **-olus**, **-olum** diminutive, small, little; as petiole (L)

oo- egg; as oocyte (Gr)

-osis process of change, condition, disease; as metamorphosis (Gr)

ovi- egg; as oviparous (L)

par-, **para-** beside; as parapodium (Gr)

peri- about, all around; as pericardium (Gr)

-phase condition, appearance, stage; as anaphase (Gr)

-plasma living substance; as cytoplasm (Gr)

-plast a formed body; as chloroplast (Gr)

-pod, **-pode** foot; as pseudopod (Gr)

poly- many, much; as polymorph (Gr)

post- after, behind; as posterior (L)

pre- in front of, before; as premolar (L)

pro- primitive, in front, before; as procambium (Gr)

proto- first, earliest; as protozoan (Gr)

-phore bearer; as chromatophore (Gr)

-phyll leaf; as chlorophyll (Gr)

-phyte plant; as metaphyte (Gr)

re- back, again; as reflex (L)

soma-, **-some** body; as somatoplasm, centrosome (Gr)

sperma-, **spermato-** sperm, seed (when referring to plants); as spermatogenesis (Gr)

stoma-, **-stome** mouth; as stomate, cyclostome (Gr)

sub- under, below; as subcutaneous (L)

super- over, above; as superior (L)

supra- over, above; as supracaudal (L)

sym-, **syn-** together; as symbiosis, synthesis (Gr)

-theca case, receptacle; as spermatheca (L)

tri- three; as triploid (Gr)

-tropic responding to a stimulus; as phototropic (Gr)

-ule, **-ula**, **-ulus**, **-ulum** small, little; as granule (Gr)

uro- tail; as urodele (Gr)

vita- life; as vitamin (L)

xero- dry; as xerophyte (Gr)

xyl-, **xylo-** wood; as xylem (Gr)

-zoa, **zoo-**, **-zoon** animal; as zoology (Gr)

APPENDIX C

Textbooks and Other References

Biology Textbooks

Alters, S. *Biology—Understanding Life*. St. Louis: Mosby, 1996

Arms, K., and P. S. Camp. *Biology*, 4th ed. Philadelphia: Saunders College Publishing, 1995

Audesirk, T., and G. Audesirk. *Biology—Life on Earth*, 4th ed. Englewood Cliffs, New Jersey: Prentice Hall, 1996

Avilla, V. L. *Investigating Life on Earth*, 2nd ed. Boston: Jones and Bartlett, 1995

Brum, G., et al. *Biology—Exploring Life*, 2nd ed. New York: Wiley, 1994

Campbell, N. A., J. B. Reece, and L. G. Mitchell. *Biology*, 5th ed. Menlo Park, California: Addison Wesley Longman, 1999

Curtis, H., and N. S. Barnes, *Invitation to Biology*, 6th ed. New York: Worth, 1998

Mader, S. *Biology*, 5th ed., Dubuque, Iowa: W.C. Brown, 1996

McFadden, C., and W. Keeton. *Biology—Exploration of Life*. New York: Norton, 1995

Morgan, J., and M. Carter. *Investigating Biology*, 2nd ed. Redwood, California: Benjamin/Cummings, 1996

Purves, W. K., and G. H. Orians. *Life: The Science of Biology*, 4th ed. Sunderland, Massachusetts: Sinauer and Freedman, 1996

Raven, P. H., and G. B. Johnson, *Biology*, 4th ed. Dubuque, Iowa: W.C. Brown, 1996

Solomon, E. P., et al. *Biology*, 4th ed. Philadelphia: Saunders College Publishing, 1996

Starr, C., and B. McMillan. *Human Biology*. Belmont, California: Wadsworth, 1995

Wallace, R., et al. *Biology—The Science of Life*, 4th ed. New York: HarperCollins, 1996

Other References

Adams, F., and G. Laughlin. *The Five Ages of the Universe—Inside the Physics of Eternity.* New York: Free Press, 1999

Alberts, B. D., et al. *Molecular Biology of the Cell*, 3rd ed. New York: Garland, 1996

Bertrand, Y. *Earth from Above.* New York: Harry N. Abrams, 1999

Blackburn, G., and M. Gail, *Nucleic Acids in Chemistry and Biology*, 2nd ed. New York: Oxford University Press, 1996

Brown, D., and P. Rothery, *Models in Biology, Mathematics, Statistics, and Computing.* New York: Wiley, 1993

Crichton, M. *The Lost World.* New York: Alfred A. Knopf, 1995

Cummings, M. R. *Human Heredity Principles and Issues*, 5th ed. St. Paul, Minnesota: West, 2000

Elseth, G., and K. Baumgardner, *Principles of Modern Genetics.* New York: West, 1995

Harris, C. L. *Concepts in Zoology*, 2nd ed. New York: HarperCollins, 1996

Hine, R. *The Facts on File Dictionary of Biology.* New York: Checkmark Books, 1999

Kalthoff, K. *Analysis of Biological Development.* New York: McGraw-Hill, 1996

Kormandy E. *Concepts of Ecology*, 4th ed. New York: HarperCollins, 1996

Karp G. *Cell and Molecular Biology.* New York: Wiley, 1996

Lawlor, D. *Photosynthesis.* 2nd ed. Essex, England: Longman Scientific and Technical Publications, 1993

Lear, L. *Rachael Carson: Witness for Nature.* New York: Henry Holt & Co., 1997

Madigan, M., et al. *Brock—Biology of Microorganisms*, 8th ed. Englewood Cliffs, New Jersey: Prentice Hall, 1996

Martini, F. *Fundamentals of Anatomy and Physiology*, 3rd ed. Englewood Cliffs, New Jersey: Prentice Hall, 1996

Miller, G. T. *Living in the Environment.* New York: Wadsworth, 1996

Nester, E., et al. *Microbiology: A Human Perspective.* Dubuque, Iowa: W.C. Brown, 1996

Parker, S. P. (Editor). *Dictionary of Bioscience.* New York: McGraw-Hill, 1997

Phillips, K. *Tracking the Vanishing Frog—An Ecological Mystery.* New York: St. Martin's Press, 1994

Ridley, M. *Genome: The Autobiography of a Species in 23 Chapters.* New York: HarperCollins, 2000

Russel, P. J. *Genetics,* 4th ed. New York: HarperCollins, 1996

Salisbury, F., and C. Ross. *Plant Physiology,* 4th ed. Belmont, California: Wadsworth, 1992

Selander, R. K., et al. *Evolution at the Molecular Level.* Sunderland, Massachusetts: Sinauer, 1991

Simmons, J. *The Scientific 100: A Ranking of the Most Influential Scientists, Past and Present.* New York: Citadel Press, 1996

Singer, M., and P. Berg. *Genes and Genomes: A Changing Perspective.* Mill Valley, California: University Science Books, 1991

Spence, A., and E. Mason. *Human Anatomy and Physiology,* 4th ed. New York: West, 1992

Starr, D. *Blood: An Epic History of Medicine and Commerce.* New York: Alfred A. Knopf, 1998

Stern, K. R. *Introducing Plant Physiology,* 7th ed. Dubuque, Iowa: W.C. Brown, 1997

Taiz, L., and E. Zeiser. *Plant Physiology.* Redwood City, California: Benjamin/Cummings, 1991

Tortora, G., et al. *Microbiology, An Introduction.* Redwood City, California: Benjamin/Cummings, 1995

Voet, D., and J. Voet, *Biochemistry,* 2nd ed. New York: Wiley, 1995

Walker, D. *Energy, Plants, and Man.* Mill Valley, California: University Science Books, 1992

Winfree, A. T. *The Geometry of Biological Time.* Berlin: Springer-Verlag, 1990

APPENDIX D
Glossary

Abaxial Dorsal, facing away from the stem.

Abdomen In mammals, the part of the body containing most of the viscera except for the heart and lungs.

Abscissic acid Growth-inhibiting plant hormone; frequently antagonizes other plant growth hormones such as auxins or gibberellins by inhibiting nucleic acids and protein synthesis.

Abscission layer Base of leaf stalk in woody plants where the parenchyma cells become separated from each other because of disappearance of the middle lamella prior to leaf fall.

Acetylcholine A neurohumor, secreted by neurons.

Aconitase An enzyme that catalyzes the conversion of citric acid to cisaconitic acid and isocitric acid during the Krebs cycle.

Acrosome The forward tip of an animal sperm. Contains enzyme that allows sperm cell to enter the egg cell.

Actin One of the two major proteins of muscle.

Active site The locus on the surface of an enzyme where the substrate adheres and where catalysis occurs.

Active transport The movement of a substance across the cell membrane against a concentration gradient, that is, from a region of low concentration to a region of high concentration. Active transport involves the expenditure of energy.

Adaptation A special structure, physiological process, or behavior that makes an organism more fit to survive and reproduce.

Adaptive radiation The evolutionary division of a single species into several species adapted to divergent forms of life.

Adaxial Referring to a leaf surface, facing the stem.

Adenosine triphosphate (ATP) A triply phosphorylated organic compound that serves as an energy bank for cells.

Adrenergic Referring to a motor nerve fiber that secretes adrenaline (epinephrine) and noradrenaline (norepinephrine).

Afferent neurons Nerve cells that carry impulses to the central nervous system.

Aleurone layer A specialized cell layer just between the seed coat and the endosperm in grass seeds, synthesizing hydrolytic enzymes stimulated by gibberellin.

Algae A wide range of photosynthetic plants belonging to several phyla. Most live in the water; most are unicellular, but some are multicellular.

Allele One form of a gene, different from other forms or alleles of the same gene.

Allopatric speciation Speciation brought about when an ancestral population becomes separated by a geographical barrier.

Allosteric Of an enzyme, having two binding sites, of an effector molecule which does not have the same structure as any of the enzyme's substrate.

Alternation of generations The alternation of haploid and diploid phases in a sexually reproducing organism. In animals, the haploid stage consists only of gametes. In plants, the haploid phase may be the more dominant phase (as in fungi and mosses). In higher plants, the diploid phase is the more prominent.

Alveolus A small saclike cavity; the alveoli constitute the blind pouches of the lung.

Amino acid An organic compound having the general formula $H_2N—CHR—COOH$, where R is an alkyl group. An amino acid is so named because it has both a basic amino group, $—NH_2$, and an acidic carboxyl group, $—COOH$.

Ammonotelic An organism in which the final product of the breakdown of protein compounds is ammonia.

Amoeba, Ameba Any one of a large number of different animal-like protists belonging to the subphylum (of Protozoa) Sarcodina, characterized by its ability to change shape frequently through the extension of cytoplasm called pseudopodia.

Amphibian A member of the vertebrate class Amphibia, such as a frog, toad, or salamander, that spends at least part of its life cycle in the water.

Anaerobic Referring to respiration without the use of oxygen.

Anaphase The stage in cell division at which the separation of homologous chromosomes occurs. Anaphase lasts from the moment of first separation to the time at which the chromosomes move to the poles.

Angiosperm Flowering plants; seeds contained in an ovary.

Anion A negatively charged ion which moves toward the positive electrode (anode).

Anisogamy The presence of two dissimilar motile gametes.

Annelid A member of the phylum Annelida; one of the coelomate and segmented invertebrates (as the earthworm or the sandworm).

Annual A plant whose life cycle is completed in one growing season.

Anterior Toward the front end, toward the head end.

Anther The pollen-bearing portion of the stamen of a flower.

Antheridium The multicellular structure that produces the male gamete in nonseed plants.

Anticodon A "triplet" of nucleotides in transfer RNA that is able to pair with a complementary triplet in messenger RNA (a codon), thus arranging transfer RNA to the proper site on the messenger RNA.

Apex The tip or highest point of a structure, as the apex of a growing stem or root.

Apical meristem The actively growing tissue at the tip of a stem or root.

Archegonium The multicellular structure that produces eggs in liverworts and mosses.

Archenteron The cavity appearing in the early embryo during the gastrula stage. It ultimately becomes the gut cavity.

Arthropod A member of the phylum Arthropoda of invertebrates (as crustaceans, insects, arachnids).

Asexual reproduction A form of reproduction, such as budding or binary fission, that does not involve the fusion of gametes.

ATP See adenosine triphosphate.

ATP synthase The enzyme that makes possible the synthesis of ATP from ADP and inorganic phosphate in mitochondria and chloroplasts.

Auricle The thin-walled chamber, also called "atrium," of the heart which receives blood and passes it to the ventricle for pumping.

Autonomic nervous system The system, made up of sympathetic and parasympathetic subsystems, that controls involuntary functions such as digestion, heartbeat, and breathing.

Autosome A chromosome that is not a sex chromosome.

Autotrophic Capable of manufacturing organic nutrients from inorganic materials. Most plants, blue-green algae, and many bacteria are autotrophs.

Auxin In plants, a chemical substance that regulates growth by affecting cell elongation.

Axon Extension of the neuron which can carry impulses; is often the longest and least branched process of the cyton and usually carries impulses away from the cell body of the neuron.

Bacteriophage One of a group of viruses that infect bacteria and cause their disintegration. Normally present in sewage and in body products.

Bacterium A prokaryote, that is, a cell with a simple DNA molecule not contained in a nuclear membrane. It has a cell wall and does not carry on photosynthesis.

Bilateral symmetry The condition in which the right and left sides are divided exactly down the axis and are mirror images of each other.

Binomial Having two names; as in the binomial nomenclature of biology, which gives the name of the genus followed by the name of the species.

Biogeochemical cycle The cycle of particular materials, such as water or carbon, from the physical environment into organisms and back out again, in an unending sequence.

Bioluminescence The production of light by biochemical processes in an organism.

Biomass The total weight of living organisms found in a given area.

Biome A major portion of the living environment of a particular region, such as a beech-maple forest or a grassland, characterized by its distinctive vegetation, and surviving within the local climate.

Biota All of the organisms found in a given area.

Biotic Pertaining to any aspect of the life and characteristics of entire populations and ecosystems.

Blastocoel The hollow central cavity of a blastula.

Blastodisk The embryo-forming portion of an egg with discoidal cleavage usually appearing as a small disk on the upper surface of the yolk mass.

Blastula An early stage in animal embryology; a hollow ball of cells surrounding a central cavity.

Budding Asexual reproduction in which a new organism grows from the body of the parent organism and eventually detaches itself.

Buffering A process by which a system resists changes in pH.

Calorie The amount of heat required to raise the temperature of one gram of water by one degree Celsius ($1°C$). (A kilocalorie is the amount of heat needed to raise the temperature of a kilogram of water by $1°C$.)

Calyx The external, usually green or leafy part of a flower, consisting of sepals.

Capillaries Very small tubes; the smallest blood-carrying vessels of animals.

cAMP See cyclic AMP.

Carbohydrates Organic compounds with the general formula $(CH_2O)_n$. Sugars, starches, and celluloses are examples.

Carbonic anhydrase An enzyme that joins carbon dioxide and water to form carbonic acid.

Carboxylic acid An organic acid containing the carboxyl group: —COOH.

Cardiac Pertaining to the heart and its functions.

Carotenoid A type of photosynthesis accessory pigment related to the xanthophylls and carotenes.

Carrier-facilitated diffusion Transport of a substance across the plasma membrane by carrier molecules but without energy. This process cannot cause the net transport of a substance from a region of low concentration to a region of high concentration. This is a form of passive transport.

Carrying capacity In ecology, the largest number of organisms of a given species that can be maintained indefinitely in a particular part of the environment.

Catalyst A chemical substance that speeds up a reaction without itself being used up in the overall course of the reaction. Enzymes are biologic catalysts.

Cation A positively charged ion which moves toward the negative electrode (cathode).

Cellulose A straight-chain polymer of glucose molecules secreted by plants and used as a structural supporting material.

Cell wall A relatively rigid structure composed of cellulose that encloses the cells of plants. The cell wall gives these cells their shape and limits their expansion in hypotonic media.

Central nervous system That part of the nervous system which consists of the brain and spinal cord of vertebrates; the chain of cerebral, thoracic, and abdominal ganglia of arthropods.

Centriole A cylindrical organelle of animal cells (and some cells of lower plants) containing microtubules in a characteristic arrangement of nine triplets. The exact role of the centriole in the formation of the spindle is not known.

Centromere The point on a chromosome to which spindle fibers attach at cell division.

Chitin The flexible organic material of the exoskeleton of insects and other arthropods, consisting of a polysaccharide-protein complex.

Chromatid Each member of a pair of sister chromosomes.

Chromatin The nucleic acid-protein complex forming the nucleoplasm, out of which chromosomes are made; carries the genes.

Chromatophore A pigment-bearing cell.

Chromosome A complex structure present in the nucleus of a eukaryotic cell, composed of DNA, RNA, and proteins, and bearing part of the genetic information of the cell.

Cilium A hairlike organelle used for locomotion by many protozoans and some algae.

Cistron The genetic unit of function, considered equivalent to a gene. Each cistron contains the genetic information for a single polypeptide chain.

Class In taxonomy, the category below phylum and above order.

Code, genetic In protein synthesis, the exact order in which amino acids are arranged in a protein, determined by the sequence of nucleotides in messenger RNA.

Codon A "triplet" of three nucleotides in messenger RNA that directs the order of a particular amino acid in a protein molecule.

Coelom The body cavity of higher animals, which is lined with epithelial tissue.

Coenocyte A cell bounded by a single plasma membrane but containing many nuclei.

Cofactor An organic molecule attached to an enzyme and necessary for its catalytic activity. Vitamins serve as cofactors or coenzymes.

Coleoptile   A pointed sheath covering the shoot of a grass seedling.

Collagen   A fibrous protein found in bone and connective tissue.

Commensalism   The form of symbiosis in which one species benefits from the association while the other is neither harmed nor helped.

Community   An ecologically integrated group of species inhabiting a given area.

Companion cell   A specialized cell found in close association with each sieve tube element in plants. It probably provides the energy for the movement of materials through the phloem.

Conifer   One of the cone-bearing trees, such as pines and firs.

Conjugation   The close association of two cells during which they exchange genetic material, as in ciliated protozoans.

Contractile vacuole   An organelle, found in protozoa, that pumps excess water out of cells.

Corepressor   A compound of low molecular weight that unites with a protein (the aporepressor) to form a functional repressor.

Corolla   All of the petals of a flower.

Cotyledon   The seed leaf of a sprouting seedling.

Covalent bond   A chemical bond resulting from the sharing of electrons between atoms.

Crista   Small shelflike structures of the inner membrane of the mitochondrion. The cristae are the sites of oxidative phosphorylation.

Crossing-over   The mechanisms by which linked genes undergo recombination. The term refers to the reciprocal exchange of parts having corresponding loci between two homologous chromatids.

Cross-pollination   The transfer of pollen from the anther of one plant to the stigma of another plant.

Cutin   Wax secreted by the plant epidermis, providing an impermeable coating on aerial plant parts.

Cyclic AMP (cAMP; cyclic adenosine monophosphate)   A compound, formed from ATP, that regulates the effects of numerous hormones in animals (second messenger).

Cytochromes   Iron-containing red proteins, molecules of the electron transfer machinery in photosynthesis and respiration.

Cytokinesis   The division of the cytoplasm of a dividing cell.

Cytokinin   A plant growth hormone playing a role in cell division.

Cytoplasm   The living contents of the cell excluding the nucleus.

Darwinism   The theory of evolution of plants and animals by natural selection, formulated by Charles Darwin. Individuals with certain characteristics have a better chance of surviving and reproducing than others.

Deciduous   A plant that sheds its leaves seasonally.

Decomposer   An organism, such as a bacterium or fungus, that converts dead organic matter into plant nutrients.

Deletion   A mutation resulting from the loss of a segment of a chromosome.

Dendrite   Any of the usually branching and short processes that carry impulses to the cell body of the neuron.

Dicot (dicotyledon)   An angiosperm plant in which the embryo produces two leaves prior to germination.

Differentiation   Process whereby originally similar cells follow different developmental pathways.

Dikaryon   An organism, usually a fungus, with paired nuclei derived from different parents. The nuclei are not fused.

Dioecious   Organisms in which the two sexes are separate so that eggs and sperms are not produced in the same individuals.

Diploid   A cell (2*n*) usually arising as a result of the fusion of two gametes, each contributing one copy of each chromosome.

DNA (deoxyribonucleic acid)   The fundamental hereditary material of all living organisms, stored in the cell nucleus and consisting of a number of nucleotides.

Dominance   The ability of one allelic form of a gene to determine the phenotype of a heterozygous individual.

Dorsal   The back of an animal, not the hind end. The upper surface of a leaf, wing.

Duplication (genetic)   A mutation resulting from the inclusion into the genome of an extra copy of a piece of a gene or chromosome.

Ecdysone   In insects, a hormone inducing molting and metamorphosis.

Echinoderm   Any of a phylum (Echinodermata) of radially symmetrical coelomate marine animals, for example, starfish, sea urchin.

Ecology   The study of the interaction of organisms with their environment.

Ecosystem   The organisms of a particular habitat together with the physical environment in which they live.

Ectoderm   The outermost of the three embryonic tissue layers. Gives rise to the skin, sense organs, nervous system.

Effector   Any organ or cell that causes movement or secretions in response to a stimulus from the environment.

Efferents   Neurons that carry impulses from the central nervous system.

Electron   One of the fundamental particles of matter that carries a negative electric charge and is normally found within orbitals surrounding the nucleus of an atom.

Endergonic reaction   A nonspontaneous chemical reaction in which free energy is absorbed from the surrounding environment.

Endocrine gland   Any ductless gland that secretes hormones into the body through the blood or lymph.

Endocytosis   Incorporation of substances into a cell by pinocytosis or phagocytosis.

Endoderm   The innermost of the three embryonic tissue layers. Gives rise to the digestive and respiratory tracts and the structures associated with them.

Endodermis   The innermost layer of cells in the cortex in stems and roots of vascular plants.

Endoplasmic reticulum (ER)   The membranous canals that wind through the cytoplasm of eukaryotic cells. Rough ER has ribosomes attached. Smooth ER is devoid of ribosomes.

Endosperm   A specialized tissue, found only in angiosperms, that contains stored food for the developing embryo.

Endotherms   Animals that have constant body temperature.

Enolase   An enzyme that converts 2-phosphoglyceric acid into phosphoenolpyruvic acid.

Entropy   A measure of the disorder or randomness in a system.

Enzyme   A globular protein which governs the rate of a given metabolic reaction.

Epidermis   In plants and animals, the outermost cell layer.

Epinephrine   The hormone secreted by the medulla of the adrenal gland, also called adrenaline. It is secreted as a result of stress and produces effects on the circulatory system and on glucose mobilization.

Epiphyte   A specialized plant that grows on the surface of other plants but does not parasitize them.

Estrus The period of maximum sexual receptivity in the female mammal. Estrus is also the time of the release of eggs in the female.

Eukaryotes Organisms whose cells contain the genetic material enclosed in a nucleus. Includes all organisms above the level of bacteria and blue-green algae.

Evergreen A plant that has leaves through all seasons.

Evolution The process of gradual change. Any genetic change in organisms from generation to generation over vast periods of time.

Exergonic reaction A spontaneous chemical reaction in which free energy is released.

Exocrine gland Any gland, such as the salivary gland, whose secretion is discharged through a duct opening on an internal or external suface of the body.

Exons The coding regions of genes.

Exoskeleton A hard covering on the outside of the animal body. The exoskeleton of arthropods has the same functions of support as the bony internal skeleton of vertebrates.

Family In taxonomy, a group of related, similar genera.

Fauna All of the animals found in a given area.

Feedback control Control of a single step of a multistep process activated by the presence or absence of a product of one of the later steps.

Fermentation The degradation of a fuel molecule to smaller molecules, resulting in the extraction of energy, without the use of oxygen; also known as anaerobic fermentation.

Flagellate A unicellular organism that propels itself by flagella.

Flagellum A long, whiplike appendage that propels unicellular organisms.

Flora All of the plants found in a given area.

Flower A shoot that contains the reproductive structures of an angiosperm; its basic parts include the calyx, corolla, stamen, and pistil.

Food chain A segment of a food web; most commonly, a simple sequence of producer, consumer, predator.

Food web A series of interrelated food chains.

Fossil The preserved remains of an organism from prehistoric times.

Gamete The mature sexual reproductive cell: the egg or the sperm.

Gametocyte The primary sex cell. The cell that divides to produce sex cells: either the eggs or the sperm.

Gametogenesis The series of cellular divisions that lead to the production of sex cells.

Gametophyte In a plant that has alternation of generations, the haploid phase that produces the gametes.

Gastropoda A class of mollusks that includes the snails, slugs, conchs.

Gastrovascular cavity The central cavity in the body of coelenterates that is used for digestion and circulation.

Gastrula A stage in embryo development exhibiting the characteristic three cell layers: ectoderm, mesoderm, and endoderm.

Gene A unit of heredity.

Gene flow The exchange of genes between different species or between different populations of the same species.

Gene pool All of the genes in a population.

Genetic drift Evolution in small populations by chance processes alone due to preservation or extinction of particular genes.

Genome The total haploid set of chromosomes found in each nucleus of a particular species.

Genotype The genetic constitution of an individual, either with respect to a single trait or with respect to a larger set of traits.

Genus A group of related, similar species.

Geotropism Downward root growth and upward shoot growth in plants in response to gravity.

Gibberellin A plant growth regulator that plays a role in the stem elongation, seed germination, and flowering of certain plants.

Gizzard A muscular organ of birds that grinds up food, sometimes with the aid of fragments of stone.

Glia Connective tissue in the vertebrate central nervous system that does not conduct nerve impulses.

Gill slits A series of openings on each side of the pharynx to the outside; always present in fish and some amphibia; embryonic in only reptiles, birds, and mammals.

Gluconeogenesis The formation of glucose by the liver from components derived from the breakdown of proteins or fats.

Haploid Having just one copy of each chromosome. This is the normal number of chromosomes in gametes (monoploid).

Hardy-Weinberg law A law stating that population gene frequencies remain constant under certain conditions: random mating and no mutations or other changes.

Hemoglobin The colored protein of the blood which carries oxygen.

Hermaphroditism The presence of both female and male sex organs in the same organism.

Heterospory The production in plants of two kinds of spores, the small microspores and the large megaspores.

Heterozygote A diploid organism having different alleles of a given gene on the pair of homologous chromosomes carrying that gene.

Holandric traits Traits transmitted by genes that are on the male chromosome.

Holdfast In many large attached algae, specialized tissue attaching the plant to its substratum.

Homeostasis A stable state, such as a constant temperature or metabolic rate, maintained by means of physiological feedback responses.

Homiobox Sequences of DNA that regulate patterns of differentiation during the embryonic development of an organism.

Homiotherm (homeotherm) An animal which maintains a constant body temperature (despite variations in environmental temperature) because of its own heating and cooling mechanisms. A so-called warm-blooded animal.

Homologue One of a pair of chromosomes having the same overall genetic makeup.

Homology A similarity between two structures due to inheritance from a common ancestor.

Homozygote A diploid organism having identical alleles of a given gene on both homologous chromosomes.

Hormone A chemical product secreted by an endocrine gland into the blood that triggers physiologic activity or organs in other parts of the body.

Hydrogenase An enzyme that can transfer electrons to H+ to generate molecular hydrogen. It can also transfer H+ from molecular hydrogen to some other electron acceptor having H+ ions.

Hydrolyze To break a chemical bond by the addition of water, —H and —OH, at the cleaved chain ends.

Hydrophobic Referring to the attraction of molecules for one another without mixing with water. Oil is a hydrophobic substance.

Hyperosmotic   Referring to a medium with a higher concentration of osmotically active molecules than is present in cells. In hyperosmotic solutions, water will flow out of cells, causing shrinkage of the cytoplasm, or plasmolysis.

Hypha   In fungi, any single filament. May be multinucleate or multicellular.

Hypoosmotic   Referring to a medium that has a lower concentration of osmotically active molecules than does a cell. In hypoosmotic media, water will flow into cells.

Imbibition   The binding of a solvent to another molecule.

Imprinting   A rapid form of learning in early life in which an animal makes a particular response only to one other animal or object.

Independent assortment   The random separation, during meiosis, of genes carried on nonhomologous chromosomes. There is random separation of the nonhomologous chromosomes as well.

Inducer   A molecule of small molecular weight which, when added to a system, causes a large increase in the level of enzyme.

Inhibitor   A molecule that bonds to the surface of an enzyme and blocks its action on its substrate molecules.

Instinct   Behavior that is inborn and unlearned; a series of reflex acts that culminates in a given activity, for example, nest building by birds.

Insulin   A hormone synthesized in the islets of Langerhans in the pancreas which regulates the level of glucose in the blood.

Interferon   A hormone of the immune system, secreted by virus-infected cells, that enables other cells to resist the virus.

Interphase   The period between successive cell divisions during which the chromosomes are very thin and elongated and the nuclear membrane is intact. This phase has been misnamed "the resting stage," but it is during this period that replication of chromosomes occurs.

Introns   Sequences of noncoding genes.

Inversion   A mutational event that leads to the reversal of the order of genes on a portion of a chromosome as though that segment had been removed from the chromosome, turned upside down, and then reattached.

Invertebrate   An animal that lacks a backbone.

In vitro   In a test tube rather than in a living organism. Said of results obtained by experiments rather than observed in life.

In vivo   In the living state rather than in a test tube.

Ion   An atom with electrons added or removed, giving it a negative or a positive charge; a charged particle.

Ionic bond   A strong chemical bond that forms between ions or groups of ions of opposite charge.

Isotonic   Referring to a solution that has the same concentration of salts as a cell, so there is no net inflow or outflow of water.

Isotope   A chemical element having the same number of protons in its nucleus as another, but differing in the number of neutrons.

Jejunum   The part of the small intestine that serves as the main absorptive region. It has a larger diameter than the rest of the small intestine.

Jugular vein   The main vein returning blood from the head to the heart.

Jurassic   A geologic period after the Cretaceous and before the Triassic period that lasted from 190 to 135 million B.C., during which dinosaurs flourished and birds made their first appearance.

Juvenile hormone   In insects, a hormone maintaining larval growth by inhibiting maturation to the adult form.

Karyokinesis The division of the nucleus of a dividing cell in the formation of daughter cells.

Karyotype A chart of chromosomes organized according to number, size, and type.

Keratin A sulfur-containing protein that is part of hard tissue, such as horn, nail, and the outermost layers of the skin.

Key In taxonomy, an arrangement of the characteristics for quickly identifying a specimen down to the species.

Kilocalorie The amount of heat required to raise the temperature of one kilogram of water one degree Celsius.

Kinetosome A cylindrical, microtubular organelle found at the base of cilia and flagella of some protozoans and tissue cells.

Krebs cycle Also known as the citric acid cycle. Sequential metabolic reactions in aerobic respiration in which acetyl CoA is completely oxidized to CO_2 by way of interconversions of oxaloacetate, citrate, ketogluterate, succinate, fumarate, and malate.

Lac operon The three structural genes that program lactose metabolism.

Lamarckism The theory of evolution of acquired characteristics proposed by Jean Baptiste de Lamarck (1744–1829).

Larva An immature stage of an animal that is markedly different in appearance from the adult, such as the caterpillar and the tadpole.

Lateral Situated at a side.

Lateral line The longitudinal line along each side of the body of fish that denotes the position of sensory cells that enable fish to perceive movement and sound in water.

Life cycle The entire span of the life of an organism.

Linkage Genes associated on the same chromosome so that they do not assort at random but are inherited together.

Lipids Fats, oils, waxes, steroids, and other large organic molecules that are some of the principal parts of living cells.

Lumen The cavity of a tubular organ, such as that of the intestine or a kidney tubule.

Lymphatics Spaces between the capillaries and the body cells which collect the lymph and return it to the bloodstream.

Lysosome A membrane-bounded inclusion found in the cytoplasm of cells. Lysosomes contain a mixture of enzymes that are capable of digesting most of the complex molecules found in the rest of the cell.

Mammal A higher vertebrate of the class Mammalia, characterized by the production of milk by the female mammary glands, skin covered by hair, a four-chambered heart, and a constant body temperature.

Marsupial Any of an order (Marsupialia) of lower mammals comprising kangaroos, wombats, opossums, and related animals that with few exceptions develop no placenta and have a pouch on the abdomen of the female containing the milk glands and serving to carry the young.

Medulla The central part of an organ or structure.

Medulla oblongata The posterior part of the brain of vertebrates merging into the spinal cord. It controls involuntary activities.

Megasporangium The special structure that produces megaspores.

Megaspore In plants, a large haploid spore that produces a female gametophyte.

Meiosis Cell division of a diploid cell to produce four haploid daughter cells. It occurs during maturation of gametes.

Meristem A plant tissue characterized by cell division through which specialized plant tissues are formed.

Mesenchyme The meshwork of embryonic or unspecialized cells developed from the mesoderm.

Mesoderm The middle of the three embryonic tissue layers. Gives rise to the skeleton, circulatory system, muscles, excretory system, and most of the reproductive system.

Messenger RNA (mRNA) A copy of one of the strands of DNA. It carries information for the synthesis of proteins.

Metamorphosis A change in body form occurring from one development stage to another, for example, a tadpole to a frog, or an insect larva to an adult.

Metaphase The stage in cell division at which chromosomes are arranged on a plane in the center (equator) of the cell. The middle stage of mitosis or meiosis.

Metazoan Any of a group (Metazoa) that comprises all animals having the body composed of cells differentiated into tissues and organs and usually a digestive cavity lined with specialized cells.

Microfilaments Two intertwined strands of actin that function in muscle contraction and changes in cell shape.

Micropyle An opening in the integument of an ovule of a seed plant through which the pollen tube penetrates to the embryo sac to reach the female gametophyte within.

Microsporangium The structure that produces the microspores.

Microspore In plants, a small haploid spore that produces a male gametophyte.

Microtubules Minute tubular structures found in centrioles, spindle apparatuses, cilia, and flagella.

Mitochondrion An organelle that occurs in eukaryotic cells in great numbers. It is the site of aerobic cellular respiration.

Mitosis Cell division leading to the formation of two daughter cells, each with a chromosome number and composition identical to that of the parent cell.

Mollusk A member of the phylum Mollusca of invertebrates, such as the snail, the clam, or the octopus.

Monerans The members of the kingdom Monera: the bacteria (Schizomycetes) and the blue-green-algae (Cyanophyta).

Monocot (monocotyledon) Any member of the angiosperm class Monocotyledoneae. Plants in which the embryo produces but a single seed leaf. Leaves of most monocots have their major veins arranged parallel to each other.

Monoecious Referring to organisms in which both sexes occur in a single individual that produces both eggs and sperms; hermaphroditic, for example, earthworms, hydras.

Morphogenesis The formation and differentiation of tissues and organs.

Morphology The study of the organic form of plants and animals.

Multicellular Consisting of more than one cell.

Mutagen A chemical agent that increases the amount or rate of mutation.

Mutation Any change in the genetic constitution of an organism.

Mutualism The type of symbiosis in which two species benefit from living in close association, for example, fungus and alga that form lichens.

Mycelium In fungi, a proliferation of hyphae.

Myoglobin The special hemoglobin found in muscle, a smaller molecule that carries less oxygen than blood hemoglobin.

Myosin A major protein of muscle. It makes up the thick filaments.

Nanometer Ten angstrom units; 10^{-3} micrometer; also known as a millimicron.

Natural selection The theory of evolution formulated by Charles Darwin that states that the groups or individuals best suited to their environment tend to survive and perpetuate the species.

Negative control The process in which a repressor functions to turn transcription off.

Nematocyst A threadlike structure produced by cells of jellyfish and other coelenterates, used to paralyze and capture prey; stinging cells.

Nematode A member of the phylum Nematoda; a roundworm.

Neoteny The reaching of sexual maturation in the larva state, for example, the axolotl, the aquatic larval stage of an American salamander.

Nephridium An organ, present in many coelomate invertebrates, that is involved in excretion and in water control. Consists of a tube of ectodermal origin, which opens to the exterior at one end.

Nephron The excretory unit of the vertebrate kidney, consisting of a glomerulus in a Bowman's capsule and uriniferous tubules.

Nerve net The network of nerve cells distributed randomly throughout the body tissues of coelenterates and echinoderms.

Neural crest A ridge of cells that become detached from the forming neural tube and wander great distances to fulfill a wide range of developmental roles. Gives rise to a variety of tissues, such as dorsal root ganglia, Schwann cells, melanophores, and cartilage of visceral arches.

Neutron A particle of matter within the atomic nucleus having a mass of 1 atomic mass unit and without electric charge.

Nitrogenase In nitrogen-fixing bacteria and blue-green algae, an enzyme which controls the stepwise reduction of atmospheric N_2 to ammonia.

Nomenclature The method of assigning names in the classification of organisms.

Nonsense mutations Mutations that substitute a codon for an amino acid to one of the codons that determine termination of translation. The gene that is produced is a shortened polypeptide that begins normally at the amino terminal end and ends at the position corresponding to the altered codon. This is also known as a chain that terminates mutation.

Notochord A thin line of cells of mesodermal origin extending along the back in the early chordate embryo that directs the formation of the neural tube.

Nuclear membrane The double membrane that encloses the nucleus of a eukaryotic cell.

Nucleic acid A long-chain molecule formed from a large number of nucleotides, that is, DNA and RNA.

Nucleolus A small, spherical body found within the nucleus of a eukaryotic cell. It is the site of synthesis of ribosomal RNA.

Nucleotide The unit of structure in a nucleic acid. A nucleotide in RNA consists of one of four nitrogenous bases linked to the sugar, ribose. In DNA, deoxyribose is present instead of ribose, and the base thymine is present instead of uracil. The sugar and nitrogenous bases are linked to phosphate.

Nucleus (1) The dense central portion of an atom, made up of protons and neutrons, with a positive charge, surrounded by negatively charged electrons. (2) A centrally located organelle of a eukaryotic cell that is bounded by a double membrane and contains the chromosome and one or more nucleoli.

Ommatidium One of the units that make up the compound eye of arthropods. It is estimated that 20,000 or more ommatidia are present in one compound eye.

Ontogeny The development of a single organism in the course of its life.

Oocyte The cell that gives rise to eggs by meiosis.

Oogenesis   Female gametogenesis; the production of egg cells.

Operator   The region of an operon that acts as the binding site for the repressor; triggers formation of mRNA.

Operculum   (1) The cover of gill slits of fish. (2) The exoskeletal plate of some gastropods (such as snails) that can close the shell opening when the animal withdraws inside.

Operon   A genetic unit of transcription, consisting of several structural genes or cistrons that are transcribed to produce a single messenger RNA molecule. The operon contains at least two transcriptional control regions: the promoter and the operator.

Organ   A formed body structure such as the eye, liver, brain, stem, or leaf, composed of different tissues and cells integrated to perform a distinct function for the organism as a whole.

Organelles   Organized structures that are found in cells, for example, ribosomes, nuclei, mitchondria, chloroplasts, cilia, contractile vacuoles.

Organic   Referring to any chemical compound that contains carbon.

Organism   Any living plant or animal.

Osmosis   The tendency of water to move from a region in which it is more concentrated to a region in which it is less concentrated. The diffusion of water through a membrane.

Ovary   The female organ in plants or animals that produces eggs.

Ovule   In plants, an organ that contains a gametophyte and within the gametophyte, an egg; when it matures, an ovule becomes a seed.

Ovum   The egg, the female sex cell.

Palisade mesophyll   One or more layers of tightly packed columnar photosynthetic cells in leaves, found just below the upper epidermis and above the spongy mesophyll.

Parasitism   The form of symbiosis in which one species lives at the expense of another but not ordinarily to the point of killing its host. Such destruction would be designated as predation.

Parenchyma   A plant tissue composed of thin-walled cells functioning either in photosynthesis or storage walls.

Parthenogenesis   The production of an organism from an unfertilized egg.

Passive transport   The movement of molecules by means of diffusion, osmosis, or carrier-facilitated diffusion without the expenditure of energy by the cell.

Pathogen   An organism, such as a virus or a bacterium, that causes disease.

Peptide bond   A linkage site in a protein chain formed by the removal of water during the joining of amino acids, —COOH to —NH$_2$. Also called an amide bond.

Perennial   A plant that lives from year to year.

Pericycle   The tissue just within the endodermis in plant roots but outside of the root vascular tissue.

Period   A division of geologic time of intermediate duration, less than an era and greater than an epoch. For example, the Triassic period lasted for about 50 million years.

Petal   A nonphotosynthetic leaf, often brightly colored, that surrounds the reproductive organs in the angiosperm flower.

Phage   A shortened form of "bacteriophage."

Phagocytosis   The engulfing of a solid particle by the formation of a pouch of cell membrane around the particle and by pinching off the pouch to form an intracellular vacuole containing the particle bounded by membrane.

Phenotype   The visible characteristics of an individual as they have developed under the combined influences of the genetic constitution of the individual and environmental factors.

Pheromone   A chemical substance used as a stimulus for a behavior response between insects or other animals of the same species.

Phloem   In vascular plants, the food-conducting tissue, which consists of sieve tubes, fibers, and other specialized cells.

Photoperiod   The duration of the periods of light and dark in a 24-hour cycle as they affect the growth and maturity of an organism.

Photosynthesis   The metabolic process by which visible light is trapped in the chlorophyll-containing tissues of plants and the energy used to synthesize carbohydrates from CO_2 and H_2O.

Phototropism   The growth of a plant in light.

Phycocyanin   A photosynthetic accessory protein pigment, greenish blue, found in blue-green algae.

Phycoerythrin   A photosynthetic accessory pigment, bright red or orange, found in the cells of red algae.

Phylogeny   The evolutionary history of a group of organisms as opposed to the development of an individual in a group.

Phylum   In taxonomy, a major category just beneath the kingdom and above the class. A group of related classes, for example, Chordata.

Physiology   The study of the functions of living organisms and their parts, such as organs, tissues, and cells.

Phytochrome   A photoreversible plant pigment, energized by red light to a chemically active form and functioning as an enzyme. In darkness, phytochrome is reconverted into the inactive form.

Phytoplankton   The photosynthetic plant portion of the plankton, consisting mostly of algae.

Pinocytosis   The uptake of liquids or very small particles into a cell by pinching off of the cell membrane to form an intracellular vacuole.

Pistil   The female sex organ of an angiosperm flower within which the ovules are borne. The pistil consists of the stigma, the style, and the ovary.

Pit   In plants, a small depression in a tracheid cell wall which is not a complete hole and which moves water and dissolved substances from one cell to the next.

Pith   In plants, unspecialized spongy tissue found within a ring of vascular tissue and probably used for storage.

Placenta   The organ, found in mammals (except marsupials), that provides for the nourishment of the fetus and the elimination of the fetal waste products by diffusion through capillaries. The placenta is formed by the union of membranes of the maternal uterine lining with membranes from the fetus.

Plankton   Microorganisms and small plants and animals passively floating or swimming at the surface of a body of water.

Plaque   A circular clear area on the surface of a nutrient agar where bacteria are growing and are infected by bacteriophage.

Plasma membrane   A double membrane that regulates the entry and exit of molecules and ions in a cell.

Plasmodesma   An open channel through which cytoplasmic strands connect two adjacent plant cells.

Plasmodium   A multinucleate mass of protoplasm, surrounded by a membrane, characteristic of the vegetative feeding stage in slime molds.

Plasmolysis   Shrinking of cell cytoplasm by loss of water in a hypertonic solution. The plasma membrane and the cytoplasm shrink and pull away from the cell wall.

Point mutation   A mutation that results from a small, localized change in the chemical structure of a gene. In genetic crosses, a point mutation behaves as though it were on a single point on a chromosome.

Polar body   A minute, nonfunctional cell produced during meiosis. The meiosis that produces the mammalian egg produces, in addition, three polar bodies, which disintegrate.

Pollen   The microspores of flowering plants that form the male gametophytes.

Pollination   The process of transferring pollen from the anther to the pistil in plants.

Polymorphism   Having several different forms. Different types of the same species, for example, human blood groups.

Polyploid   A cell in which the number of complete sets of chromosomes is greater than two.

Polyribosome   A threadlike molecule of messenger RNA and many ribosomes. The ribosomes move along the mRNA, synthesizing polypeptide chains as they proceed.

Polytene   Chromosomes that consist of many parallel, identical chromatids that remain paired through repeated duplication. Fly salivary gland chromosomes are polytene.

Population   Any group of organisms belonging to the same species and coexisting at the same time and in the same place.

Positive control   The situation in which a regulatory macromolecule is needed to activate transcription of structural genes.

Posterior   Situated behind, near the tail end.

Predator   An organism that eats other organisms.

Primate   Any of an order (Primates) of mammals including human beings, the lemur, monkey, or ape.

Primitive streak   A longitudinal thickening extending axially along the blastodisk. This appears in the early embryo of the bird or mammal, during the gastrula stage of cleavage.

Procambium   In plants, the embryonic tissue that gives rise to the first xylem and phloem.

Prokaryotes   Organisms whose chromosome material is distributed throughout the cell rather than being confined in a distinct nucleus; includes bacteria and blue-green algae.

Promoter   The site of an operon that serves as the initial binding site for RNA polymerase.

Prophage   The noninfectious bacteriophage units that are attached to the chromosomes of the host bacteria and multiply with them but do not cause the destruction of the cell.

Prophase   The first stage of cell division during which chromosomes condense to thickened compact bodies.

Protease   A proteolytic enzyme.

Protein   An extremely complex organic compound composed of amino acids with many different common side chains.

Proteolytic enzyme   An enzyme whose major catalytic function is the digestion of a protein or polypeptide chain. The digestive enzymes trypsin, pepsin, and carboxypeptidase are all proteolytic enzymes.

Proton   One of the three most fundamental particles of matter with a mass of 1 atomic unit and charge +1. Found in the atomic nucleus.

Protonema   The hairlike growth form in the early stage of the development of the moss gametophyte.

Protoplasmic streaming   The circulation in the contents of a eukaryotic cell. This process of cyclosis allows transport of materials over distances that are too great to allow mixing by diffusion alone.

Protoplast   A cell that would normally have a cell wall but from which the wall has been removed by enzymatic digestion or by special growth conditions. Equivalent to what a zoologist calls a cell.

Protozoa A group of single-celled animals, including the flagellates, amoebas, and ciliates.

Pseudopodium A temporary, flowing extension of the cell body of the amoeba that is used for locomotion, attachment to surfaces, or engulfing particles.

Pupa In higher insects, the inactive developmental stage of the life cycle between the larva and the adult.

Q_{10} Temperature coefficient. A measure of the increase in the rate of a process for each 10°C rise in temperature.

Quadrant One square meter of vegetation chosen at random for ecological study.

Radial symmetry The condition in which two halves of a body are mirror images of each other regardless of the angle of the cut, providing the cut is made along the central axis. This is a characteristic of coelenterates and echinoderms.

Recapitulation The repetition of stages of evolution of a species (phylogeny) in the stages of development in the individual organism (ontogeny). The "recapitulation principle" expressed by E. H. Haeckel is "Ontogeny recapitulates phylogeny."

Receptacle In a flowerng plant, the end of the stem upon which all of the various flora parts are attached.

Recessiveness The character or gene that is hidden by the contrasting dominant allele.

Recombinant A chromosome carrying, at two or more sites, genes that were originally carried on two separate chromosomes; offspring emerging from a genetic cross that carry such a recombinant chromosome.

Reduction In photosynthesis, gain of electrons by the addition of hydrogen. Most reductions lead to the packaging of chemical energy which can be released later by an oxidation reaction.

Reflex An automatic, inborn action, involving only a few neurons, in which motor response swiftly follows a sensory stimulus.

Regulator gene A gene that contains the information for coding a regulatory macromolecule, usually a repressor protein.

Repressor A protein molecule coded by a site on a regulator gene. The repressor may bind to a specific operator and prevent transcription of the operon.

Respiration (1) Breathing. (2) Cellular respiration: the oxidation of glucose to produce energy. The oxidant in the respiration of all higher organisms is molecular oxygen gas.

Reversion A mutation that restores a wild-type phenotype to a mutant. A throwback.

Rhizoids Hairlike extensions of cells in mosses, liverworts, and a few vascular plants that serve the same function as roots and root hairs in higher plants.

Rhizome An underground plant stem that runs horizontally and functions in vegetative propagation or storage.

Ribonucleic acid (RNA) A nucleic acid consisting of a large number of nucleotides, each of which contains ribose sugar and four bases, including uracil, and a phosphate radical.

Ribose A sugar of chemical formula $C_5H_{10}O_5$: one of the building blocks of nucleic acid.

Ribosome A small organelle that is the site of protein synthesis.

Root pressure In plants, the movement of water into the stem from the roots driven by osmotic pressure.

Sap A solution of nutrients and minerals in plants that passes from roots to stems to leaves.

Saprophyte A nongreen plant that obtains its nutrients directly from dead organic matter.

Segregation The separation of alleles, or of homologous chromosomes, into different gametes during meiosis. Each daughter cell produced by meiosis contains one or the other member of the pair of chromsomes found in the mother cell.

Self-pollination The transfer of pollen from the anther to the stigma of the same plant.

Sepal The green leaves enclosing the bud stage of a flower and comprising the calyx.

Sex chromosomes Those chromosomes, inherited differently in the two sexes, that determine the sex of an individual. In females, XX; in males, XY.

Sex-linkage A term used with reference to genes linked on the sex chromosomes or conditions transmitted by a gene on an X or Y chromosome.

Sieve cell A specialized cell found in the phloem tubules through which carbohydrates move from the leaves to the rest of the plant.

Sieve tube A column of sieve cells in the phloem specialized to conduct carbohydrates from the leaves to other plant parts; found principally in flowering plants.

Social insects Colony-forming insects that show a division of labor among the members. Examples include ants, termites, bees, and wasps.

Sodium pump A system, requiring ATP, that transports sodium against its concentration gradient from the inside of cells to the outside. An example of active transport.

Somite One of a series of blocks into which a vertebrate embryo is segmented longitudinally (flanking the notochord in two strips). Each somite is innervated by one ventral root nerve. Therefore, the nerves, the spinal cord, and associated muscles are segmented. Also present in arthropods and annelids.

Sonar Echolocation. The emitting and receiving back, as echoes, of high-pitched sounds to determine location and distance of objects. This ability is possessed by bats, dolphins, and porpoises.

Spawning The direct release of sex cells (eggs and sperm) into the water by fish and amphibians in reproduction; external fertilization.

Speciation The process by which new, discrete interbreeding units or subgroups are formed in a population.

Species The unit of classification of organisms so closely related that they can interbreed and produce viable offspring.

Specificity site That particular site on the surface of an enzyme where a compatible site of the substrate binds. This helps the enzyme select the proper substrate.

Sperm A male reproductive cell; the male gamete.

Spermatocyte The cell that gives rise to the sperm. The primary sex cell.

Spermatogenesis Male gametogenesis leading to the production of sperm. Also referred to as maturation of sperm.

Spermatozoon A mature male germ cell.

Sphincter A ring of muscle which can close an opening, for example, the pyloric sphincter, which controls the passageway from the stomach to the intestine.

Spindle fibers The fibers extending from the metaphase chromosomes to the poles of a dividing cell and aiding in the movement of chromosomes at cell division.

Spongy mesophyll In leaves, a layer of loosely packed photosynthetic cells spaced for gas diffusion and found between the palisade layer and the lower epidermis.

Spontaneous generation (abiogenesis) The now discredited idea that life can arise from nonliving matter. Redi, Spallanzani, and Pasteur disproved the possibility of spontaneous generation.

Spontaneous reaction A chemical reaction that will proceed on its own without any outside help. Need not be rapid. For example, complex nutrients, starches, proteins, and fats will hydrolyze spontaneously without enzymes over long periods of time.

Sporangium In plants, any reproductive structure within which one or more spores is formed.

Spore An asexual reproductive cell capable of developing into an adult plant without the fusion of gametes. In plants there is an alternation of generations in which haploid spores develop into gametophytes, diploid spores into sporophytes. In nonnucleated cells, a spore is a resistant cell of decreased metabolism that can survive unfavorable conditions.

Stamen A male unit of a flower, composed of an anther, which bears the pollen, and a filament, which is a stalk supporting the anther.

Steady state (homeostasis) An unchanging condition of cell metabolism.

Stele Primary vascular tissue; primary phloem and primary xylem that extends throughout the column of the primary plant body.

Stigma The part of the pistil, at the top of the style, that is structured to receive pollen and on which pollen germinates.

Stoma A small pore in the plant epidermis that permits gas exchange. Each stoma is bounded by a pair of guard cells whose osmotic status regulates the size of the opening.

Strobilus A cone or multiple fruit of a gymnosperm.

Structural gene A gene that encodes the basic structure of an enzyme.

Style The stalk bearing the stigma. A stalk of the pistil in flowering plants.

Subspecies A subdivision of a species. Usually defined more narrowly as a geographic race: a population or series of populations occupying a discrete range and differing genetically from other geographic races of the same species.

Substrate The molecules on which an enzyme exerts catalytic action. The base on which an organism lives.

Succession In ecology, the gradual and predictable series of changes in species composition or organisms from the time of colonization to the maturation of the final, stable climax community.

Symbiosis The living together of two or more species of organisms in a prolonged and intimate ecological relationship.

Symmetry In biology, the property that two halves of an object are mirror images of each other.

Sympatric Referring to populations whose geographic ranges overlap but which do not interbreed.

Synapsis The highly specific paralleled alignment of homologous chromosomes during the first division of meiosis.

Systematics The science of classification of organisms. Taxonomy.

Taxis The movement of an organism or a cell in a particular direction in response to a stimulus. The movement may be toward or away from the stimulus.

Taxonomy The science of classification of organisms.

Telophase The final phase of mitosis or meiosis during which chromosomes become thin and elongated, the nuclear membrane reforms, and nucleoli reappear in the new nuclei.

Territory A preempted area from which an animal or group of animals bars other members of the same species by aggressive behavior or feigned aggressiveness.

Testis The male gonads. The organ that produces the male sex cells.

Tetrad A group of homologous chromosomes formed during prophase of meiosis, each consisting of two chromatids and each joined at a centromere.

Tissue   A group of similar, specialized cells that work together to perform a given function. Many tissues form an organ.

Trachea (windpipe)   The tube that carries air to the bronchi of the lungs of land-dwelling vertebrates. A cartilagenous tube.

Tracheid   A long tubular conduction and support cell found in the xylem of vascular plants. It has tapering ends and pitted walls.

Transcription   The synthesis of mRNA; a strand of DNA is used as the template and transfers genetic information to the mRNA.

Transduction   The transfer of genes from one bacterium to another using a virus which acts as the carrier of the genes.

Transfer RNA (tRNA)   A relatively small RNA that functions in transferring amino acids into polypeptide chains. Each tRNA is specific for an activated amino acid.

Transformation   The transfer of genetic information from one bacterium to another, accomplished by extracting DNA from bacteria of one genotype and transferring it to bacteria of another genotype.

Transition mutation   A mutation in which a purine (A or G) replaces a purine or a pyrimidine (T or C) replaces a pyrimidine. This is a protein base substitution mutation.

Translation   The synthesis of a polypeptide made up of a particular sequence of nucleotides.

Translocation   The transfer of a part of a chromosome to a different part of a homologous or nonhomologous chromosome.

Transpiration   The loss of water through the leaves of a plant.

Transversion mutation   A mutation in which a purine (A or G) replaces a pyrimidine (T or C).

Trichocysts   Threadlike organelles that are discharged from protozoans, especially ciliates, in response to an irritation stimulus.

Trochophore   The free-swimming larvae of annelids and mollusks. A wheel-like band of cilia encircles the middle. This larval form establishes an evolutionary relationship between the two groups.

Trophic level   The position of a species in a food chain measured in steps away from the primary producers.

T-system   A set of tubules that extend from the cell membranes of muscle cells into the interior of muscle fibers, ending in blind sacs called terminal cisternae. The T-tubules transmit impulses to the sacs, which release calcium ions in response.

Turgor pressure   The pressure in a plant cell as a result of the osmotic concentration of the cell.

Unicellular   Consisting of a single cell, for example, a paramecium.

Ureotelic organism   An organism that synthesizes urea as a waste product of nitrogen metabolism. Examples include mammals, adult frogs and toads, cartilagenous marine fish.

Uricotelic organism   An organism that excretes uric acid as a nitrogenous waste, which serves as a water-conserving mechanism. Examples include grasshoppers, birds, land-dwelling reptiles.

Uterus   A specialized portion of the female mammal reproductive system which receives the fertilized egg and maintains it through embryo and fetal stages until birth.

Vacuole   A space in the cytoplasm enclosed by a membrane and filled with liquid. Some vacuoles are digestive vesicles; some are storage areas.

Vascular   Referring to organs and tissues that conduct fluid. Examples are blood vessels in animals, and phloem and xylem in plants.

Ventral   The bellyside or underside of an animal; the undersurface of a leaf or wing.

Vertebrates   Animals with a backbone (spine or spinal column). The individual segments of the backbone are known as vertebrae. The nerve cord is enclosed in the backbone. The major groups of vertebrates are fishes, amphibians, reptiles, birds, and mammals.

Vessel In angiosperms, a tubular element associated with the xylem and its water-conducting functions.

Villi Finger-shaped microscopic projections that extend from the membranous walls of the intestine.

Virus An ultramicroscopic particle that can reproduce only inside living cells. Viruses do not contain organelles equivalent to those of a cell.

Vitamins Chemical molecules that function as coenzymes or parts of coenzymes and that are utilized by cells for metabolic regulation.

Wall pressure The opposing pressure exerted by the plant cell wall in response to turgor pressure.

Wild-type A standard or reference-type genetic specimen of the natural environment, as opposed to the atypical mutant.

Xylem The woody tissue that conducts water and minerals up from the roots through the stem and leaves. Xylem consists of tracheids, vessels, fibers, and other highly specialized cells.

Yolk The stored food in animal eggs, rich in protein and lipid.

Zoology The study of animals.

Zooplankton The animal portion of the plankton.

Zoospore Any swimming spore or protist. May be diploid or haploid.

Zygote The cell created by the union of two gametes, in which the gamete nuclei are also fused. A fertilized egg.

Index